Diversity and Non-integer Differentiation for System Dynamics

Diversity and Non-integer Differentiation for System Dynamics

Alain Oustaloup

Series Editor
Bernard Dubuisson

iSTE

WILEY

First published 2014 in Great Britain and the United States by ISTE Ltd and John Wiley & Sons, Inc.

ISTE Ltd
27-37 St George's Road
London SW19 4EU
UK

www.iste.co.uk

John Wiley & Sons, Inc.
111 River Street
Hoboken, NJ 07030
USA

www.wiley.com

Library of Congress Control Number: 2014934408

British Library Cataloguing-in-Publication Data
A CIP record for this book is available from the British Library
ISBN 978-1-84821-475-0

MIX
Paper from responsible sources
FSC® C013604

Printed and bound in Great Britain by CPI Group (UK) Ltd., Croydon, Surrey CR0 4YY

Table of Contents

Acknowledgments

If my acknowledgments are naturally intended for the CRONE team, I would particularly like to thank François Levron for his contributions to the mathematical aspects and Stéphane Victor for his major contribution both to the translation and to the achievement of the book.

Preface

Non-integer differentiation does not escape to the slogan "different operator, different properties and performances". This is indeed the concise formula that is likely to explain the "why" of this operator, especially as most of its properties and performances favorably distinguish not only the operator itself but also the models that use it.

It is true that we have established non-integer models that overcome the mass-damping dilemma in mechanics and the stability-precision dilemma in automatic control; the technological achievements associated with these models have been made possible thanks to an adequate synthesis of the non-integer differentiation operator.

It is indeed the idea to synthesize non-integer differentiation (in a medium frequency range) through a recursive distribution of passive components, of transitional frequencies or of zeros and poles, which is at the origin of the non-integer differentiation operator real time use and, therefore, of both analogical and numerical applications that arise from it. As for the corresponding dates, the synthesis as we have led it has been developed by stages and thus proposed and experimented in the 70s for half-integer orders, in the 80s for real non-integer orders and in the 90s for complex non-integer orders.

The first technological applications of this operator (henceforth usable in real time) and notably the first application in 1975 of a "half-integer order controller" to the frequency control of a continuous dye laser, have widely contributed to take the non-integer differentiation out of the mathematician's drawers and to give rise to new developments likely to enrich the theoretical corpus of circuits and systems.

In this way France has been the first country to experience a interest in non-integer differentiation, this renewal having been well relayed thanks to the dynamism of the foreign scientific communities, at the European level as well as at the international level.

In this context, the French institutions have encouraged research in this field through the acknowledgement of major scientific advances and the support of initiatives or actions aiming to favor the synergies between the different themes and between the academic and industrial components, the university–industry partnership having indeed been nationally rewarded by the *Association Française pour la Cybernétique Economique et Technique* (AFCET) '95 Trophy distinguishing the CRONE suspension as the best technological innovation. Concerning the acknowledgements, let us cite the selection of the CRONE control as a "striking fact" of the *Centre National de la Recherche Scientifique* (CNRS) in 1997 and as "Flagship Innovation" of Alstom in 2000 (Hanover and Baden Baden International Fairs, 2000), a silver medal of the CNRS in 1997 and the Grand Prix Lazare Carnot 2011 of the Science Academy (founded by the Ministry of Defense). Concerning the supports, let us cite the actions financially supported by the CNRS and the Ministry of Research: the edition of "La commande CRONE" (Hermès, 1991) with the exceptional help of the ministry; the International Summer School "Fractal and hyperbolic geometries, fractional and fractal derivatives in engineering, applied physics and economics" (Bordeaux, 1994); the national project of the CNRS, "Non-integer differentiation in vibratory insulation" (1997–1999); the colloquium "Fractional Differential systems" (Paris, 1998); the launching in 1999 of the thematic action of the Ministry of Research "Systems with non-integer derivatives"; the launching in 2004 of the International Federation of Automatic Control (IFAC) Workshop "Fractional Differentiation and its Applications" through the first Workshop FDA '04 (Bordeaux, 2004) with S. Samko as chairman of the International Programme Committee; the magisterial lecture "From diversity to unexpected dynamic performances" initiated by the French Science Academy (Bordeaux, 5 January 2012).

But this academic support also found a guarantee in the industrial support brought by a strong partnership with major companies such as the Peugeot Société Anonyme (PSA) Peugeot-Citroën, Bosch (Stuttgart) and Alstom, such a partnership indeed led to a high number of patents and technological transfers that have widely proved the industrial interest of non-integer approaches.

Alongside the shared efforts to inscribe these approaches in the realist frame of the university–industry relations, our efforts have never stopped being shared with those of the international scientific community to develop the best relations and collaborations within this community. Without aiming for exhaustiveness, let us cite the involvement of European countries involved in the diffusion, promotion and animation within the community: the research group "Fracalmo", which originates from *Fra*ctional *cal*culus *mo*deling, started with a round table discussion in 1996 during the 2nd International Conference "Transform methods and special functions" held in Bulgaria; the journal "Fractional Calculus and Applied Analysis" (FCAA) started in 1998 with V. Kiryakova as managing editor; the survey on the "Recent history of fractional calculus" (*Communications in Nonlinear Science and Numerical Simulation*, 2010) at the initiative of J.T. Machado who desired to make an inventory of the major documents and events in the area of fractional calculus that had been produced or organized since 1974; the symposium "Fractional Signals and Systems" (FSS) launched by M. Ortigueira in 2009 in Lisbon, then held in 2011 at Coimbra (Portugal) and in 2013 at Ghent (Belgium). Both with in and out side of Europe, let us also recall the various events of the Workshop FDA after its launching at Bordeaux in 2004 (under the aegis of IFAC): Porto (Portugal) in 2006; Ankara (Turkey) in 2008; Badajoz (Spain) in 2010; Nanjing (China) in 2012; Grenoble (France) in 2013 and Catania (Italy) in 2014 (under the aegis of IEEE).

As the founder of the CRONE team, which counts today about 10 permanent researchers, I recognize this team to have always escorted me in the federative actions, launched within the national or international scientific community, to energize and harmonize research on both theoretical and applicative aspects of non-integer differentiation.

Introduction

If beyond integer differentiation many people indistinctly talk about fractional, non-integer or generalized differentiation, it seems through database analysis that fractional qualifying is used more in titles, whereas non-integer differentiation is used more in texts, with generalized differentiation appearing to be less used. So, why does non-integer qualifying benefit from our preference in this book?

In the formulation of generalized differentiation, the differentiation order is integer or non-integer, the generalized indeed including the integer and the non-integer. But the specificity of our book is certainly on the non-integer, notably through the link of non-integer differentiation with the recursivity and fractality that diversity covers. Besides, the real or complex non-integer differentiation synthesis, which is major in our contribution, is exclusively a matter for the non-integer, hence our preference. As for the fractional itself that excludes the irrational, it cannot be preferred to the non-integer that does include the irrational and the non-integer rational.

Beyond the originality and the interest of the book's contents that the thematic nature confers, the originality of the contents' structuration results from the author's will to offer readings of the book at several speeds.

To that effect, the book is structured so as to offer to the reader several reading levels corresponding to an increasing level of complexity and/or of specialization. The first three chapters have indeed a more general character whereas the next three chapters are more specialized, notably Chapters 4 and 5 that respectively deal with the CRONE suspension and the CRONE control (CRONE being the French abbreviation of *Commande Robuste d'Ordre Non Entier*, in English, *non-integer order robust control*).

Thus, the reader who only requires a consciousness-raising to the non-integer approach and to the most striking results on the subject can limit his reading to Chapter 1. The specificity of this chapter is indeed its approach, which is both very general to constitute an overview on the non-integer approach and very targeted to enable a simplified presentation, even an educational example, liable to quickly sensibilize the reader to the thematic interest and thus to give him the desire to go further in his reading.

The reader, who is concerned about the proofs and notably about a deepening of the proofs briefly led in Chapter 1, can partially or fully read the chapters and appendices cited in Chapter 1 with a link enabling then to directly find the expected supplement.

Finally, the reader eager to benefit from the subtleties acquired by the author in 40 years of sustained researche in the non-integer field, is invited to browse the whole book, including the Appendices, which constitute mini-chapter, complementary at the physics level as well as the mathematical level.

Chapter 1 largely contributes to the originality of the book. It indeed results from the magisterial conference presented in the framework of the *Grand Prix Lazare Carnot 2011* of the French Academy of Sciences. On the form, given the various disciplines within this academy, our contribution is a simplified presentation at the limit of scientific popularization. On the content, through a *structured approach of diversity*, this chapter unquestionably offers a framework, on the one hand, to the introduction of non-integer derivative as a modeling tool, on the other hand, to the use of such a modeling form to put into light dynamic performances (and notably of damping) unsuspected in an integer approach of mechanics and automatic control. The non-integer approach indeed enables us to overcome the mass-damping dilemma in mechanics and consequently the stability-precision dilemma in automatic control. Furthermore, the metallic and hydropneumatic versions of the CRONE suspension are the subject of a unified presentation through various forms of diversity leading to non-integer differentiation. To illustrate our strategy through a study system, we have studied the relaxation of water on a porous dyke. Such as obtained, the model of the water-dyke interface and the one governing the water relaxation are non-integer models of order between 0 and 1 for the interface and of order between 1 and 2 for the relaxation, these models being valid in a *medium frequency range*.

Chapter 2 proves and validates the *damping robustness* of the water relaxation on a porous dyke, the various study stages going from the object to its performances and to their experimental verification. The first stage (section 2.2) consists of starting off with a recursive parallel arrangement of serial RC cells to obtain a non-integer differentiation as a model of the water-dyke interface. Obtaining a differentiation non-integer order results from a smoothing of the admittance Bode asymptotic diagrams. The second stage (section 2.3) consists of using the dynamics fundamental principle to change from the non-integer derivative so obtained to a non-integer differential equation as a model governing water relaxation. After the analytical determination of the relaxation, which shows damping robustness in the time domain and constitutes the third stage (section 2.4), the fourth stage (section 2.5) illustrates damping robustness in *operational domain* and in the *frequency domain.* Lastly, the fifth stage (section 2.6) consists of experimentally verifying damping robustness through an electronic circuit made of operational amplifiers. This circuit is achieved in such a way that its transmittance respects the non-integer differential equation, which governs the water relaxation on a porous dyke.

Chapter 3 deals with non-integer differentiation, its memory and its synthesis. Section 3.2 presents, *in discrete time*, non-integer differentiation through the extension, to the non-integer case, of the generic form of integer differentiation. Section 3.3 presents, *in continuous time*, non-integer differentiation from repeated integer integration via non-integer integration. Thus presented in discrete time then in continuous time, non-integer differentiation is the subject, in section 3.4, of a study of its properties in a *sinusoidal steady state.* In this operating state and as regards kinematic magnitudes, therefore in terms of position, speed and acceleration, the non-integer derivative of position takes into account "position and speed" or "speed and acceleration" whether the differentiation order n is "between 0 and 1" or "between 1 and 2". Section 3.5 is devoted to the memory phenomenon associated with non-integer differentiation. The discrete form of the non-integer derivative as presented in section 3.2, directly shows that the function to be differentiated is taken into account through its values at all the past instants. Non-integer differentiation thus introduces a *memory notion* such that the past *attenuation* or *accentuation* is imposed by the differentiation order. As this memory notion is not without evoking a *subtle form of memory*, an investigation trail is proposed by considering an aspect of the human memory studied in cognitive psychology. Section 3.6 deals with the synthesis of non-integer differentiation through a *realistic synthesis* turning on a *non-integer differentiator bounded in frequency*. This synthesis is an *approximation* of the differentiator ideal version through a *(limited) recursive distribution of countable zeros and poles*. If the real non-integer differentiator synthesis is based on a recursive distribution of countable real zeros and poles, the complex non-integer differentiator synthesis is based on a recursive distribution of complex zeros and poles, which can also be counted [OUS 00].

Chapter 4 deals with an application of non-integer differentiation in the (automotive) vehicle suspension area, namely the CRONE suspension, CRONE being in this case the French acronym of *Comportement Robuste d'Ordre Non Entier*. The various synthesis stages, from concept to practical achievement, are developed and performance tests turning on prototypes validate the theoretical expectations. The CRONE suspension is presented through two versions in conformity with the hydropneumatic and metallic technologies used for its achievement: a *hydropneumatic version* that results from the transposition, in vibratory insulation, of the porous dyke interpretation as defined in Chapter 1 and Appendix 2; a *metallic version* that is obtained from the usual (or traditional) suspension by replacing the order 1 traditional dashpot with a *non-integer order dashpot*. As regards stability degree, robustness tests are carried out for different parametric states of the usual and CRONE suspensions. These tests turn successively on the frequency and step responses of the two suspensions and also the roots of their characteristic equations. For the CRONE suspension, they well reveal the robustness of the *resonance ratio* of the frequency response, the *first overshoot* of the step response and the *damping ratio* stemming from the characteristic equation roots.

Chapter 5 shows that the transposition of damping robustness (Chapter 2) in automatic control is at the origin of the initial approach of the CRONE control. Indeed, its first principles stem from the interpretation of the model of water relaxation on a porous dyke in conformity with two legitimate interpretations, which, respectively, define the first generation CRONE control through a *controller phase locking* and the second generation CRONE control through an *open loop phase locking*. The change from second to third generation CRONE control results from the generalization of the vertical template in conformity with two generalization levels: in the first generalization level, the vertical template is replaced by a straight line segment of any direction, called a generalized template, the direction of which is given by that of the main axis of the uncertainty domain calculated at the open loop unit gain frequency or the closed loop resonance frequency in tracking; in the second generalization level, the generalized template is replaced by a set of generalized templates, called multi-templates, which leads to a curvilinear template generalizing the rectilinear template formed by the generalized template, the tangent at each point of the curvilinear template being the main axis of the corresponding uncertainty domain.

Chapter 6 studies the close link between recursivity and non-integer differentiation through several recursive electric networks. Thus, section 6.2 presents an exhaustive study of an indefinite recursive parallel arrangement of series RC cells. Section 6.3 is devoted to the study of a recursive arborescent arrangement of gamma RLC cells, the arborescence being obtained according to a *bifurcation iterative process* in conformity with the lung intern structure seen under the respiratory angle. Led in such a way, this study presents the advantage reducing the arrangement so obtained to a *recursive cascade arrangement of gamma RLC cells*. Section 6.4 proposes a unified study of indefinite recursive parallel arrangements of RL, RC and RLC cells. The originality of this study results from methods that are said to be heuristic as they are founded on *heuristic assumptions* that, in this case, enable us to quickly obtain exact results of remarkable simplicity. Section 6.5 deals with eight recursive arrangements of RC and RL cells obtained from four possible combinations of a resistance and a capacitance and from four possible combinations of a resistance and an inductance. It is about the two arrangements stemming from a recursive parallel arrangement of series RC or RL cells, the two arrangements stemming from a recursive series arrangement of parallel RC or RL cells, the two arrangements stemming from a recursive cascade arrangement of gamma RC or RL cells, and also the two arrangements stemming from a recursive cascade arrangement of gamma CR or LR cells. Section 6.6 shows how to reparameterize the cells to define a unit gain frequency of the indefinite recursive parallel arrangements of series RC and RL cells. Section 6.7 deals with the energy stored by a non-integer differentiator then by a non-integer integrator. Two dual energetic approaches are led through an indefinite recursive parallel arrangement of series RC cells for the non-integer differentiator and an indefinite recursive series arrangement of parallel RL cells for the non-integer integrator.

Chapter 1

From Diversity to Unexpected Dynamic Performances

In the context, this chapter takes inspiration from the magisterial conference presented in the framework of the *Grand Prix Lazare Carnot 2011* of the French Academy of Sciences. In its form, given the various disciplines within this academy, our contribution is a simplified presentation, almost in layman's terms, at the limit of scientific popularization. The content, through a structured approach of diversity, our main contribution is putting the wrong mass-damping dilemma into mechanics and stability-precision dilemma into automatic control.

1.1. Introduction

This chapter offers a framework to the introduction and the use of the *non-integer* derivative.

On the one hand, this framework allows us to introduce the non-integer derivative as a modeling tool, particularly in the modeling of a porous face and the water relaxation on such a face. On the other hand, from such a *non-integer* modeling, we are able to solve a physics problem which is without solution so far, which is in this case the mass-damping dilemma in mechanics, to which the stability-precision dilemma in automatic control corresponds.

It turns out that this framework finds its essence and its adequation in a *structured approach of diversity*. Without a precedent on the matter, this approach that is inspired by various natural forms of diversity (biological among others) leads, without high mathematical developments in this chapter, to *dynamical performances (and notably damping) unexpected* in an “integer” approach of mechanics and automatic control.

In concrete terms and particularly technologically, there exist as many technological solutions as diversity forms. Therefore, a declension of these forms is proposed here, each of them being the subject of a structural definition, an illustration through a biological example and, eventually, a transposition to vibratory isolation through various technological solutions in terms of *robust suspension* (damping independent of mass):

– A first diversity form is *a multitude of different elements of same action level.* This is the most perceptible elementary form.

It is the case of an element arrangement that is borrowed from *striated muscle fibrous structure or porous face alveolar structure* (whose pores are of different sizes).

It is particularly the case of the *hydropneumatic version* of the CRONE suspension that uses different springs and different dashpots, this version having been implanted on “Citroën” vehicles, the BX (in 1995), the XM (in 2000) and the C5 (in 2010).

– A second diversity form is *a multitude of identical or similar elements of different action level.*

It is the case of an element arrangement that is borrowed from the meshed structure of similar neuron classes (cerebellum or cerebral cortex and motor neurons of the spinal cord). If the neurons are similar, they occupy different positions in the space.

– A third diversity form is *a multitude of different elements of different action level.* This form defines the highest diversity degree.

It is the case of an element arrangement that is borrowed from lung arborescent structure. The deeper the layers, the smaller the branches:

- and the greater their resistance as the restriction effect increases;

- and the smaller their capacitance as the tank effect decreases.

– Last but not least, a fourth diversity form, of different nature, is *a unique element continuously variable* that presents *a multitude of different values* on a time horizon.

It is the case of a variable device that is borrowed from an aorta transverse movement.

It is particularly the case of the controlled dashpot that presents a continuously variable oil laminating section and that equips the metallic version of the CRONE suspension, this version having been implemented on a Peugeot vehicle, the 406 (in 1998).

1.2. An issue raising a technological bottle-neck

A problem that is well known to mechanical engineers and test pilots (car, plane, etc.) consists of *a stability degree that decreases when carried mass increases*. In other words, the higher the car load, the greater the oscillation through a damping decrease.

This problem, still unsolved by the mechanics integer approach, defines a genuine *technological bottleneck*, which is called "*mass-damping interdependence*" or, more precisely, "*mass-damping dilemma*", so far an unavoidable dilemma, which well expresses that *an increase of mass is accompanied by a damping decrease.*

The Citroën 2 CV in civilian domain and the Rafale in military domain are particularly concerned by this phenomenon: the 2 CV, as it is little damped and very light, 570 kg without a load; the Rafale, as its weight varies from 9.5 tons without a load to 24 tons in full load through kerosene and arming.

As an excellent illustration, let us consider a suspension of type mass-spring-dashpot (Figure 1.1) and the bodywork response to a step displacement of the wheel for different masses (Figure 1.2).

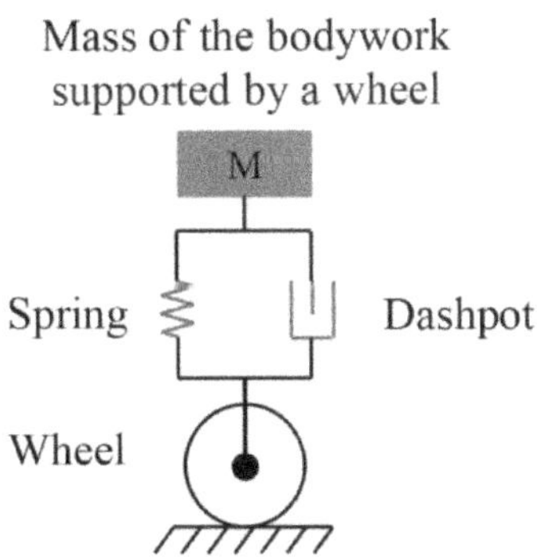

Figure 1.1. *Suspension of type mass-spring-dashpot*

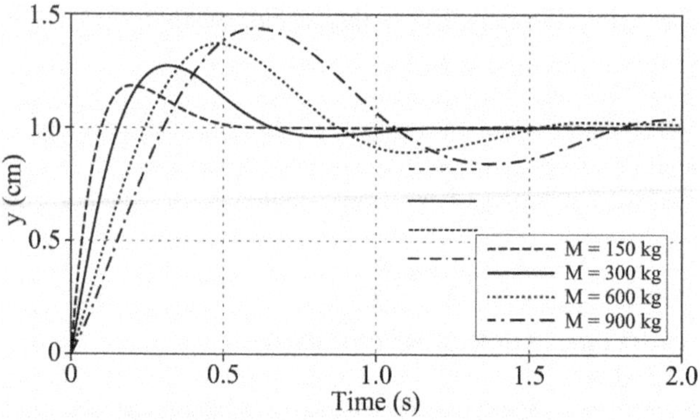

Figure 1.2. *Bodywork response*

In terms of performances, the *damping ratio*, which measures *damping* by *the decrease rate of the successive overshoots*, admits an expression of the form (Appendix 1):

$$\zeta = \frac{K_0}{\sqrt{M}}, \qquad [1.1]$$

which shows well (with the presence of the mass at the denominator) that *the damping ratio decreases when the mass increases.*

1.3. An aim liable to answer to the issue

How to answer to such an issue by an unprecedented approach in the matter? It can be answered by seeking a *structural damping* determined by the *structure* of a system independently of the system *parameters* (notably of the *mass* in our study context).

To better target this aim, we can already assume that a *complex structure* (like the one of a *high-dimensional multi-branch network* (Figure 1.3)) is, through its number of components, *liable to reduce the effect of each of the components (and then of the associated parameters).*

It is true that the higher the number of components, the less the action of each of them on the global behavior. We will indeed see, later, that *if multiplicity is not always sufficient, it is nevertheless always necessary.*

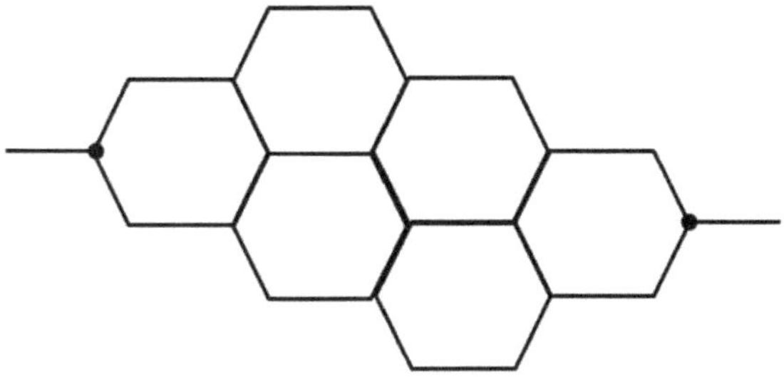

Figure 1.3. *High dimension multi-branch metwork*

1.4. A strategy idea liable to reach the aim

How to reach the aim of a damping exclusively linked to the structure?

This aim can be reached by seeking a *structure* founded on the *idea of diversity* compatible with the one of a *complex structure.*

1.4.1. *Why diversity?*

The reason for diversity is that this strategy is inspired by the interest of *disorder* taken in the sense of a natural or artificial disorder that borrows from the diversity.

Among the sources on the disorder interest, the following are of importance:

– *the damping nature of highly disturbed media,* observed by the dyke builders in the 17th Century through experimental studies of water relaxation on costal or fluvial dykes;

– *the stabilizing nature of disorder*, noticed by Prigogine, winner of the Nobel Prize in Chemistry in 1977, through his researches on instability and chaos in thermodynamics;

– *the anarchic distribution of regular blocks to artificially recreate disorder*, personally observed on the Socoa jetty in the 1980s, which has enabled us to measure human's will to recreate, according to his means, disorder.

1.4.2. *What does diversity imply?*

Through multiple different elements, *diversity* associates *multiplicity* and *difference*.

To avoid any extrapolation, we insist on specifying that the considered elements are (Figure 1.4) geometrical elements (or patterns) or system elements (or components).

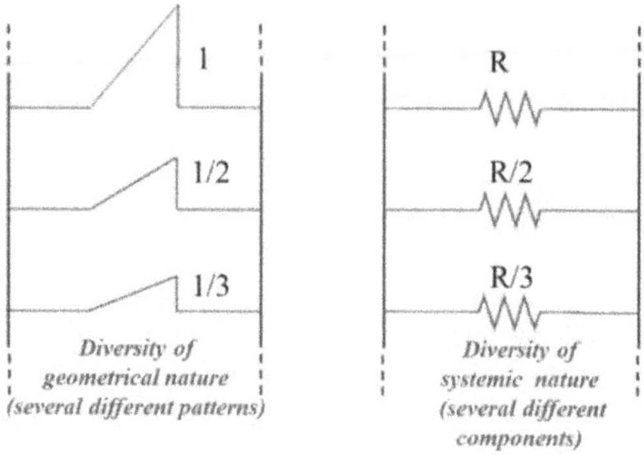

Figure 1.4. *Different paterns and different components*

1.5. On the strategy itself

1.5.1. *The study object*

To support the proceedings of our strategy, we have chosen a natural study system, though stylized, as for the representation. This study object (or system) is certainly no stranger to the damping properties of the *highly disturbed (or uneven) dykes*, notably *those forming air pockets compressible by water advance.*

This leads us to say that the study object (Figure 1.5) is *motion water* of mass *M* which (horizontally) relaxes on a *porous dyke* whose *pore localization is here reduced to the face* (Appendix 2).

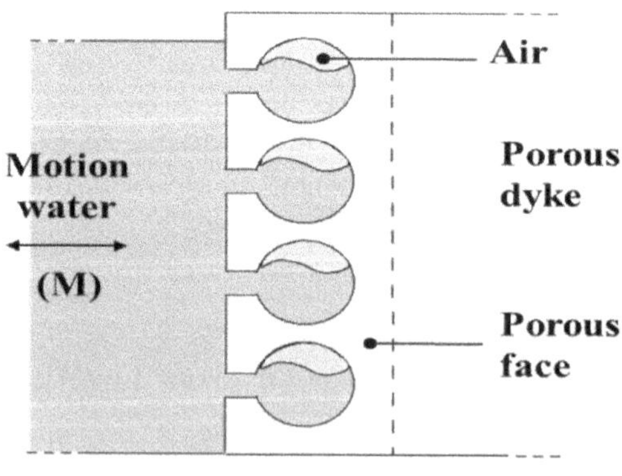

Figure 1.5. *Study object*

It is then a *porous face* replying, besides, to multiplicity by its pore number that achieves the *water-dyke interface.*

No assumption is made on the pores as for their identity or difference. They will be particularized afterwards.

1.5.2. *A pore: its model and its technological equivalent*

1.5.2.1. *The model*

Through its alveolar nature, a pore presents an *orifice* and a *cavity* (Figure 1.6):

– the *orifice*, through the restriction it offers at the water flow, serves as hydraulic resistance or energy dissipater or *dashpot*;

– the *cavity*, through the compression of air it imprisons, serves as pneumatic capacitance or energy tank or *spring*.

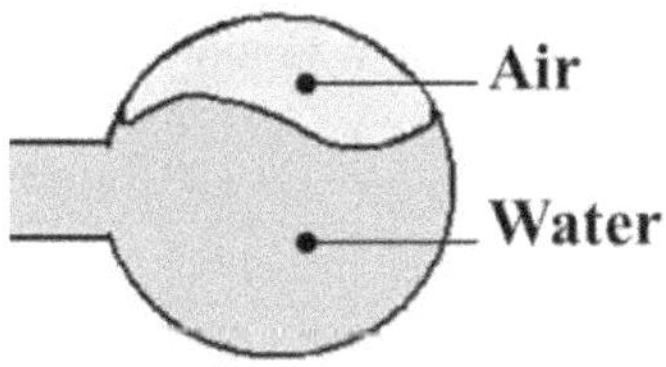

Figure 1.6. *A pore*

It is then a *resistance-capacitance cell* that represents the pore hydropneumatic model (Figure 1.7).

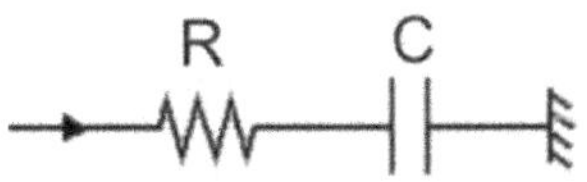

Figure 1.7. *Hydropneumatic model of a pore*

1.5.2.2. *The technological equivalent*

In the automobile industry, the technological equivalent of a pore (as configured in Figure 1.6) is unquestionably the *Citroën suspension sphere* (Figure 1.8) in which *nitrogen replaces air (for higher stability)* and *oil replaces water (for higher viscosity and stability).*

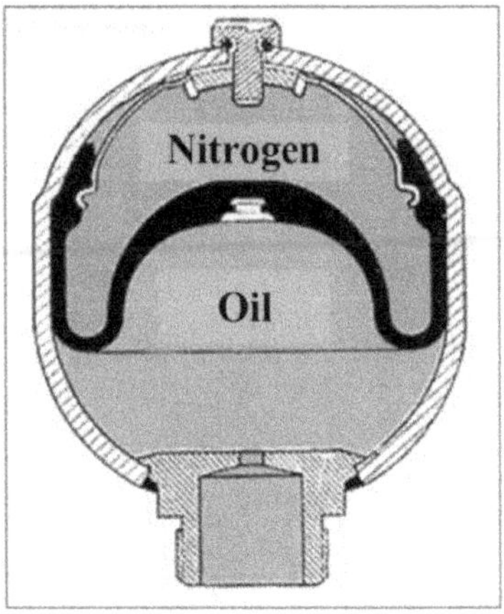

Figure 1.8. *Suspension sphere*

Of *mass-spring-dashpot type* like the suspension metallic version (therefore of the *same structure*), the hydropneumatic version presents *a damping ratio of the same form*, namely:

$$\zeta = \frac{K_1}{\sqrt{M}}, \qquad [1.2]$$

the mass appearing in the same manner at the denominator.

1.5.3. *Case of identical pores*

A first step consists of considering identical pores, the aim being to study the *structuring effect* of *multiplicity*.

As all the pores are identical, each of them absorbs the same flow (Figure 1.9).

Given the *additivity of flows* absorbed by each of the pores:

– a set of N pores (globally) absorbs a flow N times greater than the one absorbed by one pore;

– which conveys that the "*resistance*" that this set of N pores provides to the water motion *is N times lesser than a pore one*.

This leads us to say that *the passage from one pore to N pores* shifts the hydropneumatic model from a resistance-capacitance cell (Figure 1.10(a)) to a new resistance-capacitance cell (Figure 1.10(b)) whose *resistance is N times lesser than a pore one, namely R/N,* and whose *capacitance is N times greater than a pore one, namely NC.*

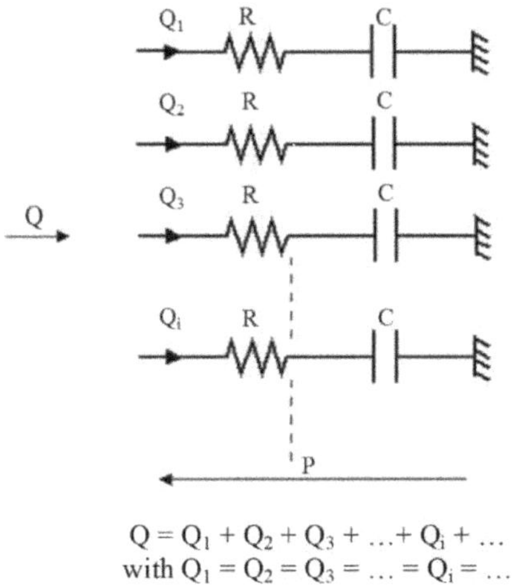

Figure 1.9. *Hydropneumatic model for identical pores*

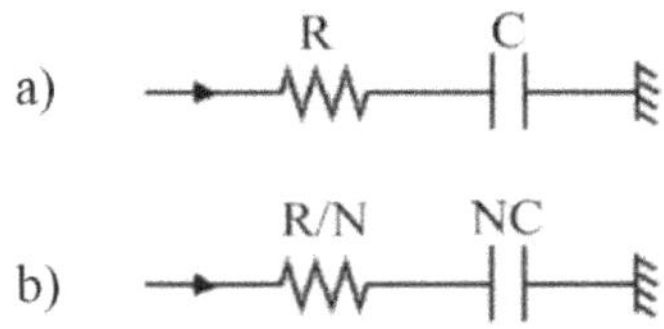

Figure 1.10. *Passage from one pore to Npores*

Actually, as the new model keeps the *same structure* (through a *unique reparameterization*), *no new structure is induced by the only pore multiplicity,* so well that *the mass appears again in the same manner in the new damping ratio expression* obtained *by replacing R with R/N and C with NC*, namely:

$$\zeta = \frac{K_2}{\sqrt{M}}. \qquad [1.3]$$

1.5.4. *Case of different pores*

A second step consists of considering the different pores, the aim being to study the *structuring effect* of *difference* associated with *multiplicity.*

As all the pores are different, the porous face answering now both to multiplicity and difference, a preamble is necessary to express as it suits the difference, this preamble concerning somewhat singular differences.

1.5.4.1. *On differences coming from regional heritage*

1.5.4.1.1. Differences of technological origin

Bordeaux's bottle and glass capacities present (Figure 1.11):

– a ratio of 2 between two consecutive bottles;

– a ratio of 1.5 between two consecutive glasses;

– ratios which, by construction, are the same independently of the considered consecutive bottles and glasses.

Figure 1.11. *Bottles and glasses*

1.5.4.1.2. A difference of natural origin

The bubbles grown into a Sauternes bottle reveal (Figure 1.12):

– a ratio of 1.6 between the diameters of two consecutive bubbles;

– a ratio that, up to measurement errors, is the same for whatever consecutive bubbles taken into account.

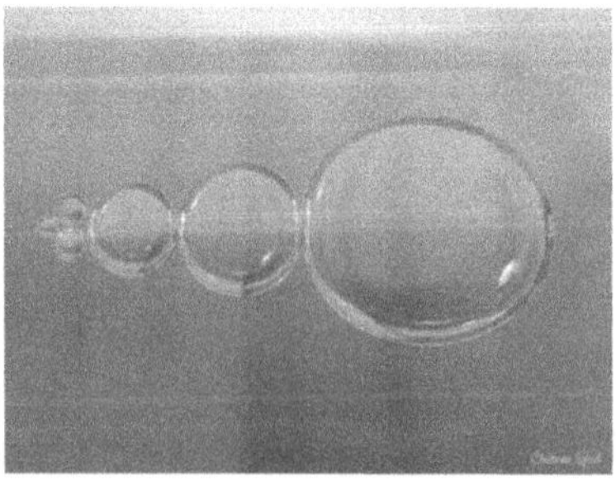

Figure 1.12. *Bubbles inside a bottle*

1.5.4.1.3. How is difference expressed?

In each of the cases, the *difference from one element to another* is expressed by a *constant ratio*: 2 for bottles, 1.5 for glasses and1.6 for bubbles.

Thus, such a *difference expression by a constant ratio* is the one kept to *quantify* (or *measure*) the difference (or more precisely *the same difference from one element to another*).

It is true that a *constant ratio,* equivalent to the *ratio* of a geometric progression or the *scaling factor* of a fractal representation, is liable to ensure, on one hand, *a regular variation of the elements* and, on the other hand, *a large range of elements* (ranging from the highest to the smallest or vice versa).

1.5.4.2. *Transposition to the study object*

Thus expressed, the difference is introduced at the porous face level (Figure 1.13):

– by differentiating the pores (through their orifices and their cavities);

– so that *the resistances and the capacitances that characterize them present* (between two consecutive pores);

- *a constant ratio* α *between the resistances,*

- *and a constant ratio* η *between the capacitances,*

– α and η being greater than a unit and called *recursive factors* in our works (knowing that recursivity can be defined as a recurrent application, which could be infinitely repeated, and characterized by a recurrence relation independent of rank).

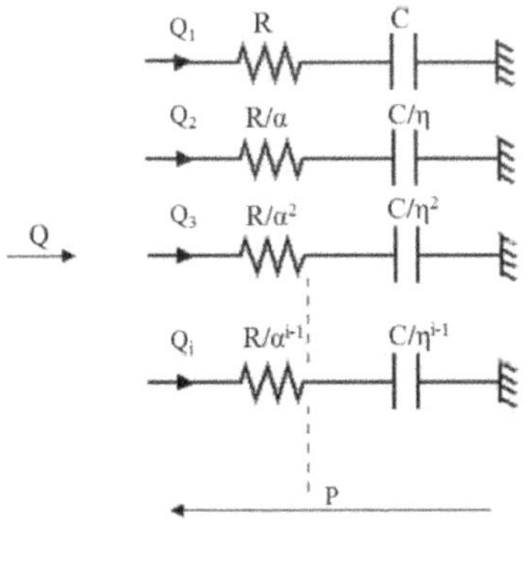

$Q = Q_1 + Q_2 + Q_3 + \ldots + Q_i + \ldots$
with $Q_1 \neq Q_2 \neq Q_3 \neq \ldots \neq Q_i \neq \ldots$

Figure 1.13. *Hydropneumatic model for different pores*

1.6. From physics to mathematics

1.6.1. *An unusual model of the porous face*

The porous face presents (Chapter 2) an admittance, $Y(j\omega)$, whose Bode asymptotic diagrams form (Figure 1.14):

– for the gain, a regular sequence of identical stair steps;

– and for the phase, a regular sequence of identical crenels.

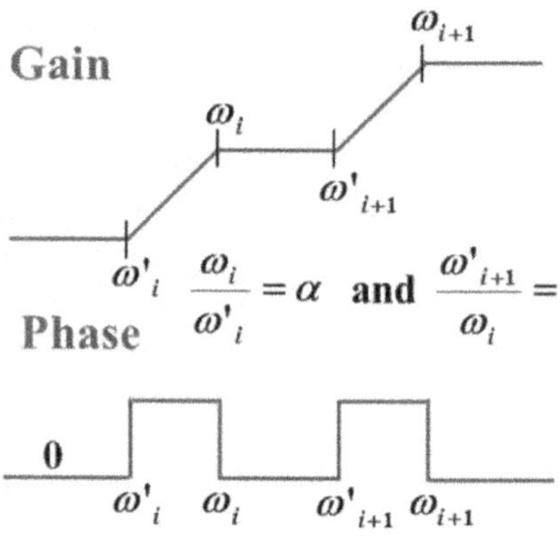

Figure 1.14. *Bode asymptotic diagrams*

The constant ratio α between the resistances is found again between the extreme transitional frequencies of a ramp. The constant ratio η between the capacitances is found again between the extreme transitional frequencies of a stage.

But how to exploit these diagrams knowing that *the transitional frequencies are too close for the Bode diagrams to approach the asymptotes?*

The intuitive idea to say the least is that the Bode diagrams can only take the *average* of the asymptotic diagrams (Figure 1.15), thus expressing the *idea of a smoothing of the steps for the gain and the crenels for the phase.*

The smoothing of the steps that constitute the gain asymptotic diagram can be made possible by a straight line, called *gain smoothing straight line*, of a slope less than 20 dB/dec, namely 20 m dB/dec *with m between 0 and 1.*

The smoothing of the crenels that constitute the phase asymptotic diagram can be made possible by a straight line, called *phase smoothing straight line*, of ordinate less than $\pi/2$, namely $m\pi/2$ *with* m *between 0 and 1*, equal to the phase asymptotic variation average.

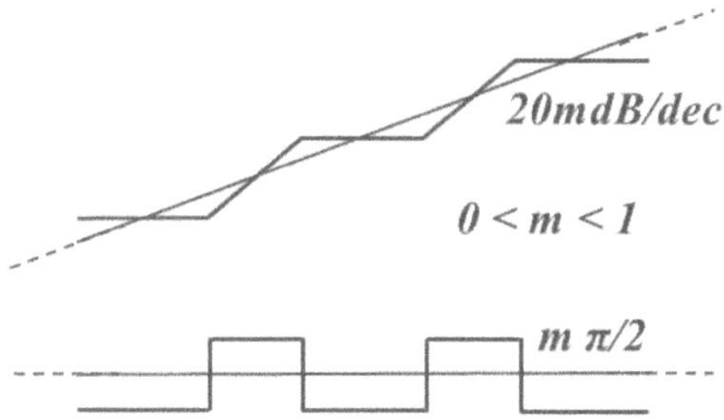

Figure 1.15. *Smoothing of the Bode asymptotic diagrams*

1.6.1.1. *A smoothing remarkable of simplicity: the one of crenels*

The uniform spread on the cyclic interval $[\omega'_i, \omega'_{i+1}]$ of the area under a crenel (Figure 1.16) is expressed by the identity of the hachured surfaces:

$$m\frac{\pi}{2}(\log\alpha + \log\eta) = \frac{\pi}{2}\log\alpha, \qquad [1.4]$$

from which m is directly deduced, namely:

$$m = \frac{\log\alpha}{\log(\alpha\eta)}. \qquad [1.5]$$

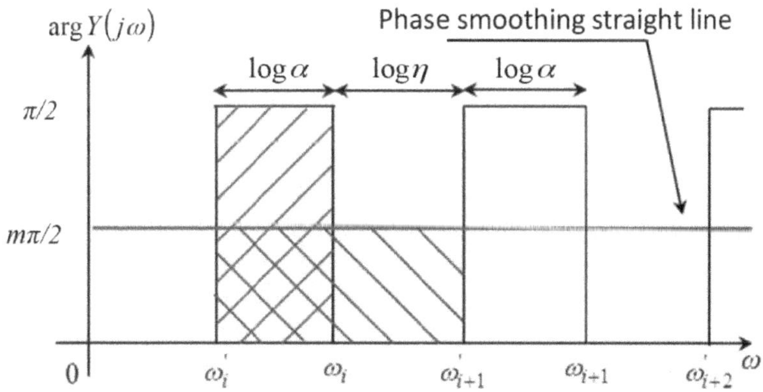

Figure 1.16. *Smoothing of the crenels*

1.6.1.2. *A non-integer derivative as a smoothing result*

The admittance in $j\omega$ that defines the smoothing admits an expression of the form:

$$Y(j\omega)=\left(j\frac{\omega}{\omega_0}\right)^m, \text{ namely} \frac{Q(j\omega)}{P(j\omega)}=\left(j\frac{\omega}{\omega_0}\right)^m, \qquad [1.6]$$

from which one draws the *symbolic equation*:

$$Q(j\omega)=\frac{1}{\omega_0^m}(j\omega)^m P(j\omega), \qquad [1.7]$$

which, in time domain, admits *the concrete (or original) equation*:

$$Q(t)=\frac{1}{\omega_0^m}\left(\frac{d}{dt}\right)^m P(t), \qquad [1.8]$$

a relation that expresses that *non-integer derivative* models the porous face.

The model parametric *compacity* (or *parsimony*) is proven by two parameters, the *differentiation non-integer order m* and the *unit gain frequency* ω_0.

1.6.1.3. *An original heuristic verification of differentiation non-integer order*

The admittance in s of an indefinite recursive parallel arrangement of series RC cells (section 6.4.2), namely

$$Y(s)=\sum_{i\in\mathbb{Z}}\frac{\alpha^i}{R}\frac{s}{s+\lambda^i\omega_0} \text{ with } \lambda=\alpha\eta \text{ and } \omega_0=\frac{1}{RC}, \qquad [1.9]$$

becomes, for $i=j+1$:

$$Y(s) = \sum_{j\in\mathbb{Z}}\frac{\alpha^{j+1}}{R}\frac{s}{s+\lambda^{j+1}\omega_0}$$

$$= \alpha\sum_{j\in\mathbb{Z}}\frac{\alpha^j}{R}\frac{s}{s+\lambda^j\lambda\omega_0} \qquad [1.10]$$

$$= \alpha \sum_{j \in \mathbb{Z}} \frac{\alpha^j}{R} \frac{\lambda^{-1} s}{\lambda^{-1} s + \lambda^j \omega_0},$$

namely, from $Y(s)$ expressed by the sum on i:

$$Y(s) = \alpha Y\left(\lambda^{-1} s\right). \qquad [1.11]$$

Ks^m being relative to the smoothing and $X(s)$ being relative to the gain and phase undulations around the smoothing straight lines ($X(s)$ tending toward 1 when λ tends toward 1), we attribute to the admittance $Y(s)$ a general expression of the form:

$$Y(s) = Ks^m X(s). \qquad [1.12]$$

This *heuristic hypothesis* is put into the general relation that $Y(s)$ verifies, namely

$$Y(s) = \alpha Y\left(\lambda^{-1} s\right). \qquad [1.13]$$

Then we obtain:

$$Ks^m X(s) = \alpha K\left(\lambda^{-1} s\right)^m X\left(\lambda^{-1} s\right), \qquad [1.14]$$

or:

$$X(s) = \alpha \lambda^{-m} X\left(\lambda^{-1} s\right), \qquad [1.15]$$

or even, in frequency domain:

$$X(j\omega) = \alpha \lambda^{-m} X\left(j \lambda^{-1} \omega\right), \qquad [1.16]$$

or also, knowing that the period $\log \lambda$ of the gain and phase undulations in a semi-logarithmic representation is translated by

$$X(j\omega) = X\left(j\lambda^{-1}\omega\right):$$ [1.17]

$$\alpha\lambda^{-m} = 1,$$ [1.18]

namely:

$$\log\alpha - m\log\lambda = 0,$$ [1.19]

from which we can immediately derive:

$$m = \frac{\log\alpha}{\log\lambda},$$ [1.20]

a result that well verifies the expression of the non-integer order m arising from the smoothing operation (section 1.6.1.1).

The exactitude and the obtaining simplicity of this result, despite the ignorance of K and $X(s)$ (except the periodicity of $X(s)$), prove the significance and originality of this heuristic method.

1.6.2. *A just as unusual model governing water relaxation*

From Newton's second law, we can write the differential equation (Figure 1.17):

$$M\frac{dV(t)}{dt} + F(t) = 0,$$ [1.21]

where $V(t)$ is the speed of the water mass M and $F(t)$ is the dyke reaction force.

If S is the flow section,

$$V(t) = Q(t)/S \text{ and } F(t) = P(t)S,$$ [1.22]

hence, the new differential equation with two variables $Q(t)$ and $P(t)$:

$$\frac{M}{S^2}\frac{dQ(t)}{dt} + P(t) = 0.$$ [1.23]

As

$$Q(t) = \frac{1}{\omega_0^m} \left(\frac{d}{dt} \right)^m P(t), \qquad [1.24]$$

this differential equation becomes a differential equation of *non-integer order* $n = 1 + m$ *between* 1 *and* 2, namely:

$$\tau^n \left(\frac{d}{dt} \right)^n P(t) + P(t) = 0, \text{ with } \tau = \left(\frac{M}{S^2} \frac{1}{\omega_0^m} \right)^{\frac{1}{n}}. \qquad [1.25]$$

Two parameters still prove the model parametric *compacity* (or *parsimony)*: the *differentiation non-integer order*, $n = 1 + m$, and the *transitional time constant*, $\tau = \tau(M)$.

The non-integer approach thus enables us to replace an infinite integer order model with a finite non-integer order model (this concerning both the porous face and the study object).

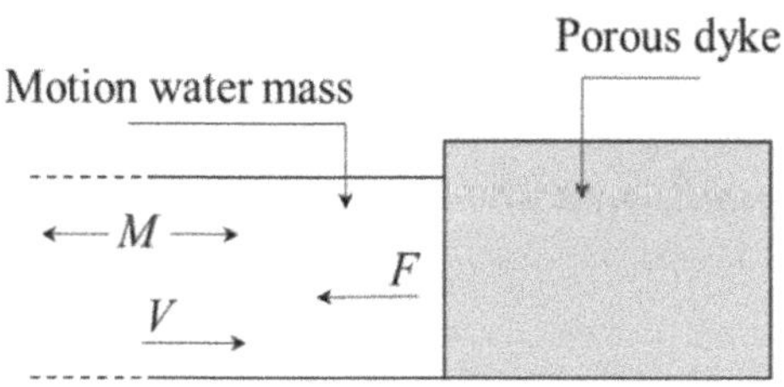

Figure 1.17. *Study system*

1.6.3. *What about a non-integer derivative which singles out these unusual models?*

Introduced as a modeling tool (for the porous face and for the study system), *a non-integer derivative* (or *non-integer differentiation operator*, $D^n = (d/dt)^n$), which owes its singularity to the *non-integer order* n:

– is an operator resulting from the will to ensure the differentiation order *continuity*;

– whose *idea and first definitions date back to the beginning of the 19th Century*;

– but whose genuine applications (in the sense of the real time) only date back to the second half of the 20th Century.

It is true that we drew this operator from mathematician drawers by proposing, from the 1970s, a *frequency synthesis*, or more precisely a *synthesis in a medium frequency range* (section 3.6), allowing *the real-time use of this operator* for:

– *half-integer orders*, from the 1970s;

– *real non-integer orders*, from the 1980s;

– *complex non-integer* orders, from the 1990s.

1.6.3.1. *On the sinusoidal state of the operator of order* $n \in [0,2]$

In *sinusoidal steady state* and in terms of *kinematic magnitudes*, therefore in terms of *position, speed* and *acceleration*, the non-integer derivative of position takes into account the *position and speed* or *speed and acceleration* whether the differentiation order n is between 0 and 1 *or* 1 and 2 (section 3.4).

1.6.3.1.1. $0 \le n \le 1$

The derivative of order $n \in [0,1]$ of the position $x(t)$ determines a "*hybrid magnitude*", which is *neither a position nor a speed* (except for $n=0$ and $n=1$), but which borrows from *both the position* $x(t)$ *and the speed* $v(t)$ according to a *linear combination* of the form:

$$\underset{n\in[0,1]}{x^{(n)}(t)} = \omega^n \left(\cos n\frac{\pi}{2} \right) x(t) + \omega^{n-1} \left(\sin n\frac{\pi}{2} \right) v(t). \qquad [1.26]$$

1.6.3.1.2. $1 \le n \le 2$

The derivative of order $n \in [1,2]$ of the position $x(t)$ determines a "*hybrid magnitude*", which is *neither a speed nor an acceleration* (except for $n=1$ and $n=2$), but which borrows from *both the speed* $v(t)$ *and the acceleration* $\gamma(t)$ according to a *linear combination* of the form:

$$\underset{n\in[1,2]}{x^{(n)}(t)} = \omega^{n-1} \left(\cos(n-1)\frac{\pi}{2} \right) v(t) + \omega^{n-2} \left(\sin(n-1)\frac{\pi}{2} \right) \gamma(t). \qquad [1.27]$$

1.6.3.2. *On the impulse state of the operator of order* $n \in \,]0,1[$

Knowing that an indefinite recursive parallel arrangement of series RC cells achieves the admittance Ks^n up to gain and phase undulations (section 1.6.1.3, where $n=m$), it is easy to exploit the admittance $Y(s)$ of such an arrangement (section 6.4.2) by rewriting it in the form:

$$Y(s)=\sum_{i\in\bullet}\frac{\alpha^i}{R}\left[1-\frac{\lambda^i\omega_0}{s+\lambda^i\omega_0}\right] \text{ with } \lambda=\alpha\eta \text{ and } \omega_0=\frac{1}{RC}, \qquad [1.28]$$

and then by expressing the corresponding original that defines the *impulse response*:

$$y(t)=\sum_{i\in\bullet}a_i\delta(t)-\sum_{i\in\bullet}b_ie^{-\omega_it}u(t), \qquad [1.29]$$

$$\text{with } a_i=\frac{\alpha^i}{R},\ b_i=(\alpha\lambda)^i\frac{\omega_0}{R} \text{ and } \omega_i=\lambda^i\omega_0, \qquad [1.30]$$

where $\delta(t)$ and $u(t)$ denote the unit impulse and step functions.

This response is split into two distinct sums:

– *an indefinite recursive series of Dirac impulses at instant* $t=0$, *of recursive factor* $a_{i+1}/a_i=a$;

– *an indefinite recursive series of decreasing exponentials for* $t>0$, *of recursive factor along the amplitude,* $b_{i+1}/b_i=\alpha\lambda$, *and along the time,* $\omega_{i+1}/\omega_i=\lambda$.

The *relaxation* defined by the impulse response for $t>0$ therefore results from an infinity of *first-order stable aperiodic modes* (of time constants $\tau_i=1/\omega_i$ between zero and infinity) among which are:

– *infinitely fast modes* (τ_i tending toward zero);

– and *infinitely slow modes* (τ_i tending toward infinity), which give a *long memory* character to the operator D^n;

– thus giving this operator an *unquestionable interest* in the domain of general physics:

- its *indimensional nature* by taking into account an *infinity of dynamics*;

- and its parametric *compacity* (or *parsimony*) through the *differentiation non-integer order*, which enables us, on its own, to take into account the whole of these dynamics.

The relaxation limit when λ tends toward 1 turns out to be proportional to the function (section 6.2.6)

$$\frac{t^{-n-1}}{\Gamma(-n)}, \qquad [1.31]$$

where $\Gamma(-n)$ denotes the gamma function and n denotes the differentiation non-integer order such as

$$n=\frac{\log\alpha}{\log\lambda}. \qquad [1.32]$$

This relaxation in t^{-n-1} is clearly different from the basic relaxation in $e^{-t/\tau}$, thus expressing its *paradoxical character* in the subject.

1.6.3.3. *An original heuristic verification of time non-integer power*

This section that deals with the relaxation $y(t)$ through the non-integer power γ of the time variable t is a *dual version* of section 1.6.1.3, which deals with the operational admittance $Y(s)$ through the non-integer power (or differentiation non-integer order) n of the operational variable s.

The relaxation of an indefinite recursive parallel arrangement of series RC cells, defined by its impulse response for $t>0$, namely

$$y(t)=-\sum_{i\in\mathbb{Z}}(\alpha\lambda)^i\frac{\omega_0}{R}e^{-\lambda^i\omega_0 t} \text{ with } \lambda=\alpha\eta \text{ and } \omega_0=\frac{1}{RC}, \qquad [1.33]$$

becomes, for $i=j+1$:

$$\begin{aligned} y(t) &= -\sum_{j\in\mathbb{Z}}(\alpha\lambda)^{j+1}\frac{\omega_0}{R}e^{-\lambda^{j+1}\omega_0 t} \\ &= -(\alpha\lambda)\sum_{j\in\mathbb{Z}}(\alpha\lambda)^j\frac{\omega_0}{R}e^{-\lambda^j\lambda\omega_0 t} \\ &= -(\alpha\lambda)\sum_{j\in\mathbb{Z}}(\alpha\lambda)^j\frac{\omega_0}{R}e^{-\lambda^j\omega_0(\lambda t)}, \end{aligned} \qquad [1.34]$$

namely, from $y(t)$ expressed by the sum on i:

$$y(t)=(\alpha\lambda)y(\lambda t). \qquad [1.35]$$

kt^{γ} being relative to the smoothing and $x(t)$ being relative to the gain and phase undulations around the smoothing straight lines ($x(t)$ tending toward 1 when λ tends toward 1), we attribute to the relaxation $y(t)$ a general expression of the form:

$$y(t) = kt^{\gamma} x(t). \qquad [1.36]$$

This *heuristic hypothesis* is put into the general relation that $y(t)$ verifies, namely

$$y(t) = (\alpha\lambda) y(\lambda t). \qquad [1.37]$$

Then we obtain:

$$kt^{\gamma} x(t) = (\alpha\lambda) k (\lambda t)^{\gamma} x(\lambda t), \qquad [1.38]$$

or:

$$x(t) = (\alpha\lambda) \lambda^{\gamma} x(\lambda t), \qquad [1.39]$$

or even, knowing that the period log λ of $x(t)$ in a semi-logarithmic representation is translated by

$$x(t) = x(\lambda t): \qquad [1.40]$$

$$(\alpha\lambda) \lambda^{\gamma} = 1, \qquad [1.41]$$

namely:

$$\log(\alpha\lambda) + \gamma \log \lambda = 0, \qquad [1.42]$$

from which we immediately derive:

$$\gamma = -\frac{\log(\alpha\lambda)}{\log \lambda} = -\frac{\log \alpha}{\log \lambda} - 1 = -n - 1, \qquad [1.43]$$

a result that well verifies the expression of the time non-integer power, $-n-1$, arising from the relaxation limit when λ tends toward 1 (section 1.6.3.2).

The exactitude and the obtaining simplicity of this result, despite the ignorance of k and $x(t)$ (except the periodicity of $x(t)$), prove the significance and originality of this heuristic method.

1.7. From the unusual to the unexpected

1.7.1. *Unexpected damping properties*

1.7.1.1. *Relaxation damping insensitivity to the mass*

The characteristic equation of the differential equation governing water relaxation is of the form:

$$(\tau s)^n + 1 = 0. \qquad [1.44]$$

Due to a cut along $\mathbb{R}^-$ of the complex plane (Figure 1.18), it has only two conjugate complex roots (section 2.5.1):

$$s_0 = \frac{1}{\tau} e^{j\frac{\pi}{n}} \text{ and } s_{-1} = \frac{1}{\tau} e^{-j\frac{\pi}{n}}. \qquad [1.45]$$

What about the oscillatory mode resulting from these roots?

The *natural frequency* is given by the projection on the imaginary axis of the root with positive imaginary part, namely (Appendix 1):

$$\omega_p = \frac{1}{\tau}\sin\theta = \frac{1}{\tau}\sin(\pi - \frac{\pi}{n}) = \frac{1}{\tau}\sin\frac{\pi}{n}. \qquad [1.46]$$

When the motion water mass M increases, the time constant τ increases and the natural frequency ω_p decreases, which is in conformity with the observation: indeed, a great mass relaxes slower than a little mass (think about a truck and a car).

The *damping ratio* is given by the cosine of the half-center angle formed by the two conjugate complex roots, namely (Appendix 1):

$$\zeta = \cos\theta = \cos\left(\pi - \frac{\pi}{n}\right) = -\cos\frac{\pi}{n}. \qquad [1.47]$$

This result conveys that *the damping ratio only depends on the non-integer order n of the differential equation.*

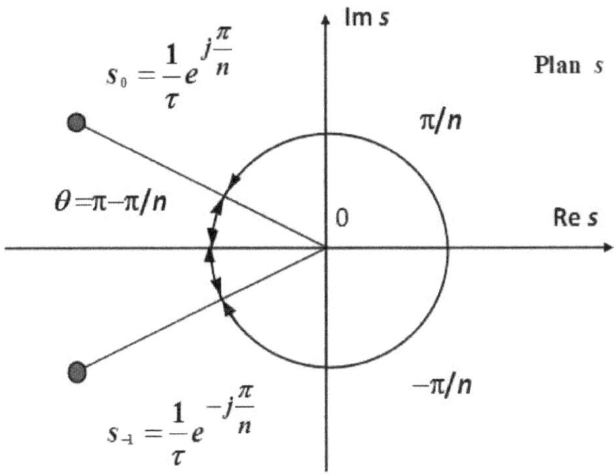

Figure 1.18. *$\theta = \pi - \pi / n$ is the half-center angle formed by the two conjugate complex roots*

Finally, knowing that $n = 1 + m$:

$$\zeta = -\cos\frac{\pi}{1+m}, \qquad [1.48]$$

where it is recalled that

$$m = \frac{\log\alpha}{\log(\alpha\eta)}. \qquad [1.49]$$

The form of this result reveals a radical change: *mass no longer appears in the expression of damping ratio.*

Moreover, through the recursive factors α and η, which express the difference from one pore to another, *the damping ratio turns out to be exclusively linked to the difference introduced in the porous face.*

To conclude, it is then *diversity that determines the damping*. This damping is *structural* because it depends only on the *structure intrinsic* ratios α and η.

1.7.1.2. *Frequency verification of the insensitivity to the mass*

The Laplace transform of the non-integer differential equation

$$\tau^n \left(\frac{d}{dt}\right)^n P(t) + P(t) = 0, \qquad [1.50]$$

allows us to write:

$$(\tau s)^n P(s) + P(s) = 0, \qquad [1.51]$$

from which we derive:

$$P(s) = -\left(\frac{1}{\tau s}\right)^n P(s) = \left(\frac{1}{\tau s}\right)^n (-P(s)), \qquad [1.52]$$

an equation translated by the functional diagram of a *free control loop* (Figure 1.19).

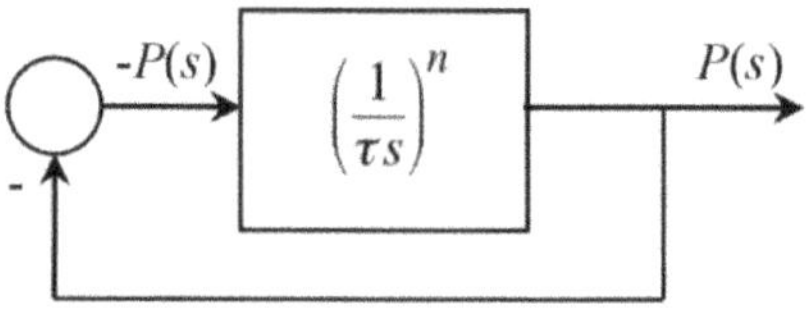

Figure 1.19. *Free control loop*

The open-loop transmittance, namely (section 2.5.2)

$$\beta(s) = \left(\frac{1}{\tau s}\right)^n = \left(\frac{\omega_u}{s}\right)^n, \qquad [1.53]$$

expresses a *non-integer integration* (of unit gain frequency $\omega_u = 1/\tau$) where the order n appears no longer as a differentiation non-integer order, but as an *integration non-integer order.*

As for the open-loop phase (Figure 1.20), namely

$$\arg \beta(j\omega) = -n\pi/2 = cte, \qquad [1.54]$$

it defines a *vertical open-loop Nichols locus, which, sliding on itself at the time of a variation of* ω_u *consecutive to a variation of the mass* M*, ensures the invariance* *of the distance to the critical point (here measured by* $\Phi_m = (2-n)\pi/2$*) and, consequently, the invariance of the corresponding damping ratio in time domain.*

Thus, due to the *invariance of the distance to the critical point*, such a frequency configuration, which puts in the wrong *mass-damping dilemma* in mechanics, also puts in the wrong *stability-precision dilemma* in automatic control.

Finally, for $n = 1.43$ (further derived from a primacy effect that concerns an aspect of human memory), the resonance ratio Q is equal to 1.28, therefore almost the optimal tuning $Q = 1.3$ (namely 2.3 dB) well known by automatic control specialists. No additional commentary will be made about this topic in order to avoid showing anthropomorphism.

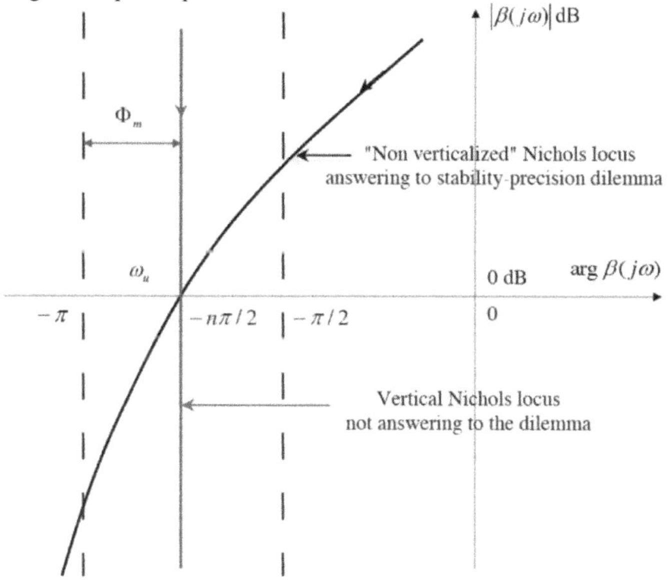

Figure 1.20. *Vertical open-loop Nichlos locus*

1.7.2. *Just as unexpected memory properties*

1.7.2.1. *Taking into account the past*

The non-integer order n derivative of the pressure $P(t)$ is calculated, at a given instant t (Figure 1.21), by the *weighted sum* of the pressures at the instants t, $t-h$, $t-2h$,...:

$$D^n P(t) = a_0 P(t) + a_1 P(t-h) + a_2 P(t-2h) + ... + a_k P(t-kh) + ..., \qquad [1.55]$$

where the coefficients a_0, a_1, a_2, ... are the *weighting coefficients* turning on the *pressure samples* $P(t)$, $P(t-h)$, $P(t-2h)$), ..., h denoting the *sampling step* (section 3.2.3).

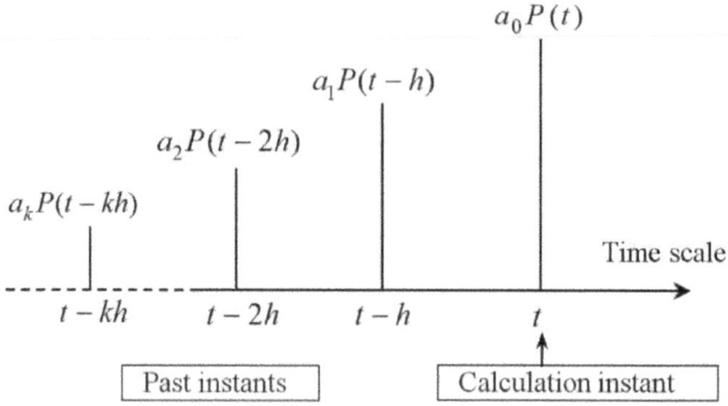

Figure 1.21. *Taking into account the past*

Through the past sampling instants $t-h, t-2h,...$, a genuine explosion is observed by *taking into account the time*:

– in the *integer case*, the time is taken at *the calculation instant* and at *the first instants of the past* (*only the first instant for an order 1 derivative*), therefore in a *local manner*;

– whereas in the *non-integer case*, the time is taken not only at *the calculation instant* but also at *all the past instants*, therefore in a *global manner*.

1.7.2.2. *Memory notion*

By taking into account the whole of the past of the function by the *weighted sum of the samples* that convey a *different weight according to the sample*, non-integer differentiation introduces a *notion of memory* (section 3.5.2) such that *the attenuation or the accentuation of the past is imposed by the differentiation order (therefore by one parameter only)*:

– *for* $n=-1$, which corresponds to an integration, *there is no weighting of the past* (Figure 1.22(a));

– *for* $n>-1$, which corresponds to less than an integration ($-1<n<0$) or to a differentiation ($n>0$), *there is an attenuation of the past* (Figure 1.22(b));

– *for* $n<-1$, which corresponds to more than an integration, *there is an accentuation of the past* (Figure 1.22(c)).

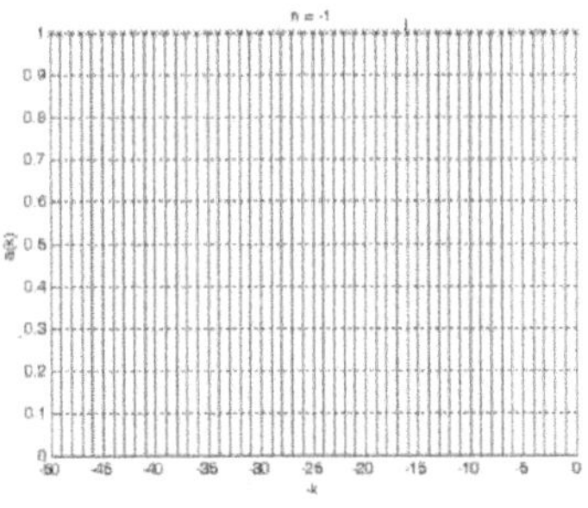

a) $n=-1$

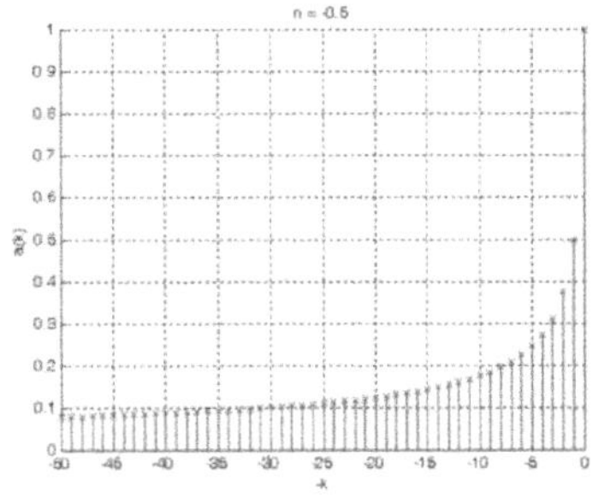

b) $n>-1$

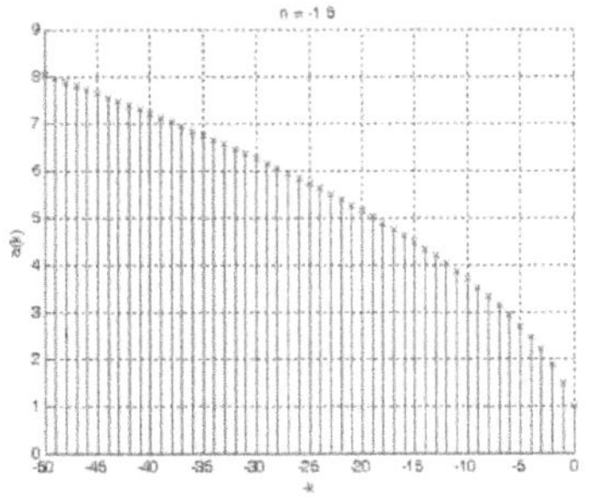

c) $n<-1$

Figure 1.22. *Weighting coefficients*

As characterized, this notion of memory tends to evoke a *subtle form of memory* through which *the recollection of an event depends on the nature of this event*:

– for usual events, *the recollection of more ancient events is less important than the recollection of more recent events*;

– on the contrary, for unusual events, *the recollection of more ancient events can be more important than the recollection of more recent events.*

1.7.2.3. *A diversion through an aspect of human memory*

1.7.2.3.1. The serial position effect

In cognitive psychology (see section 3.5.3), the study of this effect on the *recall* consists of asking a subject:

– *to recall, without a required order, a list of words*;

– *immediately after a visual presentation of this list at a certain pace.*

Drawn from the results obtained with several subjects, the *recall curve* (Figure 1.23) highlights *a best recall (or recollection)*:

– of *the first words* that are the most ancient ones (*primacy effect*);

– and of *the last words* that are the most recent ones (*recency effect*).

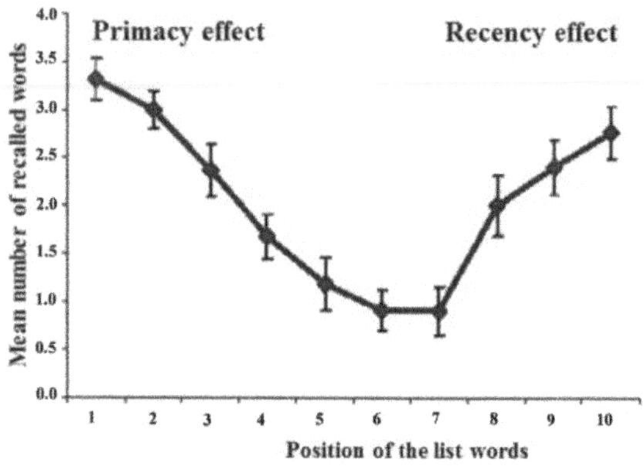

Figure 1.23. *Recall curve*

If the words at the beginning and at the end are indeed better recalled than the middle words, *the beginning ones are even better recalled than the ending ones*, expressing that *the primacy effect prevails over the recency effect.*

1.7.2.3.2. A model of the primacy effect

Over this *prevalence*, the primacy effect that actually corresponds to an *accentuation of the past* (Figure 1.24) seems to present a *curvature* in conformity with the *memory profile dictated by an integration non-integer order greater than* 1 (Figure 1.25).

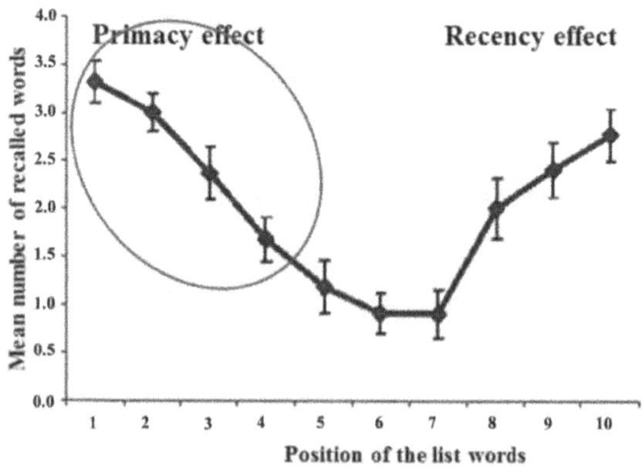

Figure 1.24. *Curvature of the primacy effect*

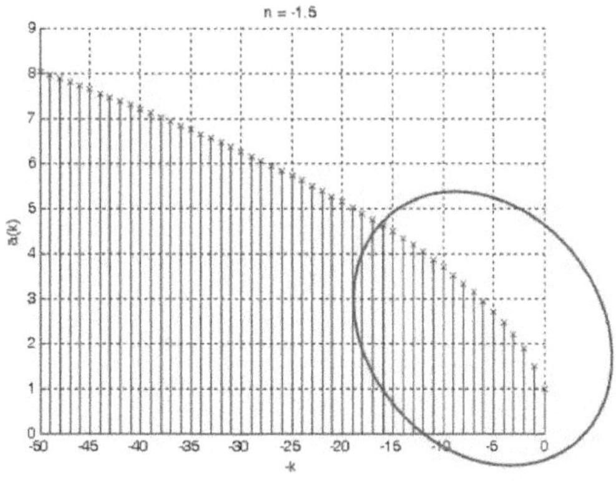

Figure 1.25. *Curvature associated with an integration non-integer order superior to one*

Hence, there is the idea to seek the *optimal non-integer order* minimizing the sum of the squares of the differences between:

– the *ordinates* of the points defining the primacy effect;

– and the *first weighting coefficients* associated with the non-integer integration (expressed up to a factor).

The result of this minimization is *an optimal order of* 1.43, which is well validated by the *corresponding normalized weighting coefficients*.

1.8. On the nature of diversity

1.8.1. *An action level to be defined*

For a given element, it is convenient to distinguish:

– its *action*, for instance the removal of an electric current;

– and its *action level*, which gives it an action more stronger (intense) than it is important.

Such an action level (Figure 1.26) is the *level* at which the element action is carried out:

– this level is definable as a *hierarchical degree* in the *organization of a system* (for instance, electric network);

– the action level of the element, indeed, depends on the *position* of the element in the organization of the system.

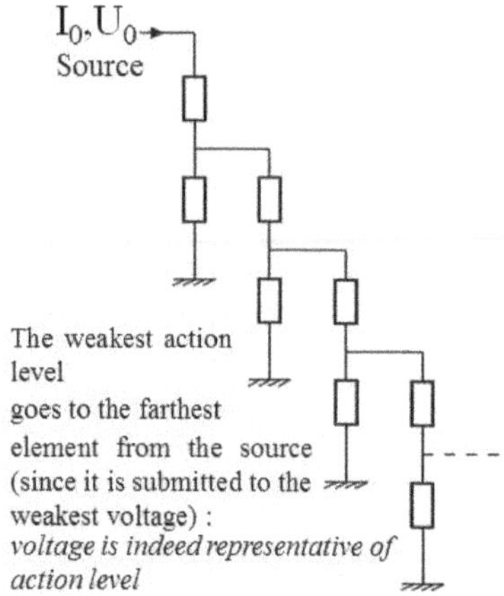

Figure 1.26. *Action level and hierarchical degree*

1.8.2. *One or several forms of diversity?*

Diversity does not have a unique form. It indeed has various forms, which call for a distinction according to the *invariance or non-variance of the elements*.

1.8.2.1. *Forms based on the invariance of the elements*

– *A multitude of different elements of same action level* (it is the case of the porous face (through its alveolar structure) or the striated muscle (through its fibrous structure));

– *A multitude of identical or similar elements of different action level* (it is the case of the thermal rod (Figure 1.27) or similar neuron classes (cerebellum or cerebral cortex and motor neurons of the spinal cord));

– *A multitude of different elements of different action level* (it is the case of the lung (through its arborescent structure (Figure 1.28)).

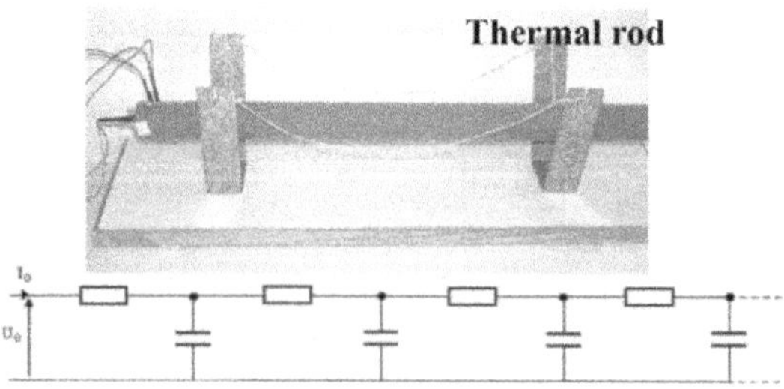

Figure 1.27. *Cascade network of identical gamma RC cells*

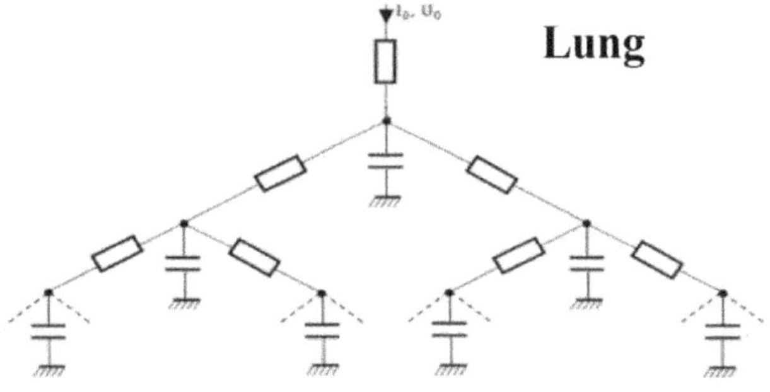

Figure 1.28. *Arborescent network of different gamma RC cells*

1.8.2.2. *A singular form based on the time variability of an element*

– *A continuously variable unique element* which presents a *multitude of different values* on a time horizon (*it is the case of the aorta*);

– Thus, defining a *technological approach of diversity* which consists of substituting (Figure 1.29) *a variable element different at each instant* for *a multitude of fixed elements identical at each instant.*

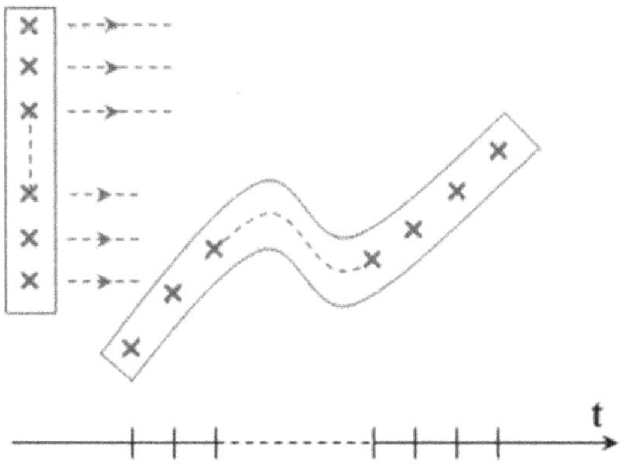

Figure 1.29. *Technological approach of diversity*

It is notably the case of the *(controlled or piloted) dashpot with variable viscous friction coefficient* $C(t)$, which is used in the *metallic version* of the CRONE suspension (Figure 1.30), and which is controlled so as to exert an effort identical to the one developed by a *non-integer order dashpot* (section 4.3).

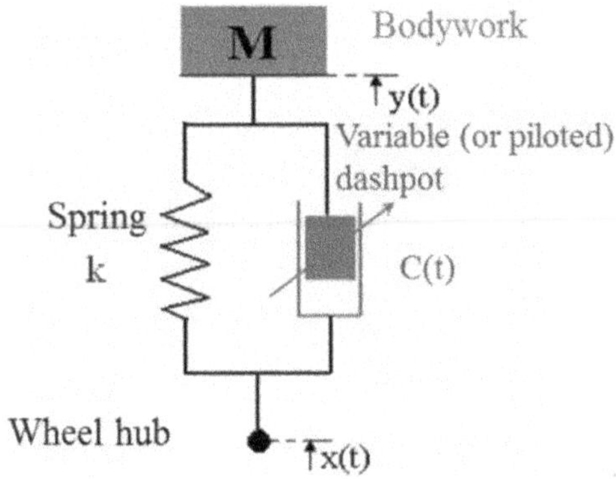

Figure 1.30. *Metallic version of the CRONE suspension*

The *controlled dashpot* defined by an effort of the form

$$F(t)=C(t)\left(\frac{d}{dt}\right)[x(t)-y(t)], \qquad [1.56]$$

where $x(t)-y(t)$ denotes the *relative displacement* of the suspension, indeed, makes it possible to achieve (or synthesize) a *non-integer order dashpot* defined by the effort

$$F(t)=C\left(\frac{d}{dt}\right)^m [x(t)-y(t)] \text{ with } 0<m<1. \qquad [1.57]$$

This achievement (or synthesis) is conditioned by the *identity of the respective efforts* that enables us to calculate the *instantaneous viscous friction coefficient* $C(t)$ from the *relative displacement* $[x(t)-y(t)]$ recorded by a position sensor, namely

$$C(t)=C\frac{\left(\frac{d}{dt}\right)^m [x(t)-y(t)]}{\left(\frac{d}{dt}\right)[x(t)-y(t)]}. \qquad [1.58]$$

Then, this value of $C(t)$ enables us to calculate the *oil laminating instantaneous section* $s(t)$.

The contribution of the corresponding collaboration with PSA is the design and achievement of the *non-integer order dashpot* (conforming to patents 90 046 13 and 95 050 84 filed in 1990 and 1995, respectively), and also its implementation on a Peugeot 406 in 1998 with the help of Coverplant company.

1.9. From the porous dyke to the CRONE suspension

To change from the porous dyke to the hydropneumatic concept of the CRONE suspension (of a quarter vehicle), it suffices (Figure 1.31):

– to rotate the dyke representation of a quarter of turn;

– to replace air with nitrogen;

– to replace water with oil;

– to set oil in motion with a piston linked to the wheel;

– to transcribe this new representation technologically with a piston and spheres (Figure 1.32).

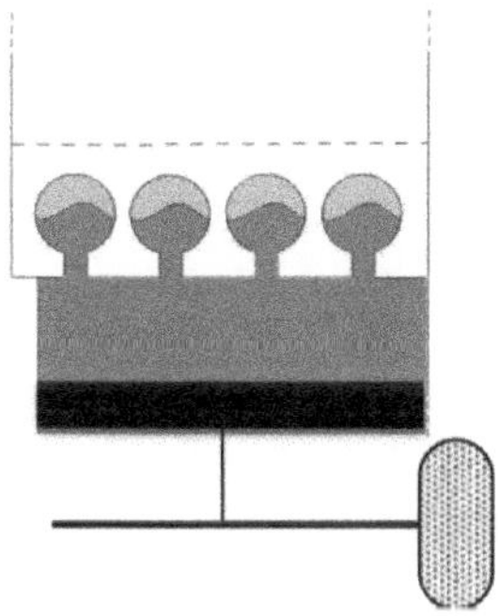

Figure 1.31. *Idea of the CRONE suspension hydropneumatic version*

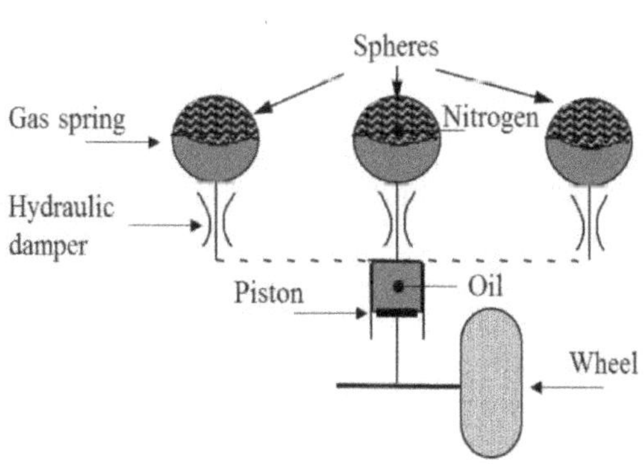

Figure 1.32. *Idea of the multishpere CRONE suspension*

To talk about the *constancy of a ratio* (between resistances or capacitances), requires:

– at least two ratios;

– therefore at least three RC cells;

– therefore at least three Citroën spheres (section 4.2.3);

– hence, the minimal version of the *multishpere CRONE suspension* (Figure 1.33).

Three different diameter holes ($Ø_1$, $Ø_2$ and $Ø_3$) ensure *three different hydraulic resistances.*

Three different calibration pressures (P_1, P_2 and P_3) ensure *three different pneumatic capacitances.*

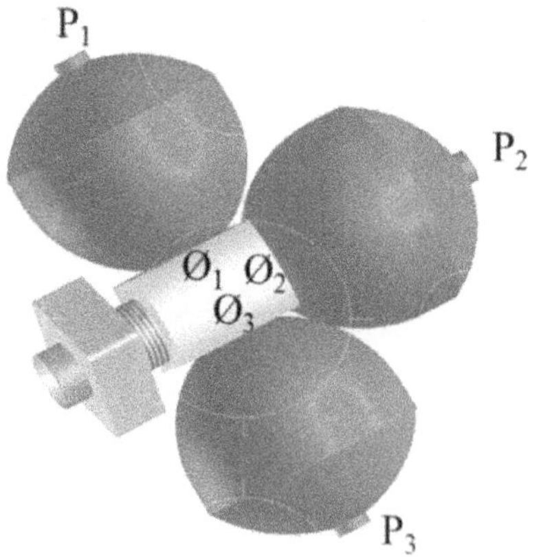

Figure 1.33. *Minimal version of the multisphere CRONE suspension*

The minimal version of the multisphere CRONE suspension as defined previously has been used successfully in the CRONE suspension implemented on a Citroën BX (Figure 1.34).

Concerning performances, Figure 1.35 gives the bodywork response to a step displacement of the wheel. This response presents the same overshoot independently of the mass. Only the natural frequency changes. In fact, the transient keeps its form up to a timescale factor (think about the accordion phenomenon).

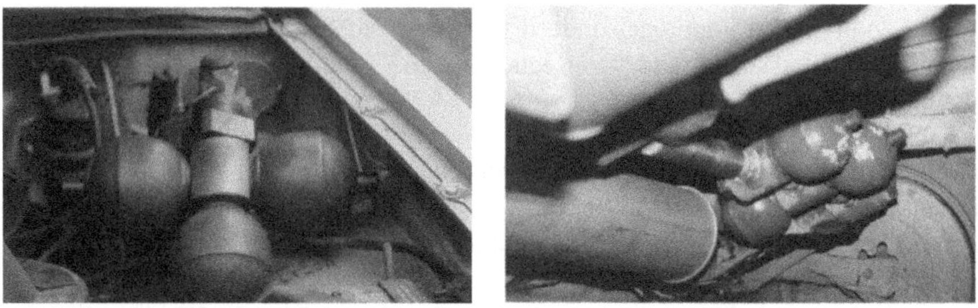

a) b)

Figure 1.34. *Citroën BX "crρned" in the front a), at the rear b) and rewarded by the AFCET'95 trophy distinguishing the best technological innovation in the University–Industry partnership (in this case PSA for the industrial component)*

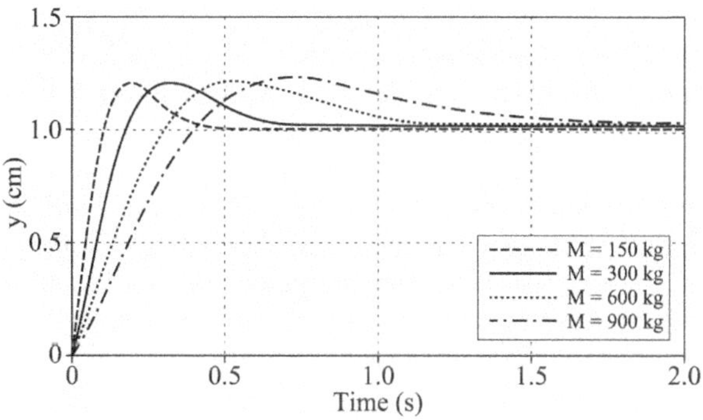

Figure 1.35. *Body work response*

1.10. Conclusion

This chapter cements the visual aids of the magisterial conference presented on 5 January 2012 in Bordeaux (France) on the occasion of the *Grand Prix Lazare Carnot 2011* of the French Academy of Sciences.

Based on a diversity-structured approach, which is notably inspired by various natural forms of diversity (biological among others), this chapter unquestionably provides a framework:

– on one hand, to the introduction of the non-integer derivative as a modeling tool;

– on the other hand, to the use of such a modeling form to put into light dynamic performances (and notably of damping) unsuspected in an "integer" approach of mechanics and automatic control.

The "non-integer" approach indeed enables us to overcome the mass-damping dilemma in mechanics and, consequently, the stability-precision dilemma in automatic control.

Furthermore, the metallic and hydropneumatic versions of the CRONE suspension are the subject of a *unified presentation* through various forms of diversity leading to non-integer differentiation.

To illustrate our strategy through a study system, we have studied the relaxation of water on a porous dyke. Obtained as such, the model of the water-dyke interface and the one governing the water relaxation are non-integer models:

– of order m between 0 and 1 for the interface;

– and of order $n = 1 + m$ between 1 and 2 for the relaxation.

As for the *validity* of these models (sections 2.2.3.4, 2.5.2, 5.2.1.2, 5.2.2.2 and A2.3.3), it is convenient to analyze the study system through the physics of the porous dyke.

Physically, a dyke cannot be made up of infinitely large and infinitely small pores. The transitional frequencies they determine can then not take values tending as well toward infinity as toward zero. Thus, they belong to a bounded frequency interval that limits the non-integer differentiation of order m and consecutively of order $n = 1 + m$ to a *medium frequency range*. More precisely, the non-integer order models so obtained are valid in a *medium frequency range belonging to the frequency interval limited by the extremal transitional frequencies of the series RC cells*. In fact, this range is the one in which these models can be representative of reality.

Moreover, the *validity range* so defined turns out to be compatible with the property according to which *the behavior at the medium frequencies is sufficient for determining the dynamics performances*:

– as well *in time domain* (reduced first overshoot, damping ratio and natural frequency);

– as *in frequency domain* (resonance ratio and resonance frequency).

1.11. Bibliography

[DOR 95] DORE P., MOREAU X., NOUILLANT M., *et al.*, Amortisseur piloté, Patent no. 9505084, INPI, 1995 (extension to Japan and USA in 1996).

[OUS 83] OUSTALOUP A., *Systèmes asservis linéaires d'ordre fractionnaire*, Masson, Paris, 1983.

[OUS 95] OUSTALOUP A., *La dérivation non entière: théorie, synthèse et applications*, Hermès, Paris, 1995.

[OUS 90] OUSTALOUP A., NOUILLANT M., Nouveau système de suspension, Patent no. 9004613, INPI, 1990.

[OUS 99] OUSTALOUP A., SABATIER J., LANUSSE P.,"From fractal robustness to the CRONE control", *Fractional Calculus and Applied Analysis (FCAA): An International Journal for Theory and Applications*, vol. 2, no. 1, pp. 1–30, January 1999.

[OUS 05] OUSTALOUP A., COIS O., LE LAY L., *Représentation et indentification par modèle non entier*, Hermès, Paris, 2005.

[OUS 06] OUSTALOUP A., "The CRONE approach: theoretical developments and major applications", Plenary lecture, *2nd IFAC Workshop on Fractional Diffenrentiation and Its Applications (FDA'06)*, Porto, Portugal, 19–21 July 2006.

[OUS 10] OUSTALOUP A., MOREAU X., "Mechanical version of the CRONE suspension", *Advances in the Theory of Control, Signals and Systems with Physical Modeling*, LNCIS 407, Springer-Verlag, Berlin-Heidelberg, pp. 99–112, 2010.

[OUS 12] OUSTALOUP A., "From diversity to unexpected dynamic performances", Keynote, *2nd IEEE International Conference on Communications Computing and Control Applications (CCCA'12)*, Marseille, France, 6–8 December 2012.

[OUS 13] OUSTALOUP A., "From diversity to unexpected dynamic performances", Plenary lecture, *International Conference on Fractional Signals and Systems FSS*, Ghent, Belgium, 24–26 October 2013.

Chapter 2

Damping Robustness

2.1. Introduction

The proof and validation of the *damping robustness* of the water relaxation on a porous dyke are presented in this chapter by a very progressive process, whose structure according to five sections develops the various study stages going from the object to its performances and to their experimental verification.

The first stage consists of starting with a recursive parallel arrangement of serial RC cells to obtain a non-integer differentiation as a model of the water-dyke interface. Obtaining a differentiation non-integer order is inscribed in a frequency approach whose originality turns on a smoothing of the admittance Bode asymptotic diagrams.

The second stage consists of using the dynamics fundamental principle to change from the non-integer derivative so obtained to a non-integer differential equation as a model governing the water relaxation. An electrical analogy is proposed through a coil loaded by a non-integer impedance.

After the analytical determination of the relaxation, which shows damping robustness in *time domain* and which constitutes the third study stage, the fourth stage illustrates damping robustness:

– in *operational domain* through two isodamping half-straight lines which can be reduced to two straight line segments;

– in *frequency domain* through a Nichols locus (in open loop) which successively forms a vertical straight line and a vertical straight line segment.

Finally, the fifth stage consists of experimentally verifying damping robustness through an electronic circuit made up of operational amplifiers. This circuit is achieved in such a way that its transmittance presents, in a medium frequency band, a denominator in conformity with the characteristic polynomial of the non-integer differential equation which governs the water relaxation on a porous dyke.

2.2. From ladder network to a non-integer derivative as a water-dyke interface model

The ladder network is nothing but the recursive parallel arrangement of series RC cells that represents the hydropneumatic model of the porous face (or the water-dyke interface). The aim of this section is to detect its non-integer nature by changing from the network to a non-integer derivative.

As directed, this change falls within a frequency approach that successively borrows from:

– a factorizing of the network admittance;

– a representation of this admittance through Bode asymptotic diagrams;

– a judicious exploitation of these diagrams.

2.2.1. *On the admittance factorizing*

As an indefinite sum of the admittances of each of the series *RC* cells, the network admittance is expressed by a series of the form:

$$Y(j\omega) = \sum_i \frac{jC_i\omega}{1 + j\dfrac{\omega}{\omega_i}} \quad \text{with } \omega_i = \frac{1}{R_i C_i}, \; R_i = \frac{R}{\alpha^i} \text{ and } C_i = \frac{C}{\eta^i}, \qquad [2.1]$$

the rank (or subscript) i being extended to the integer set $\mathbb{Z}$ in such a way that the median (or central) cell of rank 0 may be the *RC* cell, thus allowing, with finite R and C, that the ω_i cover the whole frequency domain (from zero to infinity).

Calculations directly led on this series (Chapter 6) in order to avoid a reduction to a common denominator, then explosive given the number of cells make it possible to decompose the series according to an *indefinite product* of factors of the form $\left(1 + j\omega / \omega_i'\right)\left(1 + j\omega / \omega_i\right)$, namely:

$$Y(j\omega) = K\prod_i \frac{1+j\dfrac{\omega}{\omega_i^{'}}}{1+j\dfrac{\omega}{\omega_i}}, \quad [2.2]$$

the transitional frequencies of the numerator, $\omega_i^{'}$, and those of the denominator, ω_i, being distributed in a recursive way (Figure 2.1) in conformity with the ratios:

$$\frac{\omega_{i+1}^{'}}{\omega_i^{'}} = \frac{\omega_{i+1}}{\omega_i} = \alpha\eta \text{ with } \frac{\omega_i}{\omega_i^{'}} = \alpha \text{ and } \frac{\omega_{i+1}^{'}}{\omega_i} = \eta, \quad [2.3]$$

the recursive factors α and η being greater than 1.

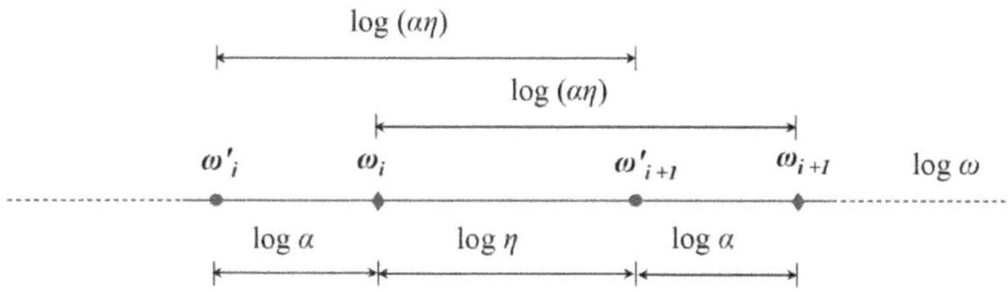

Figure 2.1. *By replacing a multiplicative term with an additive one, a logarithmic scale ensures a constant interval for a multiplication repeated by the same factor*

2.2.2. *On the asymptotic diagrams at stake*

What about the factor of rank i, $\dfrac{1+j\omega/\omega_i^{'}}{1+j\omega/\omega_i}$?

Characterized by the transitional frequencies $\omega_i^{'}$ and ω_i, such that $\omega_i^{'} < \omega_i = \alpha\omega_i^{'}$, this factor is, according to frequency areas, reduced to (Figure 2.2(a)):

– 1 for $\omega \ll \omega_i^{'}$ and ω_i, namely a gain of $20\log 1 = 0$ in dB and a nil phase;

– $j\dfrac{\omega}{\omega_i^{'}}$ for $\omega_i^{'} \ll \omega \ll \omega_i$, namely a gain of $20\log\dfrac{\omega}{\omega_i^{'}}$ in dB and a phase of $\dfrac{\pi}{2}$;

– $\dfrac{\omega_i}{\omega_i^{'}} = \alpha$ for $\omega \gg \omega_i^{'}$ and ω_i, namely a gain of $20\log\alpha$ in dB and a nil phase.

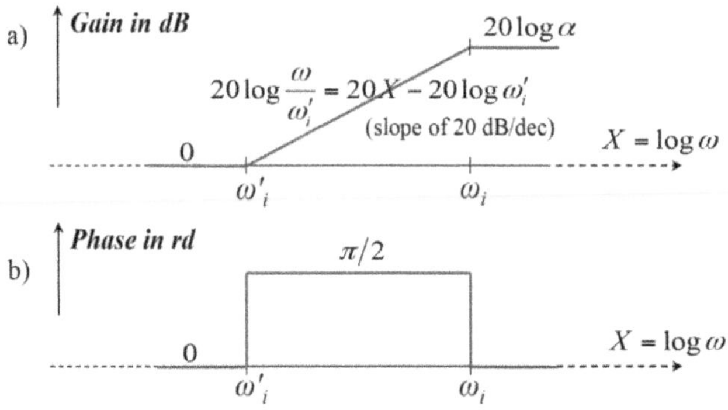

Figure 2.2. *a) A stair step as a gain asymptotic diagram and b) a crenel as a phase asymptotic diagram*

For the factor of rank $i+1$, $\dfrac{1+j\omega/\omega'_{i+1}}{1+j\omega/\omega_{i+1}}$, the asymptotic diagrams form:

– the same stair step and the same crenel;

– except for a right shift in conformity with the new transitional frequencies $\omega'_{i+1} = \alpha\eta\omega'_i$ and $\omega_{i+1} = \alpha\eta\omega_i$.

Because of the additivity of the gains (in dB) and of the phases of the factors of rank i, $i+1$,..., the asymptotic diagrams of the network admittance, $Y(j\omega)$, form (Figure 2.3):

– for the gain, a regular sequence of identical stair steps;

– for the phase, a regular sequence of identical crenels.

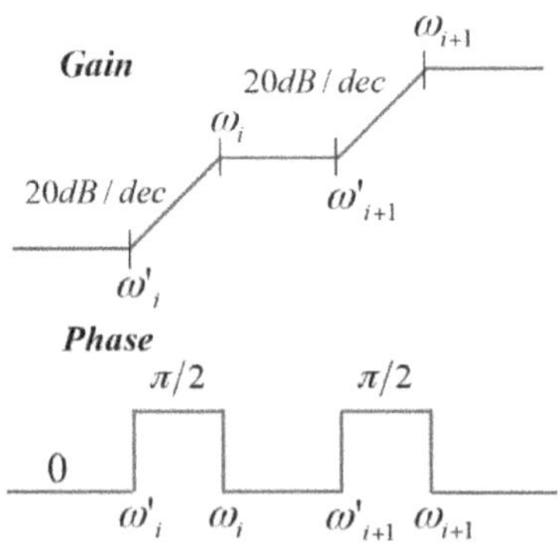

Figure 2.3. *Asymptotics diagrams of* $Y(j\omega)$

2.2.3. *On the asymptotic diagram exploiting*

The admittance asymptotic diagrams are constituted of a sequence of asymptotes (Figure 2.4) such that the Bode diagrams can approach only when these asymptotes are high enough.

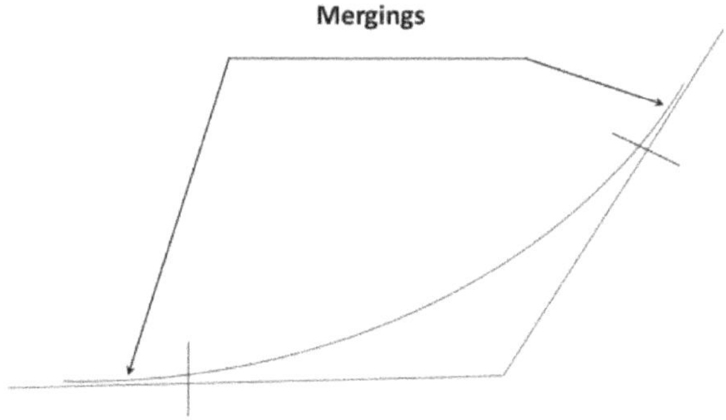

Figure 2.4. *Possible merging of a Bode diagram and an asymptotic diagram*

Now, through little-spaced transitional frequencies (because of their high density), therefore small asymptotes, our study context does not answer this condition. Thus, the Bode diagrams cannot approach these asymptotes.

The idea is then that the Bode diagrams can only take the *mean* of the asymptotic diagrams, thus expressing the idea of a *smoothing* (Figure 2.5) of stair steps for the gain and that of crenels for the phase.

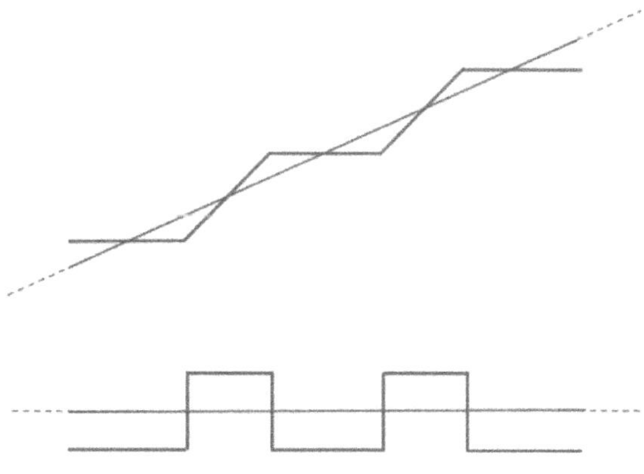

Figure 2.5. *Smoothing of the asymptotic diagrams of* $Y(j\omega)$

2.2.3.1. *Step smoothing*

The smoothing of the steps of the gain asymptotic diagram can be materialized by a straight line, called *gain smoothing straight line*, of slope less than 20 dB/dec, namely 20m dB/dec with *m* between 0 and 1 (Figure 2.6).

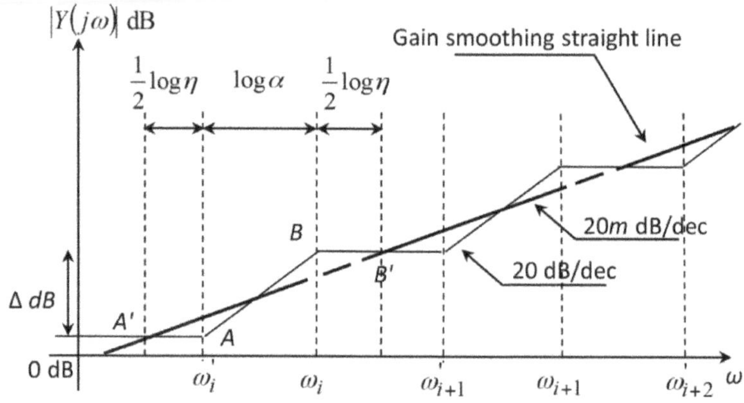

Figure 2.6. *Smoothing of the gain asymptotic diagram of* $Y(j\omega)$

ΔdB denoting the step height, the slopes of segments A'B' and AB are, respectively, given by the equations:

$$20m\ dB/dec = \frac{\Delta\, dB}{\log\alpha + \log\eta} \qquad [2.4]$$

and

$$20\ dB/dec = \frac{\Delta\, dB}{\log\alpha}, \qquad [2.5]$$

from which we draw, carrying out their ratio, the expression of *m* versus the recursive factors α and η:

$$m = \frac{\log\alpha}{\log(\alpha\eta)}. \qquad [2.6]$$

2.2.3.2. *Crenel smoothing*

The smoothing of the crenels of the phase asymptotic diagram can be materialized by a straight line, called *phase smoothing straight line*, of ordinate less than $\pi/2$, namely $m\pi/2$ with m between 0 and 1, and equal to the average of the phase asymptotic variation (Figure 2.7).

The identity of the shaded surfaces is translated by the equation:

$$m\frac{\pi}{2}(\log\alpha+\log\eta)=\frac{\pi}{2}\log\alpha, \quad [2.7]$$

from which m is directly deduced, namely:

$$m=\frac{\log\alpha}{\log(\alpha\eta)}, \quad [2.8]$$

an expression that is indeed the same as the one obtained from the gain.

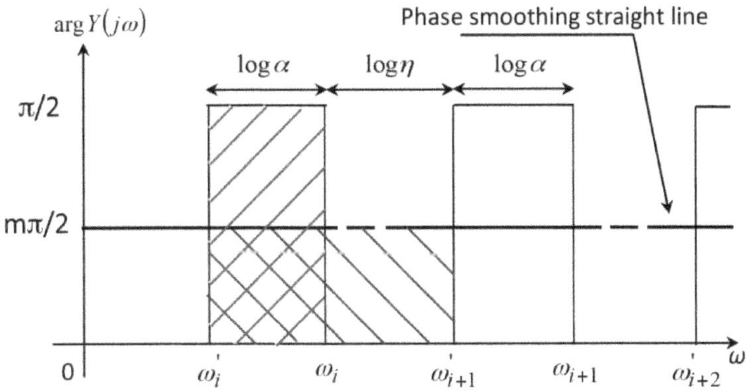

Figure 2.7. *Smoothing of the phase asymptotic diagram of* $Y(j\omega)$

2.2.3.3. *A non-integer differentiator as a smoothing result*

The gain and phase smoothing straight lines so defined (Figure 2.8) are, in fact, the Bode diagrams of a *non-integer differentiator* of frequency response:

$$Y(j\omega)=\left(j\frac{\omega}{\omega_0}\right)^m, \quad [2.9]$$

both parameters m and ω_0 being defined as follows:

– through the non-integer differentiation symbolic operator $(j\omega)^m$, m represents the differentiation non-integer order;

– ω_0 is the frequency at which the gain smoothing straight line intersects the axis 0 dB, called *unit gain frequency* or *transition frequency*.

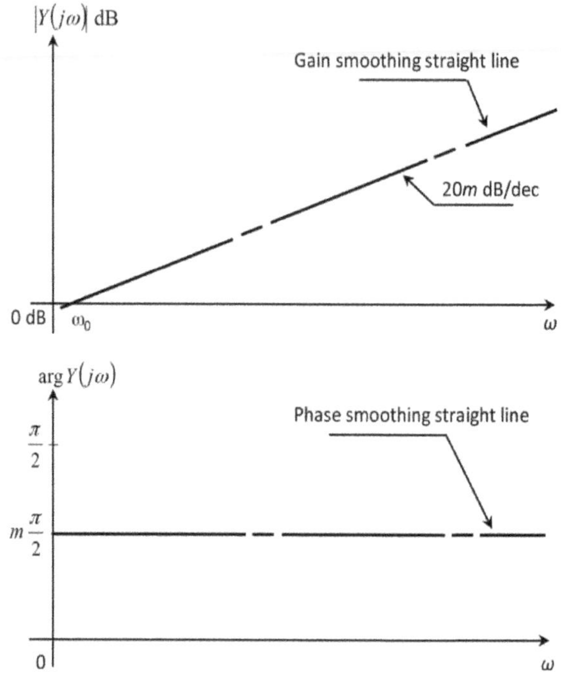

Figure 2.8. *Gain and phase smoothing straight lines of $Y(j\omega)$ as Bode diagrams of a non-integer differentiator*

2.2.3.4. *A non-integer derivative as a water-dyke interface model*

Finally, the water-dyke interface hydraulic admittance is expressed in conformity with the relation:

$$Y(j\omega) = \frac{Q(j\omega)}{P(j\omega)} = \left(\frac{j\omega}{j\omega_0}\right)^m, \qquad [2.10]$$

from which is deduced the symbolic equation:

$$Q(j\omega) = \frac{1}{\omega_0^m}(j\omega)^m \, P(j\omega), \qquad [2.11]$$

which, in time domain, admits the concrete (or original) equation:

$$Q(t) = \frac{1}{\omega_0^{\,m}} \left(\frac{d}{dt} \right)^m P(t). \tag{2.12}$$

This equation represents the (dynamic) model of the interface. It expresses that the water flow $Q(t)$ is proportional to the non-integer derivative of the pressure $P(t)$ upstream from the interface (at least in a (medium) frequency range, given that the frequency interval which regroups the whole of the ω_i and $\omega_i^{'}$ does not cover the whole frequency domain).

The non-integer approach thus permits us to replace;

– an infinite integer order model;

– by a finite non-integer order model.

2.3. From a non-integer derivative to a non-integer differential equation as a model governing water relaxation

2.3.1. *Flow-pressure differential equation*

Applying Newton's second law to the motion of water mass (Figure 2.9), here assumed not to be deformable, leads to the differential equation:

$$M \frac{dV(t)}{dt} + F(t) = 0, \tag{2.13}$$

where $V(t)$ denotes the water mass M speed and $F(t)$ denotes the dyke reaction force.

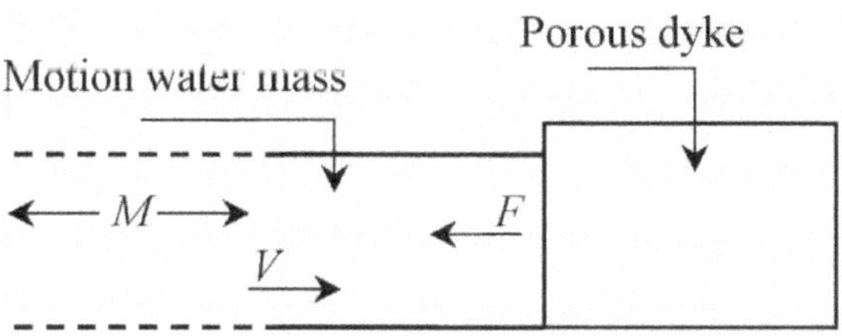

Figure 2.9. *Study system*

If S represents the water flow cross-section, the speed $V(t)$ is expressed as a function of the flow $Q(t)$ according to:

$$V(t)=\frac{Q(t)}{S}; \qquad [2.14]$$

moreover, the force $F(t)$ is expressed as a function of the pressure $P(t)$ upstream from the water-dyke interface, namely:

$$F(t)=P(t)S. \qquad [2.15]$$

Putting the expressions of $V(t)$ and $F(t)$ into the differential equation of first form [2.13], we obtain a differential equation of new form:

$$\frac{M}{S^2}\frac{dQ(t)}{dt}+P(t)=0, \qquad [2.16]$$

a differential equation of the free system, with two variables, the flow $Q(t)$ *and the pressure* $P(t)$.

2.3.2. *A non-integer differential equation as a model governing relaxation*

2.3.2.1. *Pressure as a variable of the differential equation*

Putting the expression of $Q(t)$ given by [2.12] into relation [2.16] determines a differential equation of non-integer order $n = 1 + m$ between 1 and 2 (since m is between 0 and 1), namely:

$$\frac{M}{S^2}\frac{1}{\omega_0^{\ m}}\left(\frac{d}{dt}\right)^n P(t)+P(t)=0, \qquad [2.17]$$

or, under a canonical form:

$$\tau^n\left(\frac{d}{dt}\right)^n P(t)+P(t)=0, \qquad [2.18]$$

by putting

$$\tau=\tau(M)=\left(\frac{M}{S^2}\frac{1}{\omega_0^{\ m}}\right)^{\frac{1}{n}}, \qquad [2.19]$$

which is called *transitional time constant.*

2.3.2.2. *Flow as a variable of the differential equation*

The order $-m$ derivative of the flow

$$Q(t) = \frac{1}{\omega_0^{\ m}} \left(\frac{d}{dt} \right)^m P(t), \qquad [2.20]$$

is written as:

$$\left(\frac{d}{dt} \right)^{-m} Q(t) = \frac{1}{\omega_0^{\ m}} P(t), \qquad [2.21]$$

from which is drawn:

$$P(t) = \omega_0^{\ m} \left(\frac{d}{dt} \right)^{-m} Q(t), \qquad [2.22]$$

an expression that, put into [2.18], leads to:

$$\tau^n \left(\frac{d}{dt} \right)^1 Q(t) + \left(\frac{d}{dt} \right)^{-m} Q(t) = 0, \qquad [2.23]$$

namely, through an order m differentiation:

$$\tau^n \left(\frac{d}{dt} \right)^n Q(t) + Q(t) = 0 . \qquad [2.24]$$

This differential equation, which governs the flow, has indeed the same form as the one governing the pressure (relation [2.18]).

2.3.3. *Electrical analogy*

The idea of the electrical analogy of the water relaxation on a porous dyke resulted from a discussion with Jean-Luc Thomas and Serge Poullain of AREVA-T&D in July 2003.

Given the equivalence rules between effort variables (pressure $P(t)$ equivalent to voltage $U(t)$) and between flux variables (flow $Q(t)$ equivalent to intensity $I(t)$), equations [2.16] and [2.22] can be rewritten according to:

$$\frac{M}{S^2}\frac{dI(t)}{dt} + U(t) = 0 \quad [2.25]$$

and

$$U(t) = \omega_0^m \left(\frac{d}{dt}\right)^{-m} I(t), \quad [2.26]$$

namely, by putting $L = M / S^2$:

$$U(t) = -L\frac{dI(t)}{dt} \quad [2.27]$$

and

$$U(t) = \omega_0^m \left(\frac{d}{dt}\right)^{-m} I(t). \quad [2.28]$$

These equations are the equations of the electrical circuit proposed in Figure 2.10 and made up of a self-inductance L coil loaded by a non-integer impedance of the form $(\omega_0 / s)^m$.

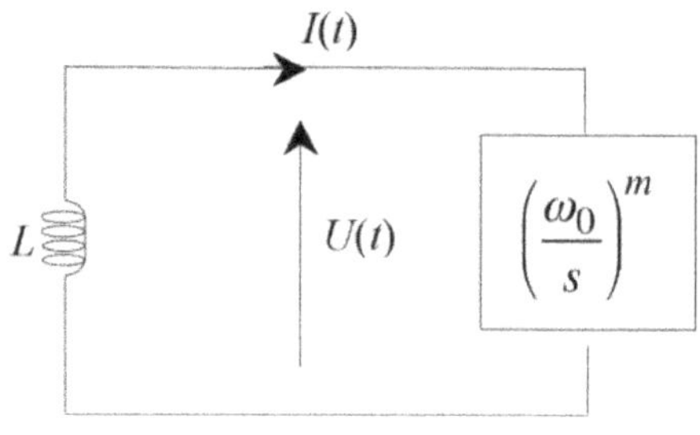

Figure 2.10. *Electrical circuit corresponding to the hydropneumatic representation defined by Figure 2.9*

2.4. Relaxation expression

The relaxation as studied in this section corresponds to the release of the kinetic energy of the water mass in the water-dyke interface, the damping being considered sufficiently important for reducing the water motion on the dyke to a single to and fro.

The study conditions are such that at time $t=0$:

– the water mass reaches the water-dyke interface with an initial speed V_0 which induces an initial flow $Q_0 = SV_0$;

– the water-dyke interface leaves an initially relaxed state through the absence of flow and pressure in the pores before the arrival of the water mass.

Thus, the *initial conditions* are:

$$Q(0^-)=Q_0 \text{ for the water mass} \qquad [2.29]$$

and

$$P(0^-)=0 \text{ for the interface,} \qquad [2.30]$$

the *initial values* then being:

$$Q(0^+)=Q(0^-)=Q_0 \qquad [2.31]$$

through the continuity of the kinetic energy of the water mass, and

$$P(0^+)=P(0^-)=0 \qquad [2.32]$$

through a nil impedance of the interface at $t=0^+$ ($P(j\omega)/Q(j\omega)=(\omega_0 / j\omega)^m=0$ for $\omega=\infty$).

Taking into account the initial values so defined permits us to write the Laplace transforms of equations [2.16] and [2.12] in the form:

$$\frac{M}{S^2}\left[sQ(s)-Q_0\right]+P(s)=0 \qquad [2.33]$$

and

$$Q(s)=\frac{1}{\omega_0^m}s^m P(s), \qquad 2.34]$$

from which is deduced, by putting into [2.33] the expression of $P(s)$ drawn from [2.34]:

$$Q(s)=\frac{\tau^n s^n}{1+\tau^n s^n}\frac{Q_0}{s}, \qquad [2.35]$$

recalling that

$$n=1+m \text{ and } \tau=\left(\frac{M}{S^2}\frac{1}{\omega_0^{\,m}}\right)^{\frac{1}{n}}. \qquad [2.36]$$

The relaxation $Q(t)$ is then given by the inverse transform of $Q(s)$ for $t>0$, namely, on this interval:

$$Q(t)=\mathcal{L}^{-1}\left[\frac{\tau^n s^n}{1+\tau^n s^n}\frac{Q_0}{s}\right], \qquad [2.37]$$

which can be rewritten as:

$$Q(t)=Q_0\tau^n\left(\frac{d}{dt}\right)^n\left\{\mathcal{L}^{-1}\left[\frac{1}{1+\tau^n s^n}\right]*\mathcal{L}^{-1}\left[\frac{1}{s}\right]\right\}, \qquad [2.38]$$

or even, given that (relation [A5.36])

$$\mathcal{L}^{-1}\left[\frac{1}{1+\tau^n s^n}\right]=\sum_{j=0}^{\infty}(-1)^j\frac{t^{n(j+1)-1}}{\tau^{n(j+1)}\Gamma(n(j+1))}u(t): \qquad [2.39]$$

$$Q(t)=Q_0\tau^n\left(\frac{d}{dt}\right)^n\left\{\left[\sum_{j=0}^{\infty}(-1)^j\frac{t^{n(j+1)-1}}{\tau^{n(j+1)}\Gamma(n(j+1))}u(t)\right]*u(t)\right\}, \qquad [2.40]$$

where $u(t)$ denotes the unit step function.

From the definition of the convolution product and for $t > 0$, relation [2.40] becomes:

$$Q(t)=Q_0\left(\frac{d}{dt}\right)^n\int_0^t\left[\sum_{j=0}^{\infty}(-1)^j\frac{v^{n(j+1)-1}}{\tau^{nj}\Gamma(n(j+1))}\right]dv, \qquad [2.41]$$

namely, through an inversion of the integral and sum signs:

$$Q(t)=Q_0\sum_{j=0}^{\infty}\frac{(-1)^j}{\tau^{nj}\Gamma(n(j+1))}\left(\frac{d}{dt}\right)^n\int_0^t v^{n(j+1)-1}dv, \qquad [2.42]$$

or, given that the order n derivative of an order 1 integral is nothing but the order $n-1$ derivative:

$$Q(t)=Q_0\sum_{j=0}^{\infty}\frac{(-1)^j}{\tau^{nj}\Gamma(n(j+1))}\left(\frac{d}{dt}\right)^{n-1}t^{n(j+1)-1}, \qquad [2.43]$$

or even, given the formula

$$\left(\frac{d}{dt}\right)^m \left[t^\alpha\right] = \Gamma(\alpha+1)\frac{t^{\alpha-m}}{\Gamma(\alpha-m+1)}: \quad [2.44]$$

$$Q(t) = Q_0 \sum_{j=0}^{\infty} \frac{(-1)^j}{\tau^{nj}\Gamma(n(j+1))}\Gamma(n(j+1))\frac{t^{nj}}{\Gamma(nj+1)}, \quad [2.45]$$

or else, finally:

$$Q(t) = Q_0 \sum_{j=0}^{\infty} (-1)^j \frac{t^{nj}}{\tau^{nj}\Gamma(nj+1)}. \quad [2.46]$$

The validation of the expression so obtained is shown in Figure 2.11, which presents, under a normalized form and for three different masses, the variation representative curves of:

– the flow $Q(t)$ analytically defined by [2.46];

– the intensity $I(t)$ stemmed from the numeric simulation of the inductive energy released by a coil in a recursive parallel arrangement of 10 series RC cells, such as (Figure 2.12):

$\alpha=1.8$; $\eta=3$; $n=1.35$; $\omega_0=0.23 rad/s$; $\tau=1.24 s$ for the nominal mass *M*; $\tau=1.5\times1.24 s$ for a mass of $1.5M$; $\tau=2\times1.24$ s for a mass of $2M$.

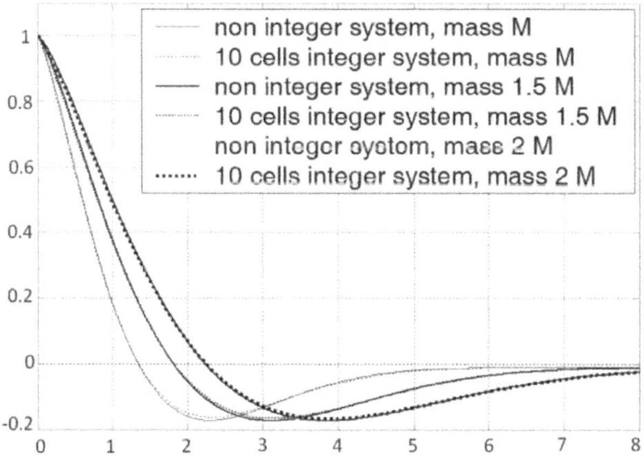

Figure 2.11. *Validation of the analytical expression of the relaxation for various masses: the relaxation shape conservation with a time scale change illustrates the damping robustness*

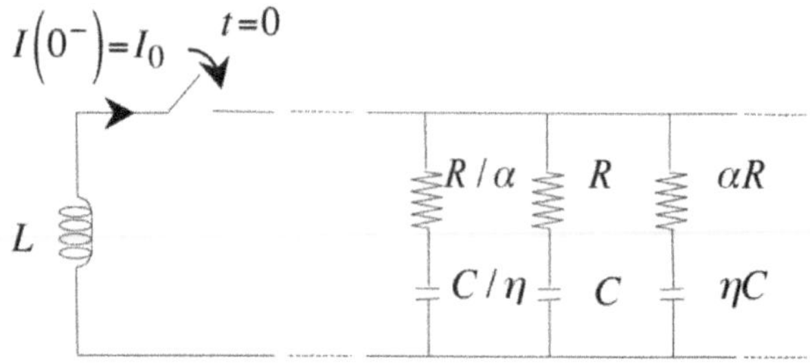

Figure 2.12. *In free state ($t>0$), the coil L releases its inductive energy $LI_0^2/2$ in the recursive parallel arrangement of series RC cells, like the water mass M releases its kinetic energy $MV_0^2/2$ in the water-dyke interface*

2.5. From a non-integer differential equation to relaxation damping robustness

The aim of this section is to demonstrate the *insensitiveness* (or *robustness*) of the relaxation damping ratio through:

– an operational approach which leads to *isodamping half-straight lines*;

– a frequency approach which leads to a *frequency template*.

These approaches constitute major contributions of our works that we have always inscribed in the framework of *fractal robustness*, defined as *damping robustness* that *fractality* ensures through *non-integer differentiation.*

Thus, demonstrated and associated with a fractal context, this robustness is validated in section 2.6 by an *experimental simulation in analogical electronics*.

2.5.1. *Operational approach*

2.5.1.1. *A multiform characteristic equation*

The characteristic equation of the differential equation

$$\tau^n \left(\frac{d}{dt}\right)^n P(t) + P(t) = 0 \qquad [2.47]$$

or

$$\tau^n \left(\frac{d}{dt}\right)^n Q(t) + Q(t) = 0\,, \qquad [2.48]$$

admits an expression of the form:

$$(\tau s)^n + 1 = 0 \,. \qquad [2.49]$$

By writing the (complex) operational variable s under the polar form

$$s = |s| e^{j\theta} \text{ with } \theta = \theta_0 + 2k\pi, \text{ where } k \text{ is an integer,} \qquad [2.50]$$

it is possible to express s^n in conformity with the relation:

$$s^n = |s|^n \, e^{jn\theta_0} \, e^{j2nk\pi}, \qquad [2.51]$$

a form that dictates the consideration of two distinct cases:

– if n is integer, $e^{j2nk\pi} = 1 \;\; \forall k$, which expresses that s^n has a single meaning, thus conveying the *uniformity* of the characteristic equation;

– if n is non-integer, $e^{j2nk\pi}$ depends on *k*, which expresses that s^n has several meanings, thus conveying the *multiformity* of the characteristic equation.

2.5.1.2. *A pair of conjugate complex roots*

To determine the roots of the characteristic equation, it is required to make this equation uniform beforehand.

To that effect, the argument of s must not describe a complete turn:

– which is possible by cutting the complex plan;

– notably along $\mathbb{R}^-$ (Figure 2.13) for two reasons:

- s^n is not defined for $s \in \mathbb{R}^-$, s^n indeed resulting from an *infinity of alternated zeros and poles, distributed in a recursive manner on* $\mathbb{R}^-$ *and infinitely close*;

- moreover, suppressing $\mathbb{R}^-$ as a belonging domain of s ensures the root *Hermitian property* through which *the complex roots are conjugate*.

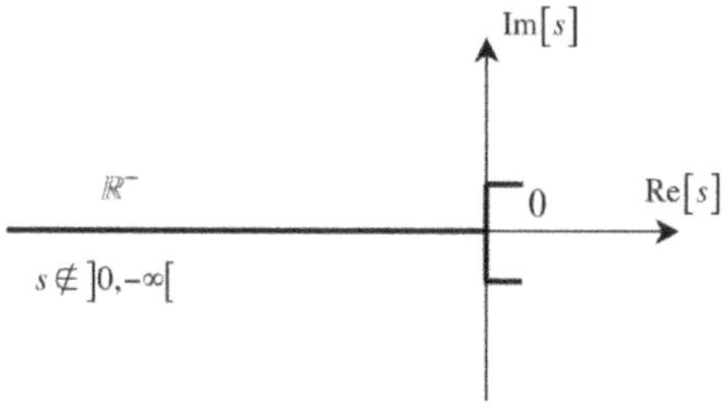

Figure 2.13. *Cutoff of the plane* s *along* $\mathbb{R}^-$: s^n *has a meaning only if* $s \in \mathbb{C} - \mathbb{R}^-$, *namely* $-\pi < \arg s < +\pi$

The roots of the characteristic equation are then the solutions of the characteristic equation made uniform, such that

$$(\tau s)^n = -1 = e^{j(\pi + 2k\pi)}, \; k \text{ integer, with } s \notin \mathbb{R}^-, \qquad [2.52]$$

namely:

$$s_k = \frac{1}{\tau} e^{j\frac{1+2k}{n}\pi}, \qquad [2.53]$$

the arguments of the s_k being constrained by the restriction $]-\pi, +\pi[$ that the cutoff of the complex plane along $\mathbb{R}^-$ imposes, namely

$$-\pi < \frac{1+2k}{n}\pi < +\pi, \qquad [2.54]$$

or

$$-1 < \frac{1+2k}{n} < +1, \qquad [2.55]$$

or even (knowing that n is positive, which allows us to multiply this double inequation by n without changing the direction of the corresponding inequalities)

$$-\frac{n+1}{2} < k < \frac{n-1}{2}. \qquad [2.56]$$

The problem thus comes down to seeking the (integer) values of k that satisfy the double inequality

$$-\frac{n+1}{2} < k < \frac{n-1}{2}, \qquad [2.57]$$

in conformity with the condition on n, namely $1 < n < 2$.

$1 < n < 2$	$1 < n < 2$
$2 < n+1 < 3$	$0 < n-1 < 1$
$-1 > -\frac{n+1}{2} > -\frac{3}{2}$	$0 < \frac{n-1}{2} < \frac{1}{2}$

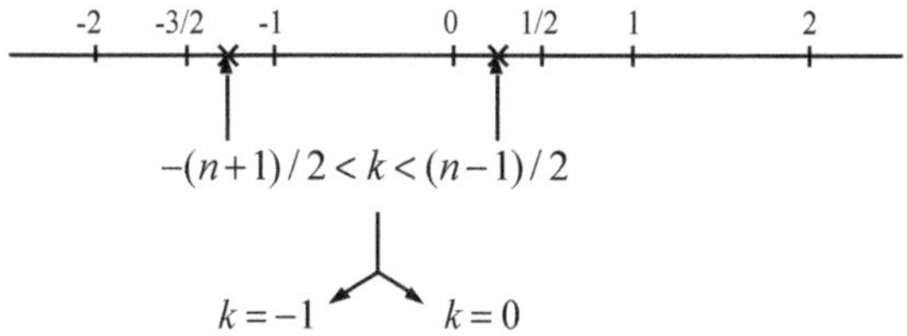

Then, there exist exclusively two roots, respectively, corresponding to $k=0$ and $k=-1$:

$$s_0 = \frac{1}{\tau} e^{j\frac{\pi}{n}} \text{ and } s_{-1} = \frac{1}{\tau} e^{-j\frac{\pi}{n}}. \qquad [2.58]$$

These roots are *complex conjugate* and they form a center angle 2θ with $\theta = \pi - \pi / n$ (Figure 2.14). They determine the oscillatory mode of the relaxation.

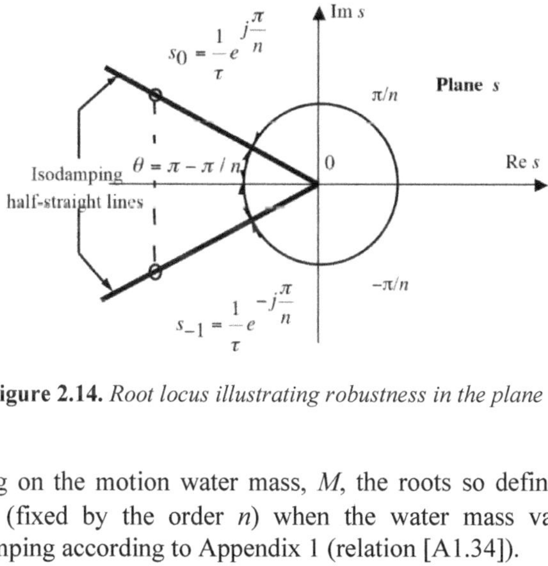

Figure 2.14. *Root locus illustrating robustness in the plane* s

τ depending on the motion water mass, *M*, the roots so defined move with a *constant angle* (fixed by the order *n*) when the water mass varies, this angle determining damping according to Appendix 1 (relation [A1.34]).

Robustness in the plane s is then illustrated by *two half-straight lines*, called *isodamping half-straight lines, forming the same angle* θ *with respect to the real axis*.

The *natural frequency* and the *damping ratio* of the oscillatory mode are directly deduced from the roots (from which the mode results) through their modulus $1/\tau$ and the half-center angle θ that they form. From Appendix 1, it is indeed known that:

– *the natural frequency* ω_p *is given by the projection on the imaginary axis of the root with positive imaginary part*:

$$\omega_p = \omega_p(n,\tau) = \frac{1}{\tau}\sin\theta = \frac{1}{\tau}\sin\left(\pi - \frac{\pi}{n}\right) = \frac{1}{\tau}\sin\frac{\pi}{n}; \qquad [2.59]$$

– *the damping ratio* ζ *is given by the cosine of the half-center angle* θ *that the pair of conjugate complex roots forms*:

$$\zeta = \zeta(n) = \cos\theta = \cos\left(\pi - \frac{\pi}{n}\right) = -\cos\frac{\pi}{n}. \qquad [2.60]$$

This last result clearly shows that the damping ratio, ζ, is:

– exclusively linked to the differentiation non-integer order, n;

– therefore independent of the motion water mass, M;

– thus allowing us to introduce the notion of *robust oscillatory mode*.

2.5.2. *Frequency approach*

The Laplace transform of the differential equation

$$\tau^n \left(\frac{d}{dt}\right)^n P(t) + P(t) = 0, \qquad [2.61]$$

admits an expression of the form:

$$(\tau s)^n P(s) + P(s) = 0, \qquad [2.62]$$

from which we can draw:

$$P(s) = -\left(\frac{1}{\tau s}\right)^n P(s). \qquad [2.63]$$

This operational equation is translated by the functional diagram of Figure 2.15, which recalls one of the free control loops (nil input $E(s)$).

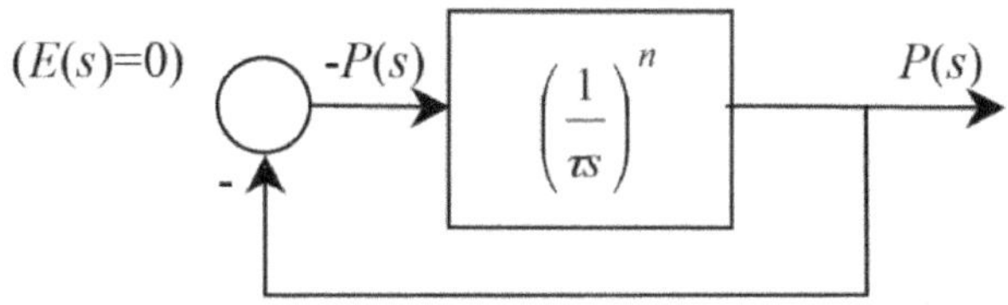

Figure 2.15. *Functional representation of a free control loop*

Because of a unit feedback, the direct chain determines an open-loop transmittance of the form:

$$\beta(s)=\left(\frac{1}{\tau s}\right)^n=\left(\frac{\omega_u}{s}\right)^n, \qquad [2.64]$$

which is the transmittance of a *non-integer integrator* of unit gain (or transition) frequency $\omega_u = 1/\tau$.

Knowing that $\arg\ \beta(j\omega) = -n\pi/2$ and $1 < n < 2$, the Nichols locus of $\beta(j\omega)$ is a *vertical straight line* of abscissa between $-\pi/2$ and $-\pi$ (Figure 2.16).

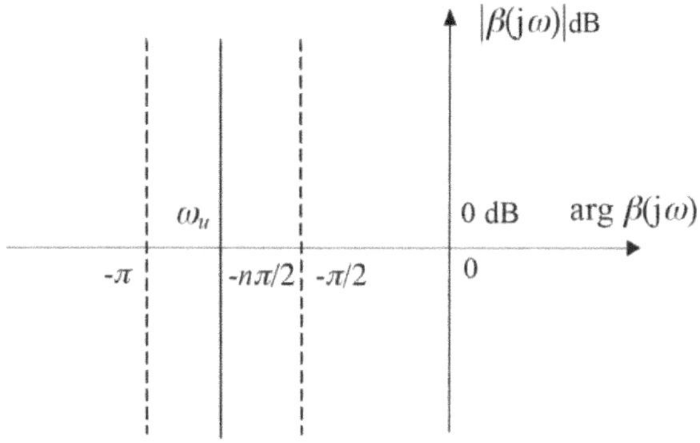

Figure 2.16. *A vertical straight line as an open-loop Nichols locus*

The form and vertical displacement of the open-loop Nichols locus illustrate robustness in the Nichols plane by ensuring the robustness of damping through the *invariance of the distance to the critical point*.

Physically, a dyke cannot be made up of infinitely large and infinitely small cavities. The transitional frequencies they determine cannot take values tending toward infinity as well as zero. Thus, they belong to a bounded frequency interval that:

– covers only partially frequency domain;

– limits the non-integer differentiation of order m (and consecutively of order n) to a medium frequency range.

Thus, the vertical straight line of abscissa $-n\pi/2$ with $n=1+m$ is reduced to a *vertical straight line segment* stretching around the unit gain frequency ω_u (Figure 2.17). This segment is called *open-loop frequency template* (or simply *vertical template* or *template*).

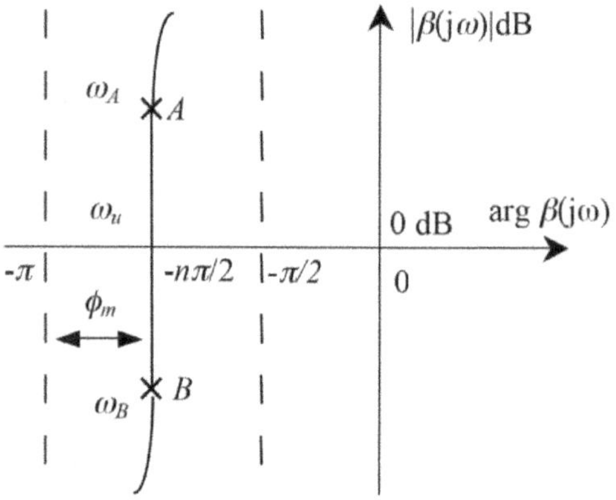

Figure 2.17. *A vertical straight line segment as an open-loop Nichols locus*

$\omega_u = 1/\tau$ depending on the motion water mass, M, the template so defined slides on itself at the time of a variation of the water mass.

Given its form, such a template's vertical displacement ensures:

– the constancy of the phase margin φ_m;

–and, consequently, that of the distance to the critical point (given the large gain margin);

– thus, conveying *damping robustness* (the robustness being all the more better as the template length is great).

2.5.3. *On the representation of robustness in a symbolic domain*

2.5.3.1. *Continuous representation in the plane s*

In frequency domain, the change from a *vertical straight line* to a *vertical straight line segment* as an open-loop Nichols locus expresses the change from an *ideal illustration* to a *more realistic illustration* of robustness (of damping). It is true that in the practical case of the porous dyke, the damping robustness can be ensured only for bounded variations of the motion water mass, *M*.

The transposition to operational domain resides in *two straight line segments forming the same angle θ in relation to the real axis* (Figure 2.18) that indeed offers, *in continuous time,* a more realistic illustration of robustness (of damping) than the ideal illustration through the isodamping half-straight lines.

These two straight line segments constitute a root locus defined by:

$$s = -\alpha \pm jm\alpha , \qquad [2.65]$$

with

$$\alpha_B \leq \alpha \leq \alpha_A \text{ and } m = cte > 0 \ (m = \tan\theta) . \qquad [2.66]$$

2.5.3.2. *Discrete representation in the plane z*

The correspondence between the planes s and z defined by the relation

$$z = e^{Ts} , \qquad [2.67]$$

enables us to represent robustness in the plane z through a root locus defined by:

$$z = e^{-T\alpha} e^{\pm jmT\alpha} , \qquad [2.68]$$

or:

$$|z| = e^{-T\alpha} \text{ and } \arg z = \pm mT\alpha , \qquad [2.69]$$

or even:

$$|z| = e^{\pm \frac{1}{m} \arg z} . \qquad [2.70]$$

This relation expresses the equation of a *logarithmic spiral* whose boundedness of α ($\alpha \in [\alpha_A, \alpha_B]$) determines *two logarithmic spiral segments* AB and $A'B'$, respectively, defined by:

$$mT\alpha_B \leq \arg z \leq mT\alpha_A \quad [2.71]$$

and

$$-mT\alpha_A \leq \arg z \leq -mT\alpha_B. \quad [2.72]$$

Thus, *in discrete time, robustness holds to a root locus formed by two logarithmic spiral segments symmetrical in relation to the real axis* (Figure 2.18).

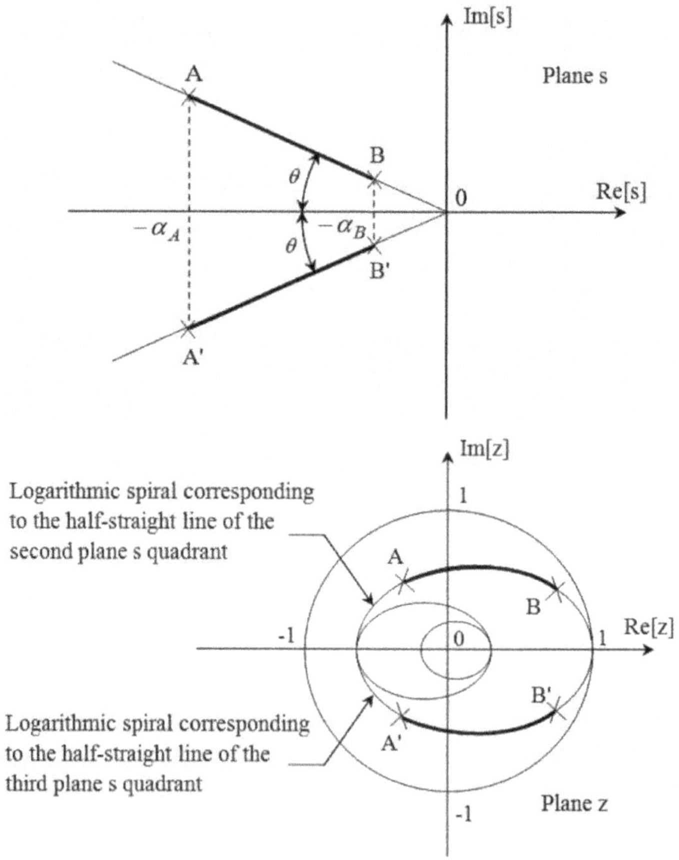

Figure 2.18. *Illustration of robustness in the planes s and z:— ideal case; — real case*

2.6. Validation by an experimental simulation in analog electronics

Damping robustness is here validated by the step response of an electronic circuit that achieves (or synthesizes):

– the most elementary transmittance (no zero at the numerator);

– whose characteristic polynomial (denominator) is no other than the one of the differential equation governing the water relaxation on a porous dyke;

– such a transmittance being, therefore, of the form:

$$F(s)=\frac{1}{1+(\tau s)^n} \text{ with } 1<n<2. \qquad [2.73]$$

2.6.1. *Simulation functional diagram*

By interpreting this transmittance as a *closed-loop synthesis transmittance* and particularly as the tracking transmittance, $F_a(s)$, of an elementary control loop of open-loop transmittance $\beta(s)$, it is possible to write, in accordance with the functional diagram in Figure 2.19:

$$F(s)=F_a(s)=\frac{S(s)}{E(s)}=\frac{\beta(s)}{1+\beta(s)}, \qquad [2.74]$$

from which we draw:

$$\beta(s)=\frac{F(s)}{1-F(s)}, \qquad [2.75]$$

namely, considering the expression of $F(s)$:

$$\beta(s)=\left(\frac{1}{\tau s}\right)^n=\left(\frac{\omega_u}{s}\right)^n, \qquad [2.76]$$

an *open-loop synthesis transmittance* which is the one of a *non-integer integrator*:

– of transition frequency $\omega_u=1/\tau$;

– and of order n between 1 and 2;

– this order being chosen in this case equal to the intermediary value 1.5 which ensures a usual phase margin of 45°.

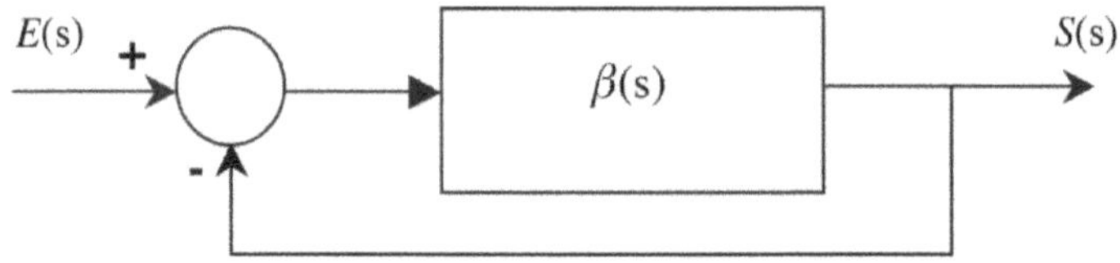

Figure 2.19. *Functional diagram (to be synthesized)*

2.6.2. *Simulation electronic circuit*

2.6.2.1. *Design*

Knowing that the closed-loop dynamic behavior (in the sense of the general form of the step response transient in tracking or regulation):

– only depends on the open-loop (frequency) behavior around the unit gain frequency ω_u;

– $\beta(s)$ can be synthesized by putting in cascade an "order 1 integrator" and an "order 0.5 integrator", respectively, presenting, at least around the frequency ω_u, an order 1 integration and an order 0.5 integration.

2.6.2.2. *Achievement*

Figure 2.20 shows the achievement diagram of the transmittance $F(s)$ in which:

– the comparator uses the structure of a differential amplifier;

– the "order 1 integrator" uses the operational amplifier reverser structure in which the feedback impedance is achieved by a capacitor;

– the "order 0.5 integrator" also uses the operational amplifier reverser structure, but in which the feedback impedance is achieved by a *recursive parallel arrangement of series RC cells*, characterized by a constant ratio α between the resistances and the capacitances of two consecutive cells:

- the ladder network having in this case seven cells whose median cell R_0C_0 is such that $\omega_u = 1/R_0C_0$;

- the recursive factor α, equaling $\sqrt{10} \approx 3.16$ given the normalized values of resistances and capacitances, α indeed having to verify the formula $\alpha^k = 10$ with k a positive integer.

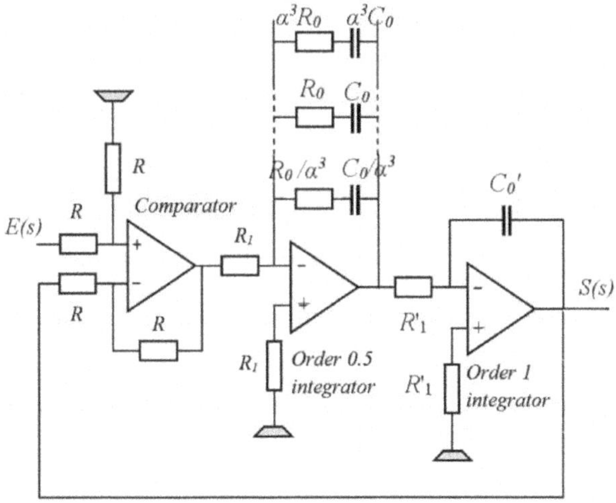

Figure 2.20. *Achievement diagram: the numerical values of the components are the following:* $R_0 = 27\,K\Omega$ *;* $C_0 = 33\,nF$*;* $f_u = 1/2\pi R_0 C_0 \approx 178\,Hz$ *;* $R = 10\,K\Omega$ *;* $R_1 = \sqrt{2}R_0 \approx 38\,K\Omega$ *(actually* $27\,K\Omega$ *in series with* $10\,K\Omega$*, namely* $37\,K\Omega$*, which is set to* $59\,K\Omega$ *for the robustness tests by adding* $22\,K\Omega$ *in series);* $R_1' = R_0 = 27\,K\Omega$ *;* $C_0' = C_0 = 33\,nF$ *;* $\alpha R_0 = 82\,K\Omega$ *;* $\alpha^2 R_0 = 270\,K\Omega$ *;* $\alpha^3 R_0 = 820\,K\Omega$ *;* $R_0/\alpha = 8.2\,K\Omega$ *;* $R_0/\alpha^2 = 2.7\,K\Omega$ *;* $R_0/\alpha^3 = 820\,\Omega$ *;* $\alpha C_0 = 100\,nF$*;* $\alpha^2 C_0 = 330\,nF$*;* $\alpha^3 C_0 = 1\,\mu F$*;* $C_0/\alpha = 10\,nF$*;* $C_0/\alpha^2 = 3.3\,nF$*;* $C_0/\alpha^3 = 1\,nF$ *; in fact, the ratios between the normalized values of the resistances and the capacitances show that* α *lightly varies from a ratio to another, in this case from 3.03 to 3.3, the average (in the geometrical sense) being, however, equal to* $\sqrt{10}$

2.6.2.3. *A greatly simplifying approximation*

By reasoning in frequency domain:

– the idea is to simplify as much as possible the ladder network at the transitional frequency of the median cell R_0C_0 arbitrarily set at ω_u;

– particularly by looking for the condition on the recursive factor α leading to the best structural reduction of the network at the frequency ω_u.

Knowing that α^2 is the ratio between the transitional frequencies of two consecutive cells, we first consider $\alpha^2 >> 1$.

In the network such as the one proposed in Figure 2.21(a):

– the top three cells, whose transitional frequencies are less than $\omega_u = 1/R_0C_0$, have then cut at the frequency ω_u and therefore present a resistive character at this frequency; hence, their equivalent resistance is $\alpha R_0 // \alpha^2 R_0 // \alpha^3 R_0$ (Figure 2.21(b));

– on the contrary, the bottom three cells, whose transitional frequencies are greater than $\omega_u = 1/R_0C_0$, have then not cut at the frequency ω_u and therefore present a capacitive character at this frequency; hence their equivalent capacitance is $C_0/\alpha + C_0/\alpha^2 + C_0/\alpha^3$ (Figure 2.21(b)).

Second, if we consider $\alpha >> 1$ (a stronger condition than the former):

– the resistance equivalent to the top three cells becomes large compared to R_0;

– the capacitance equivalent to the bottom three cells becomes small compared to C_0;

– the reduced network of Figure 2.21(b) then admits a new reduction to say the least drastic, the reduction only leading to the cell R_0C_0 (Figure 2.21(c)).

REMARK.– Somewhat rough in our study case where α is barely greater than 3, this reduction is nonetheless very useful because of the simplification it brings, especially as the error induced on the value of the frequency ω_u is not liable to affect the validity of the damping robustness tests.

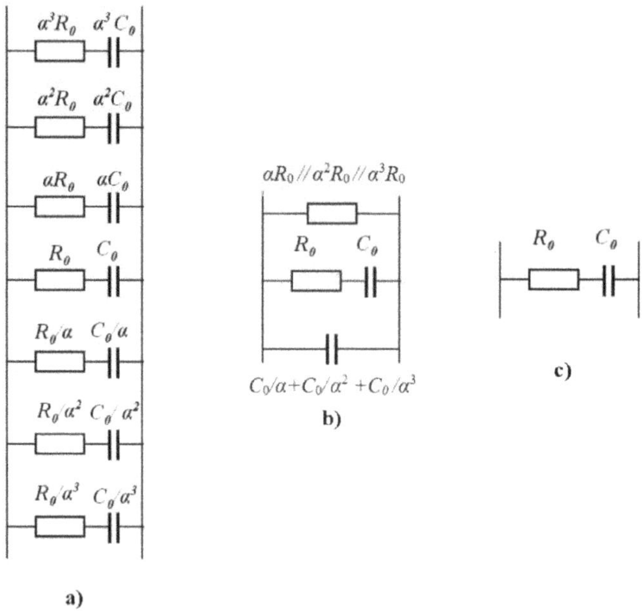

Figure 2.21. *Reduction of the ladder network at frequency $\omega_u = 1/R_0C_0$: a) network in its initial form, b) first reduction level for $\alpha^2 >> 1$ and c) second reduction level for $\alpha >> 1$*

2.6.2.4. *Modeling*

2.6.2.4.1. Order 1 integrator

The order 1 integrator frequency response is of the form:

$$I_1(j\omega) = -\frac{1}{jR_1'C_0'\omega}, \quad [2.77]$$

or, at frequency $\omega_u = 1/R_0C_0$:

$$I_1(j\omega_u) = -\frac{R_0C_0}{jR_1'C_0'}, \quad [2.78]$$

from which the gain at ω_u is drawn, which is set at 1:

$$|I_1(j\omega_u)| = \frac{R_0C_0}{R_1'C_0'} = 1, \quad [2.79]$$

hence, the relation:

$$R_1'C_0' = R_0C_0, \quad [2.80]$$

which is verified by setting, for instance:

$$R_1' = R_0 \text{ and } C_0' = C_0. \quad [2.81]$$

Given that $R_1'C_0' = R_0C_0 = 1/\omega_u$, $I_1(j\omega)$ becomes:

$$I_1(j\omega) = -\frac{\omega_u}{j\omega}, \quad [2.82]$$

namely, in operational domain:

$$I_1(s) = -\frac{\omega_u}{s}. \quad [2.83]$$

2.6.2.4.2. Order 0.5 integrator

Given the reduction of the ladder network at the frequency ω_u (Figure 2.21), the frequency response of the order 0.5 integrator at this frequency can be written as:

$$I_{0.5}(j\omega_u) = -\frac{1}{R_1}\left(R_0 + \frac{1}{jC_0\omega_u}\right), \quad [2.84]$$

namely:

$$I_{0.5}(j\omega_u) = -\frac{R_0}{R_1}\left(1 + \frac{1}{jR_0C_0\omega_u}\right), \quad [2.85]$$

or, given that $\omega_u = 1/R_0C_0$:

$$I_{0.5}(j\omega_u) = -\frac{R_0}{R_1}(1-j). \qquad [2.86]$$

The gain at ω_u which is also set at 1, namely

$$|I_{0.5}(j\omega_u)| = \frac{R_0}{R_1}\sqrt{2} = 1, \qquad [2.87]$$

then leads to the relation:

$$R_1 = \sqrt{2}R_0. \qquad [2.88]$$

The smoothing of the ladder network Bode asymptotic diagrams around ω_u leads to an admittance in $(j\omega)^{0.5}$ which makes it possible to express, around this frequency, $I_{0.5}(j\omega)$ in the form:

$$I_{0.5}(j\omega) = -\left(\frac{\omega_u}{j\omega}\right)^{0.5}, \qquad [2.89]$$

namely, in operational domain:

$$I_{0.5}(s) = -\left(\frac{\omega_u}{s}\right)^{0.5}. \qquad [2.90]$$

2.6.2.4.3. Open loop and closed loop

Because of a unitary feedback, the action chain determines an open-loop transmittance of the form:

$$\beta(s) = I_1(s)I_{0.5}(s), \qquad [2.91]$$

namely, given the respective expressions of $I_1(s)$ and $I_{0.5}(s)$:

$$\beta(s) = \left(\frac{\omega_u}{s}\right)\left(\frac{\omega_u}{s}\right)^{0.5} = \left(\frac{\omega_u}{s}\right)^{1.5}, \qquad [2.92]$$

from which, as expected, a closed-loop transmittance in conformity with the one said of synthesis is drawn:

$$F(s) = \frac{\beta(s)}{1+\beta(s)} = \frac{1}{1+\left(\frac{s}{\omega_u}\right)^{1.5}}. \qquad [2.93]$$

2.6.3. *On the simulation itself: damping robustness tests*

The variations of the motion water mass, *M*, are here simulated electronically through variations of the open-loop unit gain frequency, ω_u.

In order to *globally* vary this frequency, it suffices to act on the open-loop gain by acting on the value of a component on which it depends, for instance the resistance R_1 which is between the comparator output and the following operational amplifier reverser input.

It is, indeed, the resistance R_1 that has been modified to carry out our robustness tests. Initially equal to a nominal value of $37\ \text{K}\Omega$, it has successively been decreased and increased in conformity with the minimal and maximal values of $10\ \text{K}\Omega$ and $59\ \text{K}\Omega$, respectively.

The step responses observed on the oscilloscope screen (Figure 2.22) for the nominal, minimal and maximal values of R_1 so defined obviously show that *the transient keeps its form with a change of time scale*, thus expressing *damping robustness* and, by transposition to mechanics domain, *putting in the wrong the mass-damping interdependence*.

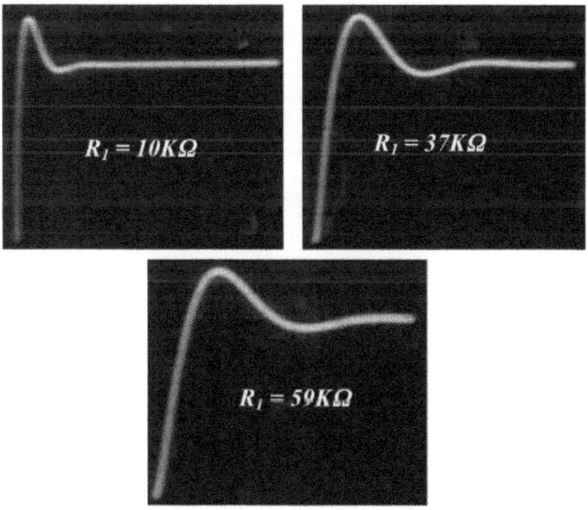

Figure 2.22. *Step responses recorded by the oscilloscope: the circuit input is a square signal of amplitude 3 V and frequency 16 Hz; the oscilloscope tunings are of 0.2 V/div for the voltage and 2 ms/div for the time*

2.7. Bibliography

[OUS 91] OUSTALOUP A., *La Commande CRONE*, Hermès, Paris, 1991.

[OUS 95] OUSTALOUP A., *La dérivation non entière: théorie, synthèse et applications*, Hermès, Paris, 1995.

[OUS 99] OUSTALOUP A., SABATIER J., LANUSSE P., "From fractal robustness to the CRONE control", *Fractional Calculus and Applied Analysis (FCAA): An International Journal for Theory and Applications*, vol. 2, no. 1, pp. 1–30, January 1999.

[OUS 05] OUSTALOUP A., COIS O., LANUSSE P., *et al.*, "A survey on the CRONE approach", Plenary lecture, *IEEE International Conference on Systems, Signals, Devices SSD*, Sousse, Tunisia, 21–24 March 2005.

Chapter 3

Non-Integer Differentiation, its Memory and its Synthesis

3.1. Introduction

Presenting non-integer differentiation through the extension, to the non-integer case, of the generic form of integer differentiation, is undoubtedly the simplest and most natural way to tackle and apprehend, *in discrete time*, the generalized differentiation concept. This is the subject of section 3.2 that uses this approach to obtain, through an immediate generalization, a very general *discrete form of derivative*, the one proposed by Grünwald in 1867.

Presenting non-integer differentiation from repeated integer integration via non-integer integration is undoubtedly the most natural and direct way to introduce, *in continuous time*, the impulse response and the transmittance of the non-integer differentiation operator. This is the subject of section 3.3 that uses this approach to successively determine, *under a continuous form*:

– repeated integer integration and non-integer integration;

– non-integer differentiation that, combining an integer differentiation operation and a non-integer integration operation, leads (according to the order of the operations) to the Riemann–Liouville derivative (1832) or the Caputo derivative (1967));

– the impulse response and the transmittance of the non-integer integration operator, from which we deduce the ones of the non-integer differentiation operator.

So, presented in discrete time and then in continuous time, non-integer differentiation is the subject of a study of its properties in *sinusoidal steady state* (section 3.4). In this operating state and with regard to kinematic magnitudes, therefore in terms of position, speed and acceleration, the non-integer derivative of position takes into account the "position and speed" or "speed and acceleration" whether the differentiation order n is "between 0 and 1" or "between 1 and 2". More precisely, taking into account these kinematic magnitudes are under the form of a *linear combination* of position and speed for n between 0 and 1, on the one hand, and of speed and acceleration for n between 1 and 2, on the other hand.

Section 3.5 is devoted to the memory phenomenon associated with non-integer differentiation. The discrete form of the non-integer derivative, as presented in section 3.2, directly shows that the function to be differentiated is taken into account through its values at all the past instants (including the calculation instant). More precisely, the whole function past being taken into account by a *weighted sum of its samples*, non-integer differentiation thus introduces a *memory notion* such that the past *attenuation* or *accentuation* is imposed by the differentiation order (therefore, by a single parameter). As this memory notion is not without the evocation of a *subtle form of memory* (through which the recollection of an event depends on the nature of this event), an investigation trail is proposed by considering (as a diversion) an aspect of the human memory: it is about the *serial position effect* studied in cognitive psychology and for which the *primacy effect* is modeled by a *behavior model* reduced to an integration non-integer order.

Finally, section 3.6 deals with the synthesis of non-integer differentiation through a *realistic synthesis* turning on a *non-integer differentiator bounded in frequency*. This synthesis is an *approximation* of the differentiator ideal version through a *(limited) recursive distribution of countable zeros and poles*. Real non-integer differentiator synthesis is based on a recursive distribution of countable real zeros and poles. Complex non-integer differentiator synthesis is based on a recursive distribution of complex zeros and poles which can also be counted [OUS 00]. The complex non-integer order is determined analytically as a function of complex recursive factors defined by the relative position of the distributions of zeros and poles. The expression so obtained is an extension to the complex number set of the expression determined graphically in the case of a real non-integer order. The real and imaginary orders are deduced and their ratio is expressed in order to reveal a specific property. An elementary stability study establishes the necessary and sufficient stability condition for the synthesized differentiator. A condition on the number of zeros or poles to be used is obtained. This number must be greater than a certain limit. The expression of this lower limit is determined analytically. This minimum is high enough to ensure the stability of the synthesized differentiator, but not enough to provide a sufficient approximation of the differentiator to be synthesized. A method of calculation is proposed which does provide a high enough number of zeros or poles. This determines the minimum number of zeros or poles to ensure an impulse response with sufficiently equivalent energy to the differentiator to be synthesized. The impulse response energy is calculated using the Aström formula [AST 70]. Our whole approach is well validated through the comparison between the impulse responses obtained for the differentiator to be synthesized and the differentiator synthesized with various numbers of zeros and poles.

3.2. From integer differentiation to non-integer differentiation

3.2.1. *Toward non-integer differentiation: the adopted approach*

A natural approach of the non-integer derivative of a causal or non-causal function $f(t)$, using a generalization of the well-known integer order derivative definition, consists of:

– expressing the order 1 derivative under an adequate form;

– also expressing the order 2 derivative under an analogous form;

– then extending to the non-integer case the resulting generic form;

– to obtain thus a generalization to the integer and non-integer case.

3.2.2. *Generic form of the order 1 and 2 derivatives*

The order 1 left derivative is defined by:

$$D^1 f(t) = \lim_{h \to 0} \frac{f(t) - f(t-h)}{h}. \qquad [3.1]$$

A discretization of t to the sampling step h, namely $t = Kh$, is translated by:

$$D^1 f(t) = \lim_{h \to 0} \frac{f(Kh) - f((K-1)h)}{h}. \qquad [3.2]$$

The introduction of the delay operator q^{-1} applicable to a concrete function and defined by

$$q^{-1} f(Kh) = f((K-1)h), \qquad [3.3]$$

allows us to write:

$$D^1 f(t) = \lim_{h \to 0} \frac{1-q^{-1}}{h} f(Kh) . \qquad [3.4]$$

A similar calculation carried out for an order 2 derivative leads us to the following equation:

$$D^2 f(t) = \lim_{h \to 0} \left(\frac{1-q^{-1}}{h} \right)^2 f(Kh) . \qquad [3.5]$$

3.2.3. *Generalization to the integer and non-integer case*

The generalization to any (integer or non-integer, real or complex) order is immediate and leads us to the definition proposed by Grünwald–Letnikov in 1867, that is to say:

$$D^n f(t) = \lim_{h \to 0} \left(\frac{1-q^{-1}}{h} \right)^n f(Kh) , \qquad [3.6]$$

namely, developing $\left(1-q^{-1}\right)^n$ by the Newton's binomial formula:

$$D^n f(t) = \lim_{h \to 0} \frac{1}{h^n} \left[\sum_{k=0}^{\infty} (-1)^k \frac{n(n-1)(n-2)...(n-k+1)}{k!} q^{-k} \right] f(Kh) , \qquad [3.7]$$

or:

$$D^n f(t) = \lim_{h \to 0} \frac{1}{h^n} \sum_{k=0}^{\infty} (-1)^k \frac{n(n-1)(n-2)...(n-k+1)}{k!} q^{-k} f(Kh) , \qquad [3.8]$$

or even, given that

$$q^{-k} f(Kh) = f((K-k)h) = f(t-kh) : \qquad [3.9]$$

$$D^n f(t) = \lim_{h \to 0} \sum_{k=0}^{\infty} (-1)^k \frac{n(n-1)(n-2)...(n-k+1)}{k!} f(t-kh) , \qquad [3.10]$$

or else, under a more condensed writing:

$$D^n f(t) = \lim_{h \to 0} \frac{1}{h^n} \sum_{k=0}^{\infty} a_k f(t-kh), \qquad [3.11]$$

by putting:

$$a_k = (-1)^k \frac{n(n-1)(n-2)...(n-k+1)}{k!}, \qquad [3.12]$$

or, using the usual symbolism:

$$a_k = (-1)^k \binom{n}{k} \text{ with } a_k = 1 \text{ for } k = 0. \qquad [3.13]$$

In the particular case of a causal (nil for $t < 0$) function $f(t)$, we have:

$f(t-kh) = 0$ for $t - kh < 0$, namely for $kh > t = Kh$, therefore for $k > K$.

The extended sum from $k = 0$ to $k = \infty$ is then reduced to the extended sum from $k = 0$ to $k = K$, namely:

$$D^n f(t) = \frac{1}{h^n} \sum_{k=0}^{K} (-1)^k \binom{n}{k} f(t-kh). \qquad [3.14]$$

3.3. From repeated integer integration to non-integer differentiation through non-integer integration

Given that the order 0 integration operator, I^0, and the order 0 differentiation operator, D^0, are such that

$$I^0(f)(t) = D^0(f)(t) = f(t), \qquad [3.15]$$

in non-integer integration or differentiation, the integration or differentiation order, real here, is strictly positive.

Regarding the structure of this section, the following developments respect the guideline used by Trigeassou [TRI 11, TRI 12].

3.3.1. *Repeated integer integration*

Let $f(t)$ be a continuous time function, real or complex, nil for $t<0$. Its successive (integer) integrals are, respectively, written as:

$$I^1(f)(t)=I(f)(t)=\int_0^t f(\theta)\,d\theta, \tag{3.16}$$

Riemann integration of f between 0 and t;

$$I^2(f)(t)=I\left(I^1(f)\right)(t)=\int_0^t I^1(f)(\theta)\,d\theta, \tag{3.17}$$

one time repeated Riemann integration of f between 0 and t;

$$I^n(f)(t)=I\left(I^{n-1}(f)\right)(t)=\int_0^t I^{n-1}(f)(\theta)\,d\theta, \tag{3.18}$$

$n-1$ times repeated Riemann integration of f between 0 and t, n being an integer greater than or equal to 1.

Let us show that, for any integer $n\geq 1$,

$$I^n(f)(t)=\int_0^t \frac{(t-\theta)^{n-1}}{(n-1)!}f(\theta)\,d\theta. \tag{3.19}$$

This formula is true for $n=1$ as, knowing that $(t-\theta)^0=1$ and $0!=1$, it then becomes:

$$I^1(f)(t)=\int_0^t f(\theta)\,d\theta. \tag{3.20}$$

Let us assume that this formula is true at the rank n, and let us show that it is also true at the rank $n+1$.

Thus, it is assumed that

$$I^n(f)(t) = \int_0^t \frac{(t-\theta)^{n-1}}{(n-1)!} f(\theta)\,d\theta, \qquad [3.21]$$

and it can be written for $n+1$:

$$\begin{aligned} I^{n+1}(f)(t) &= I\left(I^n(f)\right)(t) = \int_0^t I^n(f)(\theta)\,d\theta \\ &= \int_0^t \left(\int_0^\theta \frac{(\theta-u)^{n-1}}{(n-1)!} f(u)\,du \right) d\theta, \end{aligned} \qquad [3.22]$$

namely:

$$I^{n+1}(f)(t) = \iint_{\mathrm{D}} \frac{(\theta-u)^{n-1}}{(n-1)!} f(u)\,du\,d\theta, \qquad [3.23]$$

a double integral whose integration variables are θ and u. The integration domain is defined by

$$\mathrm{D} = \left\{(\theta,u); 0 \le u \le \theta \text{ and } 0 \le O \le t\right\} \qquad [3.24]$$

if the integration is carried out first on the variable u and then on the variable θ, or by

$$\mathrm{D} = \left\{(\theta,u); 0 \le u \le t \text{ and } \mathrm{u} \le \theta \le t\right\} \qquad [3.25]$$

if the integration is carried out the other way around.

Therefore:

$$\begin{aligned} I^{n+1}(f)(t) &= \int_0^t f(u) \left(\int_u^t \frac{(\theta-u)^{n-1}}{(n-1)!} d\theta \right) du \\ &= \int_0^t f(u) \left[\frac{(\theta-u)^n}{n!} \right]_{\theta=u}^{\theta=t} du \\ &= \int_0^t \frac{(t-u)^n}{n!} f(u)\,du. \end{aligned} \qquad [3.26]$$

The formula is then true at the rank $n+1$. Thus, by recurrence on n, this formula [3.19] is true for any non-zero positive integer n.

3.3.2. *Non-integer integration*

The generalization of the repeated integer integration to the integer and non-integer case is carried out by replacing the factorial function $(n-1)!$ with the gamma function $\Gamma(n)$ (1811), as $\Gamma(n)=(n-1)!$ for any integer $n \geq 1$, namely

$$\Gamma(n)=\int_0^{\infty} x^{n-1} e^{-x} dx, \quad n>0, \tag{3.27}$$

which enables us to obtain the Riemann–Liouville integral, said non-integer or fractional (actually n is a strictly positive real), of the function f between 0 and t, namely:

$$I^n(f)(t)=\int_0^t \frac{(t-\theta)^{n-1}}{\Gamma(n)} f(\theta) d\theta, \quad n>0. \tag{3.28}$$

This integral is nothing but the convolution product of the function $f(t)$ with the impulse response $h_n(t)$ of the non-integer integration operator, I^n, namely:

$$I^n(f)(t)=h_n(t) * f(t), \tag{3.29}$$

with

$$h_n(t)=\frac{t^{n-1}}{\Gamma(n)} u(t)=\mathcal{L}^{-1}\left[\frac{1}{s^n}\right], \tag{3.30}$$

$u(t)$ being the unit step function and $1/s^n$ the non-integer integration symbolic operator. The convolution product so defined is indeed expressed by either of the two relations:

$$I^n(f)(t)=\int_0^t \frac{\theta^{n-1}}{\Gamma(n)} f(t-\theta) d\theta \tag{3.31}$$

or

$$I^n(f)(t)=\int_0^t \frac{(t-\theta)^{n-1}}{\Gamma(n)} f(\theta)\,d\theta, \qquad [3.32]$$

this result having been the subject of an equivalence study with the Grünwald derivative in the case of a strictly negative differentiation order [OUS 95].

3.3.3. *Non-integer differentiation*

As a preamble, it is convenient to specify that in non-integer differentiation, the differentiation strategy consists of coming down to a non-integer integration through a formulation that enables to combine an integer differentiation and a non-integer integration, the chronology of these operations being at the origin of two methods of non-integer differentiation:

– the Riemann–Liouville one which consists of, first, integrating to a non-integer order and, then, differentiating to an integer order;

– the Caputo one which consists of, first, differentiating to an integer order and, then, integrating to a non-integer order.

D denoting the differentiation operator d/dt, and n the differentiation order, strictly positive real, bounded by the two consecutive integers $N-1$ and N, namely $N-1 \leq n < N$, it is possible to write:

$$D^n(f)(t)=D^{N-N+n}(f)(t), \quad n>0, \qquad [3.33]$$

a general relation which, under the assumption of nil initial conditions (as specified by the appendix (*)), is declined according to the two dual formulations:

$$D^n(f)(t)=D^N\left(D^{-N+n}(f)\right)(t) \qquad [3.34]$$

and

$$D^n(f)(t)=D^{-N+n}\left(D^N(f)\right)(t), \qquad [3.35]$$

or, knowing that the differentiation to the negative order $-N+n$, namely D^{-N+n}, is actually an integration to the positive order $N-n$, namely I^{N-n}:

$$D^n(f)(t)=D^N\left(I^{N-n}(f)\right)(t) \qquad [3.36]$$

and

$$D^n(f)(t) = I^{N-n}\left(D^N(f)\right)(t). \qquad [3.37]$$

APPENDIX (*) – If $F(s) = \mathcal{L}\left[f(t)\right]$, the Laplace transform as expressed here, namely

$$\begin{aligned} \mathcal{L}\left[D^N\left(I^{N-n}(f)\right)(t)\right] &= s^N \mathcal{L}\left[I^{N-n}(f)(t)\right] \\ &= s^N \frac{1}{s^{N-n}} F(s) \\ &= s^n F(s) \\ &= \mathcal{L}\left[D^n(f)(t)\right], \end{aligned} \qquad [3.38]$$

which involves $D^n(f)(t) = D^N\left(I^{N-n}(f)\right)(t)$, is true only if $g(t) = I^{N-n}(f)(t)$ verifies the initial conditions:

$$g(0) = g^{(1)}(0) = \cdots = g^{(N-1)}(0) = 0. \qquad [3.39]$$

Similarly, the Laplace transform as expressed here, namely

$$\begin{aligned} \mathcal{L}\left[I^{N-n}\left(D^N(f)\right)(t)\right] &= \frac{1}{s^{N-n}} \mathcal{L}\left[D^N(f)(t)\right] \\ &= \frac{1}{s^{N-n}} s^N F(s) \\ &= s^n F(s) \\ &= \mathcal{L}\left[D^n(f)(t)\right], \end{aligned} \qquad [3.40]$$

which involves $D^n(f)(t) = I^{N-n}\left(D^N(f)\right)(t)$, is true only if $f(t)$ verifies the initial conditions:

$$f(0) = f^{(1)}(0) = \cdots = f^{(N-1)}(0) = 0. \qquad [3.41]$$

Actually, in both cases, the nil initial condition necessity comes from the fact that:

– if for a strictly positive real m, one always has the relation

$$\mathcal{L}\left[I^m(f)(t)\right] = F(s)/s^m \qquad [3.42]$$

– on the contrary, for a strictly positive integer m, one uniquely has the relation

$$\mathcal{L}\left[D^m(f)(t)\right] = s^m F(s) \qquad [3.43]$$

for

$$f(0) = f^{(1)}(0) = \cdots = f^{(m-1)}(0) = 0. \qquad [3.44]$$

3.3.3.1. *Riemann–Liouville and Caputo definitions*

The first formula [3.36], relative to Riemann–Liouville, has a meaning if $f(t)$ is integrable to the order $N-n$ and if $g(t) = I^{N-n}(f)(t)$ is N times differentiable. The second formula [3.37], relative to Caputo, has a meaning if $f(t)$ is N times differentiable and if $g(t) = f^{(N)}(t)$ is integrable to the order $N-n$.

Given relation [3.28], these two formulas become:

$$D^n(f)(t) = \frac{d^N}{dt^N}\int_0^t \frac{(t-\theta)^{N-n-1}}{\Gamma(N-n)} f(\theta)\, d\theta, \qquad [3.45]$$

Riemann–Liouville derivative (1832), and

$$D^n(f)(t) = \int_0^t \frac{(t-\theta)^{N-n-1}}{\Gamma(N-n)} \frac{d^N f(\theta)}{dt^N}\, d\theta, \qquad [3.46]$$

Caputo derivative (1967).

3.3.3.2. *Impulse response of the non-integer differentiation operator*

In non-integer integration, the impulse response of the non-integer integration operator, I^n, is given by the original of the non-integer integration symbolic operator, $1/s^n$, namely (relation [3.30]):

$$\mathcal{L}^{-1}\left[\frac{1}{s^n}\right] = \frac{t^{n-1}}{\Gamma(n)} u(t) \quad \forall t \quad \text{for } n > 0. \qquad [3.47]$$

The change of n into $-n$ enables us to write:

– either

$$\mathcal{L}^{-1}\left[s^n\right]=\frac{t^{-n-1}}{\Gamma(-n)}u(t) \quad \forall t \quad \text{for } n<0, \qquad [3.48]$$

the condition $n<0$ indeed ensuring the conformity of relation [3.48] with relation [3.47];

– or

$$\mathcal{L}^{-1}\left[s^n\right]=\frac{t^{-n-1}}{\Gamma(-n)}u(t) \quad \forall t\neq 0 \quad \text{for } n>0 \text{ and } n\notin \mathbb{N}, \qquad [3.49]$$

knowing that the impulse response of the non-integer differentiation operator, D^n, or of the non-integer differentiation symbolic operator, s^n, is not defined at $t=0$ for a differentiation non-integer order strictly positive (see the synthesis of s^n through electrical components).

Concerning the gamma function as it figures in the whole of these relations, it is convenient to note that if $\Gamma(n)$ exists for $n>0$ (relation [3.47]), $\Gamma(-n)$ exists for $n<0$ (relation [3.48]) and for $n\in \mathbb{R}^+ - \mathbb{N}$ where $\mathbb{R}^+ = [0,+\infty[$ and $\mathbb{N}=0,1,2,\ldots$ (relation [3.49]).

3.4. Non-integer differentiation in sinusoidal steady state

3.4.1. *A definition of the Fresnel vector*

The Fresnel vector of a sinusoidal concrete function $f(t)=f_m\cos(\omega t+\phi)$ can be defined as the image vector of the complex quantity (or function) $F(t)$ associated with the concrete function $f(t)$ and whose real part is nothing but $f(t)$, namely (Figure 3.1):

$$F(t)=f_m \exp j(\omega t+\phi), \qquad [3.50]$$

$f(t)$ indeed verifying the equation

$$f(t)=\mathrm{Re}\left[F(t)\right]. \qquad [3.51]$$

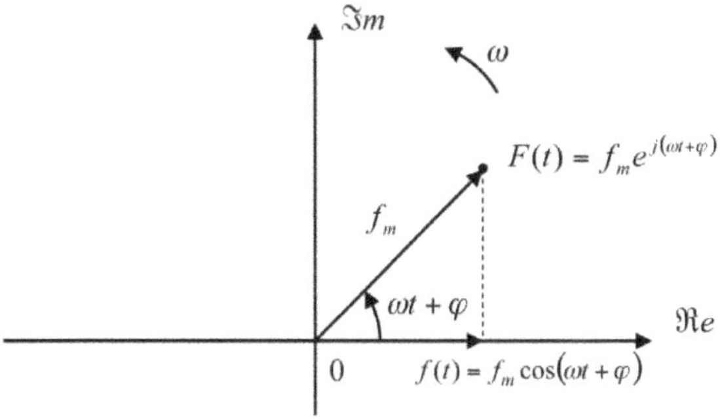

Figure 3.1. *Fresnel vector rotating (counterclockwise) at the angular speed* ω

3.4.2. *A direct application to kinematic magnitudes*

In sinusoidal steady state and in terms of position, speed and acceleration, the first *integer derivatives* of position are, respectively, written, $x(t)=x_m\cos\omega t$ defining then the position:

$$x^{(0)}(t)=\left(\frac{d}{dt}\right)^0 x(t)=x(t)=x_m\cos\omega t \qquad [3.52]$$

$$x^{(1)}(t)=\left(\frac{d}{dt}\right)^1 x(t)=v(t)=x_m\omega\cos\left(\omega t+\frac{\pi}{2}\right) \qquad [3.53]$$

$$x^{(2)}(t)=\left(\frac{d}{dt}\right)^2 x(t)=\gamma(t)=x_m\omega^2\cos(\omega t+\pi), \qquad [3.54]$$

where $v(t)$ and $\gamma(t)$ denote the speed and the acceleration.

By associating with $x(t)$, $v(t)$ and $\gamma(t)$ the complex quantities $X(t)$, $V(t)$ and $\Gamma(t)$ whose real parts are $x(t)$, $v(t)$ and $\gamma(t)$, it is possible to write:

$$X(t)=x_m\exp j\omega t, \qquad [3.55]$$

$$V(t)=x_m\omega\exp j\left(\omega t+\frac{\pi}{2}\right) \qquad [3.56]$$

and

$$\Gamma(t)=x_m\omega^2\exp j(\omega t+\pi), \qquad [3.57]$$

the corresponding Fresnel vectors being represented by Figure 3.2 in conformity with the definition given in section 3.4.1.

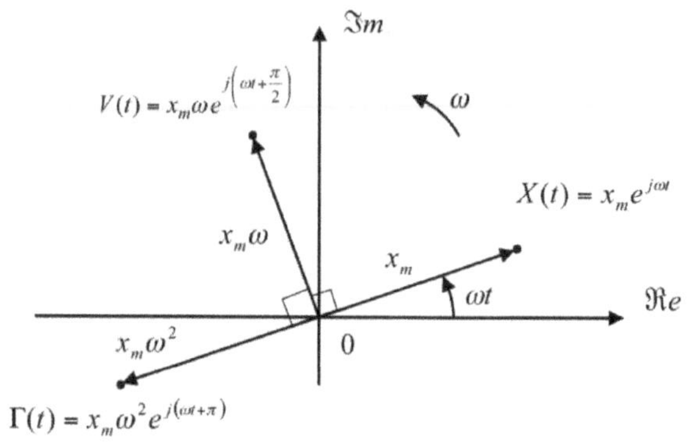

Figure 3.2. *Fresnel vectors associated with the kinematic magnitudes that are the position, the speed and the acceleration*

3.4.3. *Non-integer derivative of position*

Still defined by $x(t) = x_m \cos \omega t$, the position admits for non-integer derivative of order n:

$$x^{(n)}(t) = \left(\frac{d}{dt}\right)^n x(t) = x_m \omega^n \cos\left(\omega t + n\frac{\pi}{2}\right), \quad [3.58]$$

knowing that the frequency operator of non-integer differentiation, $(j\omega)^n$, leads:

– to multiplying the amplitude x_m of $x(t)$ by the gain ω^n of the operator;

– and to adding, to the phase ωt of $x(t)$, the phase $n\frac{\pi}{2}$ of the operator.

In a Fresnel vector representation of $x^{(n)}(t)$, it is convenient to introduce the complex quantity:

$$X_n(t) = x_m \omega^n \exp j\left(\omega t + n\frac{\pi}{2}\right), \quad [3.59]$$

such that

$$\text{Re}\left[X_n(t)\right] = x^{(n)}(t), \quad [3.60]$$

the differentiation order n being here successively considered:

– between 0 and 1 in the first case;

– between 1 and 2 in the second case.

3.4.3.1. *First case:* $0 \leq n \leq 1$

In the interval $[0,1]$, the Fresnel vector representation of $x^{(n)}(t)$ is illustrated in Figure 3.3.

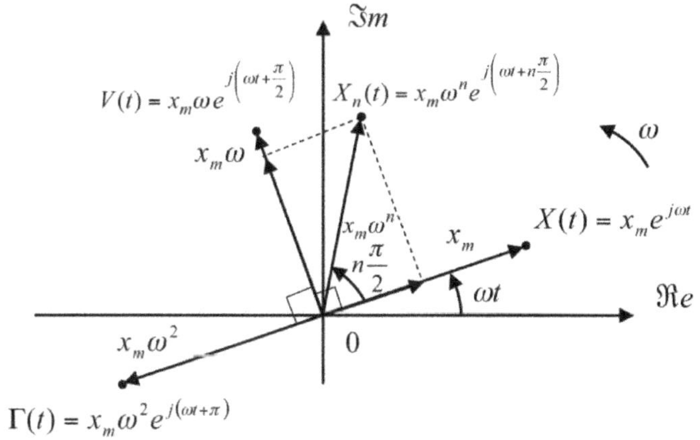

Figure 3.3. *Representation of* $x^{(n)}(t)$ *for* $n \in [0,1]$

Such a representation suggests a decomposition of the image vector of $X_n(t)$ along the directions of the Fresnel vectors associated with the position and the speed, thus enabling us to write $X_n(t)$ as:

$$X_n(t) = x_m \omega^n \left(\cos n\frac{\pi}{2}\right) e^{j\omega t} + x_m \omega^n \left(\sin n\frac{\pi}{2}\right) e^{j\left(\omega t + \frac{\pi}{2}\right)}, \qquad [3.61]$$

or:

$$X_n(t) = \omega^n \left(\cos n\frac{\pi}{2}\right) x_m e^{j\omega t} + \omega^{n-1} \left(\sin n\frac{\pi}{2}\right) x_m \omega e^{j\left(\omega t + \frac{\pi}{2}\right)}, \qquad [3.62]$$

namely:

$$X_n(t) = \omega^n \left(\cos n\frac{\pi}{2}\right) X(t) + \omega^{n-1} \left(\sin n\frac{\pi}{2}\right) V(t), \qquad [3.63]$$

from which we draw, through real part identification:

$$\underset{n\in[0,1]}{x^{(n)}(t)} = \omega^n \left(\cos n\frac{\pi}{2} \right) x(t) + \omega^{n-1} \left(\sin n\frac{\pi}{2} \right) v(t) . \quad [3.64]$$

Such a result expresses that in sinusoidal steady state, the differentiation of the position $x(t)$ to a non-integer order n between 0 and 1, namely $x^{(n)}(t)$ with $n \in [0,1]$, is a linear combination:

– of the position $x(t) = x^{(0)}(t)$;

– and of the speed $v(t) = x^{(1)}(t)$.

Particularizing the general expression of $x^{(n)}(t)$ by three differentiation orders characteristic of the interval $[0,1]$, namely $n = 0$, $n = 1/2$ and $n = 1$, leads to:

$$x^{(0)}(t) = x(t) \quad [3.65]$$

$$x^{(1/2)}(t) = \frac{1}{\sqrt{2}} \left(\omega^{\frac{1}{2}} x(t) + \omega^{-\frac{1}{2}} v(t) \right) \quad [3.66]$$

$$x^{(1)}(t) = v(t), \quad [3.67]$$

results which enable us to verify that the position $x(t)$ for $n = 0$ and the speed $v(t)$ for $n = 1$ are well recovered.

3.4.3.2. *Second case:* $1 \le n \le 2$

In the interval $[1,2]$, the Fresnel vector representation of $x^{(n)}(t)$ is illustrated in Figure 3.4. This representation suggests a decomposition of the image vector of $X_n(t)$ along the directions of the Fresnel vectors associated with the speed and the acceleration, thus enabling us to write $X_n(t)$ in the following form:

$$\begin{aligned} X_n(t) = {} & x_m\, \omega^n \left(\cos(n-1)\frac{\pi}{2} \right) e^{j(\omega t + \frac{\pi}{2})} \\ & + x_m\, \omega^n \left(\sin(n-1)\frac{\pi}{2} \right) e^{j(\omega t + \pi)}, \end{aligned} \quad [3.68]$$

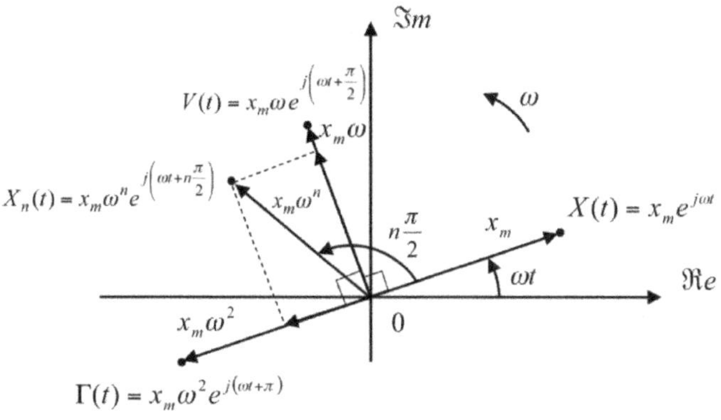

Figure 3.4. *Representation of* $x^{(n)}(t)$ *for* $n \in [1,2]$

or:

$$X_n(t) = \omega^{n-1}\left(\cos(n-1)\frac{\pi}{2}\right) x_m\, \omega\, e^{j(\omega t + \frac{\pi}{2})} + \omega^{n-2}\left(\sin(n-1)\frac{\pi}{2}\right) x_m\, \omega^2 e^{j(\omega t + \pi)}, \quad [3.69]$$

namely:

$$X_n(t) = \omega^{n-1}\left(\cos(n-1)\frac{\pi}{2}\right) V(t) + \omega^{n-2}\left(\sin(n-1)\frac{\pi}{2}\right) \Gamma(t), \quad [3.70]$$

from which we draw, through real part identification:

$$\underset{n\in[1,2]}{x^{(n)}(t)} = \omega^{n-1}\left(\cos(n-1)\frac{\pi}{2}\right) v(t) + \omega^{n-2}\left(\sin(n-1)\frac{\pi}{2}\right) \gamma(t). \quad [3.71]$$

This result expresses that in sinusoidal steady state, the differentiation of the position $x(t)$ to a non-integer order n between 1 and 2, namely $x^{(n)}(t)$ with $n \in [1,2]$, is a linear combination of:

– the speed $v(t) = x^{(1)}(t)$;

– the acceleration $\gamma(t) = x^{(2)}(t)$.

Particularizing the general expression of $x^{(n)}(t)$ by three differentiation orders characteristic of the interval $[1,2]$, namely $n=1$, $n=3/2$ and $n=2$, leads to:

$$x^{(1)}(t) = v(t) \tag{3.72}$$

$$x^{(3/2)}(t) = \frac{1}{\sqrt{2}}\left(\omega^{\frac{1}{2}}\, v(t) + \omega^{-\frac{1}{2}}\, \gamma(t)\right) \tag{3.73}$$

$$x^{(2)}(t) = \gamma(t)\ , \tag{3.74}$$

results which enable us to verify that the speed $v(t)$ for $n=1$ and the acceleration $\gamma(t)$ for $n=2$ are well recovered.

3.4.4. *Verification of the decomposition of* $x^{(n)}(t)$

First case: $0 \le n \le 1$

$$x^{(n)}(t) = \omega^{n}\left(\cos n\frac{\pi}{2}\right)x(t) + \omega^{n-1}\left(\sin n\frac{\pi}{2}\right)v(t)\ , \tag{3.75}$$

or, given the expressions of $x(t)$ and $v(t)$:

$$x^{(n)}(t) = \omega^{n}\left(\cos n\frac{\pi}{2}\right)x_m \cos\omega t + \omega^{n-1}\left(\sin n\frac{\pi}{2}\right)x_m\omega\cos\left(\omega t + \frac{\pi}{2}\right), \tag{3.76}$$

or even, knowing that

$$\cos\left(\omega t + \frac{\pi}{2}\right) = -\sin\omega t :$$

$$x^{(n)}(t) = x_m\omega^{n}\cos\omega t\cos n\frac{\pi}{2} - x_m\omega^{n}\sin\omega t\sin n\frac{\pi}{2}\ , \tag{3.77}$$

or else:

$$x^{(n)}(t) = x_m\omega^{n}\left(\cos\omega t\cos n\frac{\pi}{2} - \sin\omega t\sin n\frac{\pi}{2}\right), \tag{3.78}$$

namely, finally:

$$x^{(n)}(t) = x_m \omega^n \cos\left(\omega t + n\frac{\pi}{2}\right) . \qquad [3.79]$$

Q.E.D

Second case: $1 \leq n \leq 2$

$$x^{(n)}(t) = \omega^{n-1}\left(\cos(n-1)\frac{\pi}{2}\right)v(t) + \omega^{n-2}\left(\sin(n-1)\frac{\pi}{2}\right)\gamma(t), \qquad [3.80]$$

or, given the expressions of $v(t)$ and $\gamma(t)$:

$$x^{(n)}(t) = \omega^{n-1}\left(\cos(n-1)\frac{\pi}{2}\right)x_m \omega \cos\left(\omega t + \frac{\pi}{2}\right)$$
$$+ \omega^{n-2}\left(\sin(n-1)\frac{\pi}{2}\right)x_m \omega^2 \cos(\omega t + \pi) , \qquad [3.81]$$

or even, knowing that

$$\cos\left(\omega t + \frac{\pi}{2}\right) = -\sin \omega t , \ \cos(\omega t + \pi) = -\cos \omega t ,$$

$$\cos(n-1)\frac{\pi}{2} = \sin n\frac{\pi}{2} \text{ and } \sin(n-1)\frac{\pi}{2} = -\cos n\frac{\pi}{2} :$$

$$x^{(n)}(t) = -x_m \omega^n \sin \omega t \sin n\frac{\pi}{2} + x_m \omega^n \cos \omega t \cos n\frac{\pi}{2} , \qquad [3.82]$$

or else:

$$x^{(n)}(t) = x_m \omega^n \left(\cos \omega t \cos n\frac{\pi}{2} - \sin \omega t \sin n\frac{\pi}{2}\right), \qquad [3.83]$$

namely, finally:

$$x^{(n)}(t) = x_m \omega^n \cos\left(\omega\, t + n\frac{\pi}{2}\right). \qquad [3.84]$$

Q.E.D

3.5. On memory associated with non-integer differentiation

3.5.1. *From the local to the global by taking into account the past*

Through the function $f(t-kh)$ which introduces terms in $f(t)$, $f(t-h)$, $f(t-2h)$,..., therefore past samples, formula [3.11], namely

$$D^n f(t) = \lim_{h\to 0} \frac{1}{h^n} \sum_{k=0}^{\infty} a_k f(t-kh), \qquad [3.85]$$

shows that, contrary to integer derivative, the non-integer derivative of a function at a given instant t takes into account the values of this function at all the past instants.

Thus, if integer derivative gives a *local characterization* of the function (slope of the tangent to the curve at the instant t for the order 1 derivative), non-integer derivative turns out to give a *global characterization.*

3.5.2. *On memory notion*

Taking into account all the past of the function through a *weighted sum of the samples* that expresses (through the weighting coefficients) a different weight according to the sample, non-integer differentiation introduces a *memory notion* through a *lessening* (or *attenuation*) or a *stressing* (or *accentuation*) of the past in conformity with the weighting coefficients that are such that:

– for $n=-1$ (which corresponds to *an integration*), the past is not weighted (Figure 3.5);

– for $n>-1$ (which corresponds to *less than an integration* and to *a differentiation*), the past is lessened with weighting coefficients which keep the same sign for $-1<n<1$ and which change sign for $n>1$ (Figure 3.6);

– for $n<-1$ (which corresponds to *more than an integration*), the past is stressed with weighting coefficients which keep the same sign (Figure 3.7).

Characterized as such, this memory notion evokes a memory-subtle form through which the recollection of an event depends on the nature of this event:

– for usual events, the recollection of more ancient events is less important than the recollection of more recent events;

– on the contrary, for unusual events, the recollection of more ancient events can be more important than the recollection of more recent events.

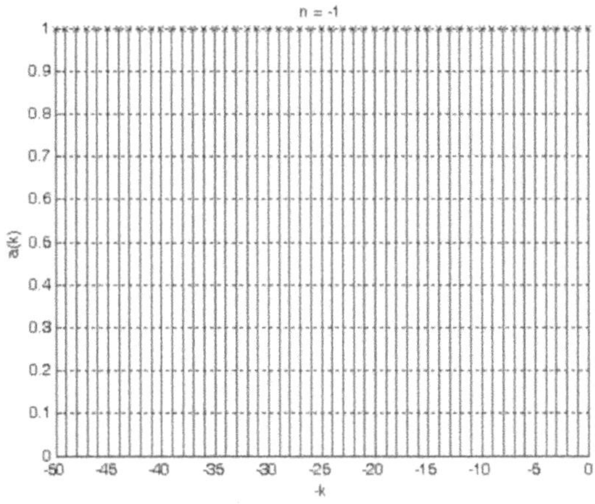

Figure 3.5. $a(k)$ *versus* k *with* $n=-1$

3.5.3. *On an aspect of human memory (an investigation trail)*

Human memory stocks attenuation and accentuation phases of the past well (even combinations of these phases). The *serial position effect* studied in cognitive psychology thus seems to be a good illustration (http://psychocognitiv67.canalblog.com/).

3.5.3.1. *Serial position effect*

The study of serial position effect on *recall* (or *recollection*) consists of:

– visually presenting, to a subject, a list of words at a certain rate;

– then asking the subject to recall *without any required order* the list of words *immediately* after the presentation of this list (case of *free recall* as opposed to *serial recall* whose order is the one of the list of words).

Stemming from the results obtained with several subjects, the *recall curve* (Figure 3.8) highlights:

– an *attenuation phase of the past* for which the recall of words is attenuated with the past;

– an *accentuation phase of the past* for which the recall of words is accentuated with the past;

– these two phases contributing to a better recall of the *first words* which are the oldest (*primacy effect*) and of the last words which are the *most recent* ones (*recency effect*).

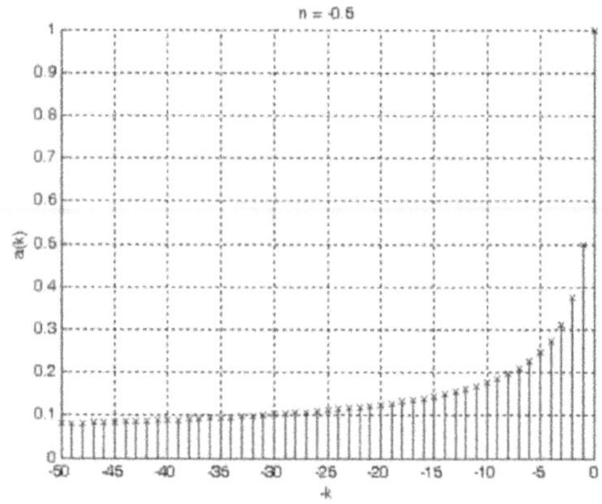

a)

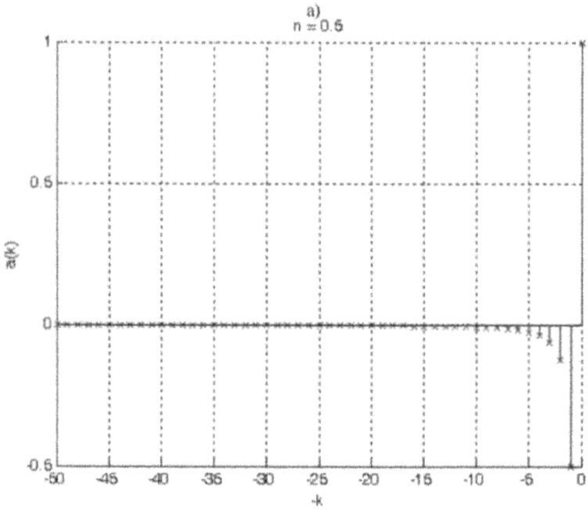

b)

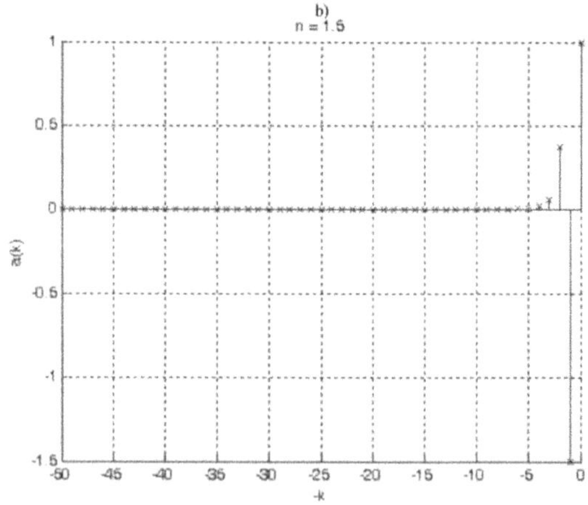

c)

Figure 3.6. *$a(k)$ versus k for different values of $n > -1$*

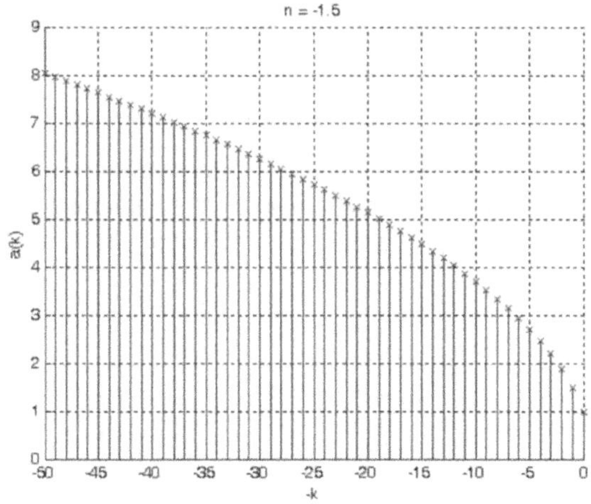

Figure 3.7. $a(k)$ *versus* k *with* $n < -1$

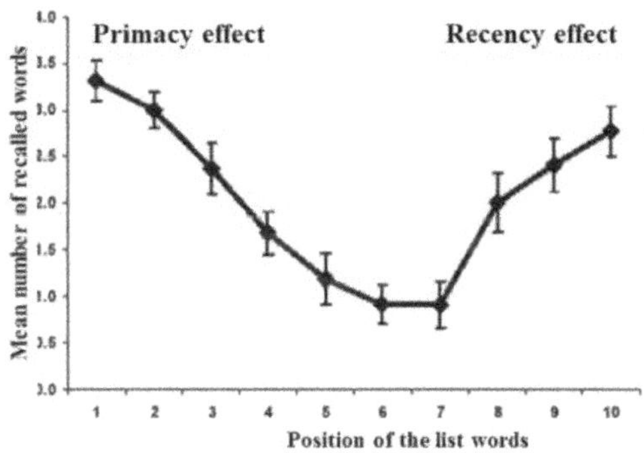

Figure 3.8. *Recall curve*

If it is true that the words at the beginning and at the end are better recalled than those at the middle:

– *those at the beginning are even better recalled*;

– expressing that *the primacy effect triumphs over the recency effect.*

According to the specialists, the primacy effect is attributed to the auto-repetition of the first words while the following ones are displayed. The fact remains that, even if the first words are the oldest, they are those which escape the most from all *routine* effects or (in other words) which benefit the most from all *newness* effects, routine and newness here evoking *the usual* and *the unusual.*

To end under an applicative angle, it is convenient to mention that these mnesic phenomena turn out to be widely used:

– notably in the diffusion of wide audience information;

– more particularly in the achievement and programming of adverts in order to obtain the best impact, their achievement favoring the beginning and the end of each advert, and their programming granting a preferential order to the first and last adverts.

3.5.3.2. *A primacy effect model*

Beyond its preponderance on recency effect, the primacy effect that actually corresponds to an *accentuation of the past* (Figure 3.9) seems to present a *curvature* in conformity with the *memory profile dictated by an integration non-integer order greater than 1* (Figure 3.10).

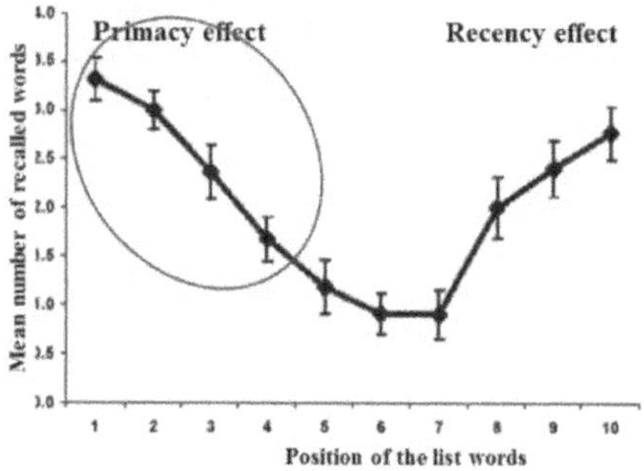

Figure 3.9. *Curvature of the primacy effect*

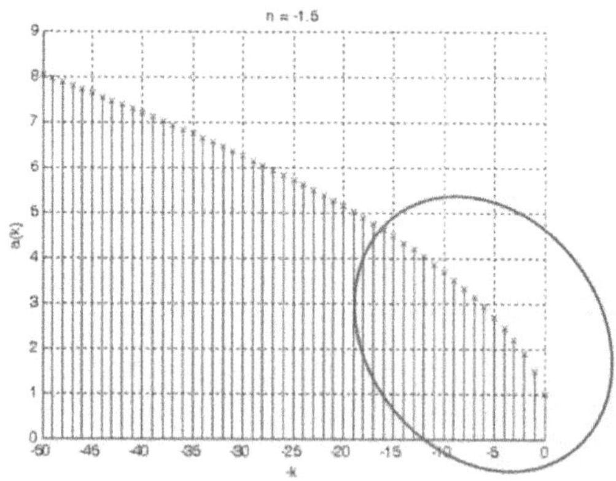

Figure 3.10. *Curvature associated with an integration non-integer order superior to one*

Hence, the idea to seek a *behavior model* through the *optimal integration non-integer order* which minimizes the sum of the squares of the differences between:

– the mnesic data stemming from the recall curve

– and the first weighting coefficients (expressed up to a factor) relative to the non-integer integration, the result of this minimization being *an optimal non-integer order of 1.434* that the *corresponding normalized weighting coefficients* are well validated as the comparison of relations [3.113] and [3.86] and also [3.114] and [3.87] shows it.

3.5.3.3. *Optimal integration non-integer order proof*

The goal of this section is to demonstrate the result so defined by determining *the optimal value of the differentiation non-integer order* n, $n<-1$, in the sense of the minimization of a quadratic criterion precisely turning:

– on one hand, on the ordinates of the points defining the primacy effect, namely (Figure 3.8)

$$y_0 = 1.66\,,\ y_1 = 2.36\,, \qquad [3.86]$$

$$y_2 = 2.97\,,\ y_3 = 3.29 \qquad [3.87]$$

– on the other hand, on the first weighting coefficients associated with the non-integer integration, a_0, a_1, a_2, a_3, expressed up to a factor A, namely

$$Aa_0 = A \text{ (as } a_0 = 1\text{)},\ Aa_1 = -An\,, \qquad [3.88]$$

$$Aa_2 = A\frac{n(n-1)}{2}\,,\ Aa_3 = -A\frac{n(n-1)(n-2)}{6}\,. \qquad [3.89]$$

The optimal couple (A,n) that globally minimizes the gaps corresponding to the whole of the approximations

$$y_0 \approx A\,,\ y_1 \approx -An\,, \qquad [3.90]$$

$$y_2 \approx A\frac{n(n-1)}{2} \text{ and } y_3 = -A\frac{n(n-1)(n-2)}{6}\,, \qquad [3.91]$$

minimizes the quadratic criterion $J(A,n)$ such that:

$$\begin{aligned} J(A,n) &= (y_0 - A)^2 + (y_1 + An)^2 + \left(y_2 - A\frac{n(n-1)}{2}\right)^2 \\ &\quad + \left(y_3 + A\frac{n(n-1)(n-2)}{6}\right)^2, \end{aligned} \quad [3.92]$$

namely:

$$\begin{aligned} J(A,n) &= (y_0 - A)^2 + (y_1 + An)^2 + \left(y_2 - \tfrac{A}{2}(n^2 - n)\right)^2 \\ &\quad + \left(y_3 + \tfrac{A}{6}(n^3 - 3n^2 + 2n)\right)^2. \end{aligned} \quad [3.93]$$

The partial derivative of J with respect to A is written as:

$$\begin{aligned} \frac{\partial J(A,n)}{\partial A} &= -2(y_0 - A) + 2(y_1 + An)n - 2\left(y_2 - \frac{A}{2}(n^2 - n)\right)\frac{(n^2 - n)}{2} \\ &\quad + 2\left(y_3 + \frac{A}{6}(n^3 - 3n^2 + 2n)\right)\frac{(n^3 - 3n^2 + 2n)}{6}, \end{aligned} \quad [3.94]$$

namely:

$$\begin{aligned} \frac{\partial J(A,n)}{\partial A} &= -2(y_0 - A) + 2(y_1 + An)n - \left(y_2 - \frac{A}{2}(n^2 - n)\right)(n^2 - n) \\ &\quad + \frac{1}{3}\left(y_3 + \frac{A}{6}(n^3 - 3n^2 + 2n)\right)(n^3 - 3n^2 + 2n). \end{aligned} \quad [3.95]$$

As for the partial derivative of J with respect to n, it is written as:

$$\begin{aligned} \frac{\partial J(A,n)}{\partial n} &= 2(y_1 + An)A + 2\left(y_2 - \frac{A}{2}(n^2 - n)\right)\left(-\frac{A}{2}\right)(2n-1) \\ &\quad + 2\left(y_3 + \frac{A}{6}(n^3 - 3n^2 + 2n)\right)\frac{A}{6}(3n^2 - 6n + 2), \end{aligned} \quad [3.96]$$

namely:

$$\frac{\partial J(A,n)}{\partial n}=2A(y_1+An)-A\left(y_2-\frac{A}{2}(n^2-n)\right)(2n-1)$$
$$+\frac{A}{3}\left(y_3+\frac{A}{6}(n^3-3n^2+2n)\right)(3n^2-6n+2). \quad [3.97]$$

$\partial J(A,n)/\partial A$ and $\partial J(A,n)/\partial n$, respectively, can be put under the form:

$$\frac{\partial J(A,n)}{\partial A}=P(n)+AQ(n) \quad [3.98]$$

and

$$\frac{\partial J(A,n)}{\partial n}=AP'(n)+A^2Q'(n), \quad [3.99]$$

by putting:

$$P(n)=2(y_1n-y_0)-y_2(n^2-n)+\frac{y_3}{3}(n^3-3n^2+2n) \quad [3.100]$$

$$Q(n)=2(1+n^2)+\frac{1}{2}(n^2-n)^2+\frac{1}{18}(n^3-3n^2+2n)^2 \quad [3.101]$$

$$P'(n)=2y_1-y_2(2n-1)+\frac{y_3}{3}(3n^2-6n+2) \quad [3.102]$$

$$Q'(n)=2n+\frac{1}{2}(n^2-n)(2n-1)+\frac{1}{18}(n^3-3n^2+2n)(3n^2-6n+2). \quad [3.103]$$

The minimization of J that results from the cancellation of the partial derivatives so determined, namely

$$\frac{\partial J(A,n)}{\partial A}=0 \text{ and } \frac{\partial J(A,n)}{\partial n}=0, \quad [3.104]$$

leads to a system of two equations:

$$P(n)+AQ(n)=0 \quad [3.105]$$

and

$$P'(n)+AQ'(n)=0, \qquad [3.106]$$

from which we draw:

$$A=-\frac{P(n)}{Q(n)} \text{ from the first equation} \qquad [3.107]$$

and

$$A=-\frac{P'(n)}{Q'(n)} \text{ from the second equation,} \qquad [3.108]$$

which enables us to express the equality of the ratios:

$$\frac{P(n)}{Q(n)}=\frac{P'(n)}{Q'(n)}, \qquad [3.109]$$

from which we draw:

$$P(n)Q'(n)-P'(n)Q(n)=0. \qquad [3.110]$$

We thus obtain a polynomial equation in n, from which one can determine the roots in n then extract (among the real roots) the one that ensures the minimal value of J, by verifying that this one is less than –1. It is true that the framework is the one of an integration corresponding to an accentuation of the past, namely $n<-1$.

It turns out that three real roots can be found, namely

$$n=-1.434\,,\ n=0.992 \text{ and } n=2.605\,, \qquad [3.111]$$

which, respectively, lead to

$$J=0.0062\,,\ J=27.725 \text{ and } n=27.961\,, \qquad [3.112]$$

and show well that $n=-1.434$ ensures the minimal value of J, namely $J=0.0062$, a result that expresses an *optimal integration non-integer order of 1.434*.

Furthermore, we get the corresponding value of A by particularizing the relation $A=-P(n)/Q(n)$ by $n=-1.434$, namely $A=1.664$.

Finally, as for the estimates of the weighting coefficients associated with the non-integer integration and expressed up to the factor A, the optimal values of n and A so obtained ($n=-1.434$ and $A=1.664$) enable us to determine them in conformity with the numerical values:

$$z_0=1.664\;;\; z_1=2.387\;; \qquad [3.113]$$

$$z_2=2.905\;;\; z_3=3.325\,. \qquad [3.114]$$

3.6. On the synthesis of non-integer differentiation

The idea of the synthesis of a non-integer differentiator is:

– an *approximation* of the differentiator ideal version;

– by a *recursive (or geometric) distribution of countable zeros and poles.*

The number of (synthesis) zeros and poles is *finite* or *infinite* depending on whether the non-integer differentiator ideal version is *frequency bounded* or *not.*

This leads us to say that a *realistic synthesis* (in the sense of a finite number of zeros and poles) can only be inscribed in the case of a *frequency bounded non-integer differentiator.* Moreover, the awareness of this context has enabled us to be at the origin of the non-integer differentiation synthesis [OUS 95, OUS 00] then of its numerous applications in engineering sciences (CRONE control, CRONE suspension,...).

In the following sections, the differentiation non-integer order is successively *real* and *complex* [OUS 00].

3.6.1. *Synthesis of a frequency-bounded real non-integer differentiator*

A *frequency bounded (or frequency-band) real non-integer differentiator*, of real non-integer order n and of transitional frequencies ω_b and ω_h distributed geometrically around the unit gain frequency ω_u ($\omega_u=(\omega_b\omega_h)^{1/2}$), is defined by a transmittance of the form:

$$D(s)=\left(\frac{\omega_u}{\omega_h}\right)^n\left(\frac{1+s/\omega_b}{1+s/\omega_h}\right)^n \text{ with } n\in\mathbb{R}, \qquad [3.115]$$

this transmittance being obtained from $D(s) = (s/\omega_u)^n$ by replacing s/ω_u with

$$\frac{\omega_b}{\omega_u}\frac{1+s/\omega_b}{1+s/\omega_h} \text{ or } \frac{\omega_u}{\omega_h}\frac{1+s/\omega_b}{1+s/\omega_h}, \qquad [3.116]$$

knowing that $\omega_u^2 = \omega_b\omega_h$.

The synthesis of such a differentiator results from an intuitive approach based on the concept of fractal through recursivity [OUS 95]. A recursive distribution of real zeros and poles is indeed used, namely (Figure 3.11):

$$D(s) = \lim_{N\to\infty} D_N(s), \qquad [3.117]$$

with

$$D_N(s) = \left(\frac{\omega_u}{\omega_h}\right)^n \prod_{k=-N}^{N} \frac{1+s/\omega_k'}{1+s/\omega_k}, \qquad [3.118]$$

where

$$\left(\frac{\omega_u}{\omega_h}\right)^n = \left(\frac{\omega_b}{\omega_u}\right)^n = \frac{1}{\alpha^{N+1/2}}; \qquad [3.119]$$

$-\omega_k'$ is a zero of rank k ; $-\omega_k$ is a zero of rank k ;

$2N+1$ is the number of zeros or poles such that

$$N = \frac{\log(\omega_N/\omega_0)}{\log(\alpha\eta)}. \qquad [3.120]$$

The recursivity of the zeros and poles is expressed by a distribution of the transitional frequencies ω_k' and ω_k which is defined by the three following relations:

$$\omega_0' = \alpha^{-1/2}\omega_u \text{ and } \omega_0 = \alpha^{1/2}\omega_u; \qquad [3.121]$$

$$\frac{\omega_{k+1}'}{\omega_k'} = \frac{\omega_{k+1}}{\omega_k} = \alpha\eta > 1; \qquad [3.122]$$

$$\frac{\omega_k}{\omega'_k} = \alpha > 0 \;;\; \frac{\omega'_{k+1}}{\omega_k} = \eta > 0 . \qquad [3.123]$$

The ratios α and η defined by [3.123], and which imply a *constant ratio* $\alpha\eta$ *between two consecutive zeros or poles* (relation [3.122]), are called *recursive factors*. For a given order of differentiation n and a given ratio μ between the transitional frequencies ω_h and ω_b, these factors only depend on parameter N and could be written as $\alpha(N)$ and $\eta(N)$. The higher N is, the weaker they are. The transitional frequencies of rank 0 (ω'_0 and ω_0) are determined in relation to ω_u using [3.121]. The others are determined from ω'_0 and ω_0 using [3.122] and [3.123] up to the lowest ω'_{-N} and the highest ω_N transitional frequencies. These frequencies are also easily expressed as functions of the transitional frequencies ω_b and ω_h of $D(s)$. The relative position of the gain asymptotic diagrams of $D(s)$ and $D_N(s)$, as illustrated in Figure 3.11(a), shows that the gain asymptote of $D(s)$ between ω_b and ω_h cuts the $D_N(s)$ steps in the middle of the plateaus of length $\log\eta$. Thus, on a logarithmic scale, ω'_{-N} and ω_N are, respectively, at $(1/2)\log\eta$ from ω_b and ω_h, namely:

$$\omega'_{-N} = \eta^{1/2}\omega_b \text{ and } \omega_N = \eta^{-1/2}\omega_h . \qquad [3.124]$$

Transmittance $D_N(s)$ of integer order $2N+1$, thus defined, which provides an approximation of $D(s)$ for finite N (the greater N is, the better the approximation), is called *synthesis transmittance.* As shown in Figure 3.11, a legitimate smoothing of its Bode asymptotic diagrams leads to those of $D(s)$.

The smoothing of the steps of the $D_N(s)$ gain asymptotic diagram between ω_b and ω_h (Figure 3.11(a)), can be represented by a straight line, called *gain smoothing straight line*, of slope less than 6 dB/oct, namely 6 n dB/oct with n between 0 and 1. This straight line coincides with the gain asymptote of $D(s)$ between ω_b and ω_h.

Δ is the height of the steps. The slopes of the gain smoothing straight line and of a riser (slopes of segments AB and $A'B'$) are, respectively, given by:

$$6n \text{ dB/oct} = \frac{\Delta \text{dB}}{\log\alpha + \log\eta} \text{ and } 6 \text{dB/oct} = \frac{\Delta \text{dB}}{\log\alpha}, \qquad [3.125]$$

from which can be drawn the expression of the non-integer order n versus the recursive factors α and η:

$$n = \frac{\log \alpha}{\log(\alpha\eta)}. \quad [3.126]$$

Smoothing the crenels of the $D_N(s)$ phase asymptotic diagram between ω_b and ω_h (Figure 3.11(b)), can be represented by a straight line, called *phase smoothing straight line*, the ordinate being less than $\pi/2$, namely $n\pi/2$ with n between 0 and 1. This straight line coincides with the phase asymptote of $D(s)$ between ω_b and ω_h.

S is the area of the shaded surfaces. The ordinate of the phase smoothing straight line and the height of a crenel are, respectively, expressed by:

$$n\frac{\pi}{2} = \frac{S}{\log\alpha + \log\eta} \text{ and } \frac{\pi}{2} = \frac{S}{\log\alpha}, \quad [3.127]$$

from which we deduce the same expression of n as the one obtained from the gain (relation [3.126]).

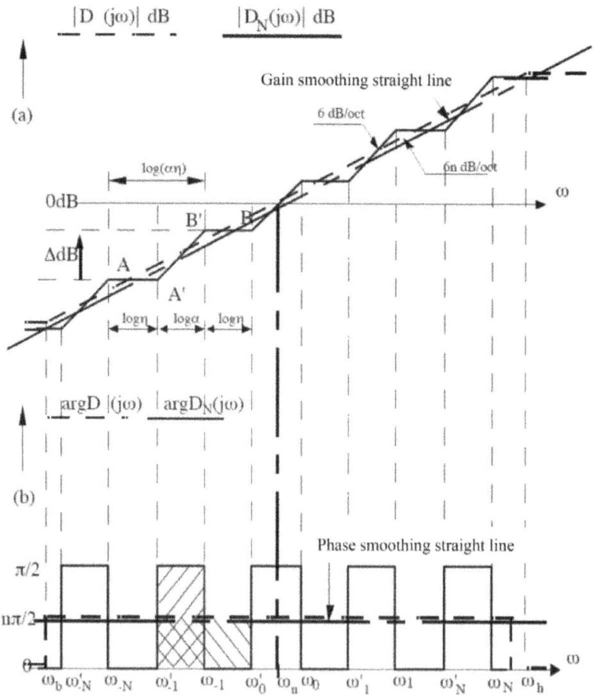

Figure 3.11. *Bode asymptotic diagrams of $D(s)$ and $D_N(s)$ for $n \in]0,1[$*

The analysis of Figure 3.11 provides the ratio of the transitional frequencies ω_h and ω_b of $D(s)$:

$$\frac{\omega_h}{\omega_b} = (\alpha\eta)^{2N+1}, \tag{3.128}$$

from which the product of the recursive factors α and η is drawn:

$$\alpha\eta = \left(\frac{\omega_h}{\omega_b}\right)^{1/(2N+1)}. \tag{3.129}$$

Furthermore, relation [3.126] makes it possible for the recursive factor α to be determined as a function of the product $\alpha\eta$:

$$\alpha = (\alpha\eta)^n. \tag{3.130}$$

The recursive factor η is deduced through its formulation under the form $\eta = (\alpha\eta)\alpha^{-1}$, namely:

$$\eta = (\alpha\eta)^{1-n}. \tag{3.131}$$

Thus, using [3.129], the recursive factors α and η can be expressed, respectively, by:

$$\alpha = \left(\frac{\omega_h}{\omega_b}\right)^{n/(2N+1)} \text{ and } \eta = \left(\frac{\omega_h}{\omega_b}\right)^{(1-n)/(2N+1)}, \tag{3.132}$$

which shows that α and η only depend on N for given n and $\mu = \omega_h / \omega_b$.

As the synthesis zeros and poles $-\omega_k'$ and $-\omega_k$ can be considered as being equivalent to the corresponding transitional frequencies ω_k' and ω_k, the zero of rank k can be written without the negative sign as:

$$\omega_k' = \omega_{-N+k+N}' = (\alpha\eta)^{k+N}\,\omega_{-N}', \tag{3.133}$$

namely, given [3.124]:

$$\omega_k' = (\alpha\eta)^{k+N}\,\eta^{1/2}\omega_b, \tag{3.134}$$

from which the pole of rank k can be deduced using [3.123]:

$$\omega_k = (\alpha\eta)^{k+N} \alpha\eta^{1/2} \omega_b . \quad [3.135]$$

Given [3.129]–[3.131], the zero and pole of rank k can be written, respectively, as:

$$\omega_k' = \left(\frac{\omega_h}{\omega_b}\right)^{(k+N+1/2-n/2)/(2N+1)} \omega_b \quad [3.136]$$

and

$$\omega_k = \left(\frac{\omega_h}{\omega_b}\right)^{(k+N+1/2+n/2)/(2N+1)} \omega_b . \quad [3.137]$$

The transitional frequencies so obtained and associated with [3.118], determine entirely the synthesis transmittance $D_N(s)$. Taking the value of N from the $2N+1$ zeros or poles used in the synthesis allows us to determine the values of the set of zeros and poles of [3.118], given the parametric data ω_u, ω_b, ω_h and n of the transmittance to be synthesized, $D(s)$.

The formulas established from Figure 3.11 may seem restrictive, as this figure is for n between 0 and 1 and, thus, for a specific distribution of zeros and poles.

Nevertheless, these formulas are valid irrespective of the range of n. They always express a *recursive distribution of zeros* and a *recursive distribution of poles*. However, these two distributions are in different relative positions depending on the range of n values. The relative displacement of these distributions makes it possible for the order n values to be scanned in accordance with relation [3.126], while respecting the other relations.

Given that the product $\alpha\eta$ is always greater than 1, relation [3.126] shows that:

– n is positive if $\alpha > 1$: the distribution of zeros and poles then starts with a zero;

– n is negative if $0 < \alpha < 1$: the distribution of zeros and poles then starts with a pole.

Figure 3.12 illustrates the validity of the formalism for $n \in \mathbb{R}$ through the Bode asymptotic diagrams of $D(s)$ and $D_N(s)$ for a range of n different from that of Figure 3.11, in this case $]2,3[$. The distribution of zeros and poles is no longer alternate from the beginning to the end. It indeed starts with three consecutive zeros and ends with three consecutive poles.

3.6.2. *Synthesis of a frequency-bounded complex non-integer differentiator*

In the particular case of real zeros and poles to which real recursive factors correspond, it is relatively easy to illustrate graphically and thus verify the identity [3.117] which defines the synthesis of a frequency-bounded real non-integer differentiator. This was in fact demonstrated in section 3.6.1.

On the contrary, in the general case of complex zeros and poles to which complex recursive factors correspond, a graphic approach cannot be used. An analytical approach is then required in order to demonstrate the validity of the synthesis transmittance $D_N(s)$ in the general case of complex parameters, notably by showing its generalization to a complex non-integer differentiation order. Its validation for such a differentiation order is carried out by calculating its limit when N tends toward infinity, and then by showing that this one is nothing but $D(s)$.

The set of relations established for a real non-integer differentiation order in section 3.6.1 is now extended to a complex non-integer differentiation order $n = a + ib$. Here, the purely imaginary unit i in the complex plane $\mathbb{C}_i$ is chosen independently of the purely imaginary unit j in the complex plane $\mathbb{C}_j$ of the operational variable $s = \sigma + j\omega$.

The synthesis transmittance $D_N(s)$ needs to be expressed under a more general form, including [3.136] and [3.137] in [3.118] with $\omega_h / \omega_b = \mu$ and $\omega_u / \omega_h = \mu^{-1/2}$ as $\omega_u = (\omega_b \omega_h)^{1/2}$:

$$D_N(s) = \mu^{-n/2} \prod_{k=-N}^{N} \frac{1 + \dfrac{q}{\mu^{\frac{k+N+1/2-n/2}{2N+1}}}}{1 + \dfrac{q}{\mu^{\frac{k+N+1/2+n/2}{2N+1}}}} \quad \text{with } n \in \mathbb{C}_i, \qquad [3.138]$$

where q is a reduced operational variable, i.e. the operational variable s normalized by dividing it by the transitional frequency $\omega_b\,(q = s/\omega_b)$.

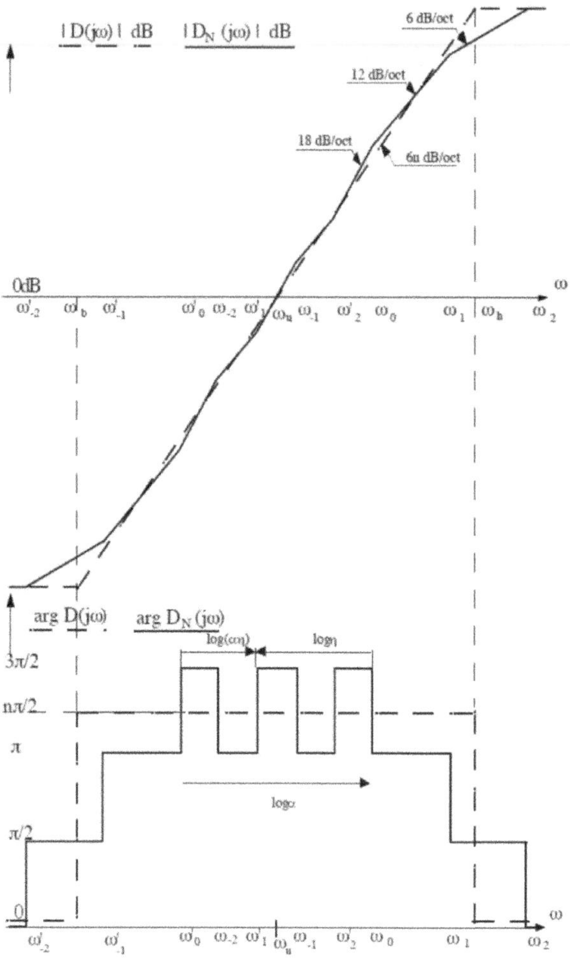

Figure 3.12. *Bode asymptotic diagrams of* $D(s)$ *and* $D_N(s)$ *for* $n \in \,]2,3[$

3.6.2.1. *Definition of the space* $\hat{\mathbb{C}}$

$D_N(s)$ is defined in a space $\hat{\mathbb{C}}$ which includes both $\mathbb{C}_i$ and $\mathbb{C}_j$. This space, such as $\hat{\mathbb{C}} = \mathbb{C}_j + i\,\mathbb{C}_j$, is a Banach algebra, whose norm generalizes the norms of $\mathbb{C}_i$ and $\mathbb{C}_j$ which are given by the modulus.

All the functions defined in $\mathbb{C}_j$ also extend in $\hat{\mathbb{C}}$. All analytic functions in $\mathbb{C}_j$ remain analytic in $\hat{\mathbb{C}}$. All convergent series in $\mathbb{C}_j$ extend in $\hat{\mathbb{C}}$ with the same circle of convergence.

In particular, we can define an exponential e^z $\forall z \in \hat{\mathbb{C}}$ and a logarithm $\ln z$ in a part of $\hat{\mathbb{C}}$.

The following properties which are valid in $\mathbb{C}_j$ remain valid in $\hat{\mathbb{C}}$:

1) $e^{z_1 + z_2} = e^{z_1} e^{z_2}$ $\forall z_1, z_2 \in \hat{\mathbb{C}}$;

2) $e^{\ln z} = z$ $\forall z$ such that $\ln z$ be defined;

3) $\ln(1+z) = \sum_{n \geq 1} (-1)^{n+1} \frac{z^n}{n}$ $\forall z \in \hat{\mathbb{C}}$ such that $\|z\| < 1$;

4) $\exists \beta > 0$ such that, if $z_1, z_2 \in \hat{\mathbb{C}}$ verify $\|z_1\| < \beta$ and $\|z_2\| < \beta$, then $\ln\left(\frac{1+z_1}{1+z_2}\right) = \ln(1+z_1) - \ln(1+z_2)$;

5) to demonstrate that $\prod_{k=-N}^{N} z_k = z$ in $z \in \hat{\mathbb{C}}$, it suffices to demonstrate that $\sum_{k=-N}^{N} \ln z_k = \ln z$.

3.6.2.2. *Demonstration of the validity of the synthesis transmittance for a complex non-integer differentiation order*

This section presents a heuristic proof by assuming that q verifies the validity conditions of the previous properties which are used below. In fact, the final result [3.148] is true $\forall q \in \mathbb{C}_j - [-\mu, -1]$, but its demonstration has not been done here due to space constraints.

In this proof, we use the properties of the space $\hat{\mathbb{C}}$ as defined in section 3.6.2.1.

The natural logarithm of $D_N(s)$ given by [3.138] allows us to replace the product sign by the sum sign:

$$\ln D_N(s) = -\frac{n}{2}\ln\mu + \sum_{k=-N}^{N}\left[\ln\left(1+\frac{q}{\mu^{\frac{k+N+1/2-n/2}{2N+1}}}\right) - \ln\left(1+\frac{1}{\mu^{\frac{k+N+1/2+n/2}{2N+1}}}\right)\right]. \quad [3.139]$$

The integer series expansion of the function $\ln(1+x)$ gives:

$$\ln D_N(s) = -\frac{n}{2}\ln\mu + \sum_{k=-N}^{N}\sum_{r=1}^{+\infty}\frac{(-1)^{r+1}}{r}\times \left[\left(\frac{q}{\mu^{\frac{k+N+1/2-n/2}{2N+1}}}\right)^r - \left(\frac{q}{\mu^{\frac{k+N+1/2+n/2}{2N+1}}}\right)^r\right]. \quad [3.140]$$

The inversion of the sum signs and the isolation of the terms dependent on k lead to:

$$\ln D_N(s) = -\frac{n}{2}\ln\mu + \sum_{r=1}^{+\infty}(-1)^{r+1}\frac{q^r}{r}\left(\mu^{\frac{-r(N+1/2-n/2)}{2N+1}} - \mu^{\frac{-r(N+1/2+n/2)}{2N+1}}\right)\sum_{k=-N}^{N}\mu^{\frac{-rk}{2N+1}}, \quad [3.141]$$

or, given that the sum over k is a geometric series of ratio $\mu^{-r/(2N+1)}$:

$$\ln D_N(s) = -\frac{n}{2}\ln\mu + \sum_{r=1}^{+\infty}(-1)^{r+1}\frac{q^r}{r}\left(\mu^{\frac{-r(N+1/2-n/2)}{2N+1}} - \mu^{\frac{-r(N+1/2+n/2)}{2N+1}}\right)\frac{\mu^{\frac{rN}{2N+1}} - \mu^{\frac{-r(N+1)}{2N+1}}}{1-\mu^{\frac{-r}{2N+1}}}. \quad [3.142]$$

For an infinite number of zeros and poles, i.e. when N tends toward infinity, the following two limits facilitate the reduction of [3.142]:

$$\lim_{N\to\infty}\left[\mu^{\frac{-r(N+1/2-n/2)}{2N+1}} - \mu^{\frac{-r(N+1/2+n/2)}{2N+1}}\right] = \mu^{-r/2}n\frac{\ln\mu}{2N+1}r \quad [3.143]$$

and

$$\lim_{N\to\infty}\left[\frac{\mu^{\frac{rN}{2N+1}}-\mu^{\frac{-r(N+1)}{2N+1}}}{1-\mu^{\frac{-r}{2N+1}}}\right]=\mu^{r/2}\left(1-\mu^{-r}\right)\frac{1}{\dfrac{\ln\mu}{2N+1}r}. \quad [3.144]$$

Relations [3.143] and [3.144] enable a simple expression of [3.142] when N tends toward infinity:

$$\lim_{N\to\infty}\ln D_N(s)=-\frac{n}{2}\ln\mu+\sum_{r=1}^{+\infty}(-1)^{r+1}\frac{q^r}{r}n\left(1-\mu^{-r}\right). \quad [3.145]$$

Developing [3.145] introduces the difference between the integer series expansions of the functions $\ln(1+q)$ and $\ln(1+q/\mu)$ according to:

$$\lim_{N\to\infty}\ln D_N(s)=-\frac{n}{2}\ln\mu+n\left[\ln(1+q)-\ln\left(1+\frac{q}{\mu}\right)\right], \quad [3.146]$$

from which is drawn:

$$\lim_{N\to\infty}D_N(s)=\left(\mu^{-1/2}\frac{1+q}{1+\dfrac{q}{\mu}}\right)^n, \quad [3.147]$$

or finally, given that $q=s/\omega_b$, $\mu=\omega_h/\omega_b$ and $\mu^{-1/2}=\omega_u/\omega_h$ and given [3.115] generalized by $n\in\mathbb{C}_i$:

$$\lim_{N\to\infty}D_N(s)=D(s). \quad [3.148]$$

Identity [3.117] has now been demonstrated with no assumption on n.

3.6.3. *Stability of the synthesis transmittance*

The stability condition of the synthesis transmittance is determined, not in the complex plane in j, $\mathbb{C}_j$, to which the operational variable $s=\sigma+j\omega$ belongs, but in the complex plane in i, $\mathbb{C}_i$, to which the synthesis zeros and poles $-\omega_k'$ and $-\omega_k$ (or ω_k' and ω_k without their signs) belong.

Given relations [3.136] and [3.137] and knowing that $n = a + ib$, these zeros and poles are written with a polar representation in the complex plane $\mathbb{C}_i$:

$$\omega_k^{'} = \rho_k^{'} e^{-i\theta} \text{ and } \omega_k = \rho_k e^{i\theta}, \qquad [3.149]$$

where:

$$\rho_k^{'} = \left(\frac{\omega_h}{\omega_b}\right)^{(k+N+1/2-a/2)/(2N+1)} \omega_b \qquad [3.150]$$

$$\rho_k = \left(\frac{\omega_h}{\omega_b}\right)^{(k+N+1/2+a/2)/(2N+1)} \omega_b \qquad [3.151]$$

and

$$\theta = \frac{b}{2(2N+1)} \ln \frac{\omega_h}{\omega_b}. \qquad [3.152]$$

Their modulus verifies a recursive distribution through [3.150] and [3.151] whereas their argument is independent of their rank (relation [3.152]).

The synthesis transmittance $D_N(s)$ can be decomposed in partial fractions according to:

$$D_N(s) = K + \sum_{k=-N}^{N} \frac{A_k}{1 + s/\omega_k}, \qquad [3.153]$$

with:

$$K = D_N(\infty) = \left(\frac{\omega_u}{\omega_h}\right)^n \prod_{k=-N}^{N} \frac{\omega_k}{\omega_k^{'}} \qquad [3.154]$$

and

$$A_k = \left[\left(1 + \frac{s}{\omega_k}\right) D_N(s)\right]_{s=-\omega_k}. \qquad [3.155]$$

The corresponding (complex) impulse response, namely

$$L^{-1}\left[D_N(s)\right] = D_N(\infty)\delta(t) + \sum_{k=-N}^{N} A_k \omega_k e^{-w_k t} u(t), \qquad [3.156]$$

where $\delta(t)$ and $u(t)$ denote the unit impulse and step, expresses the stability condition in $\mathbb{C}_i$ as it is only stable if $\mathrm{Re}\left[-\omega_k\right] < 0$.

Thus, as in the complex plane $\mathbb{C}_j$, the necessary and sufficient condition of stability in the complex plane $\mathbb{C}_i$ is that the poles must have a negative real part.

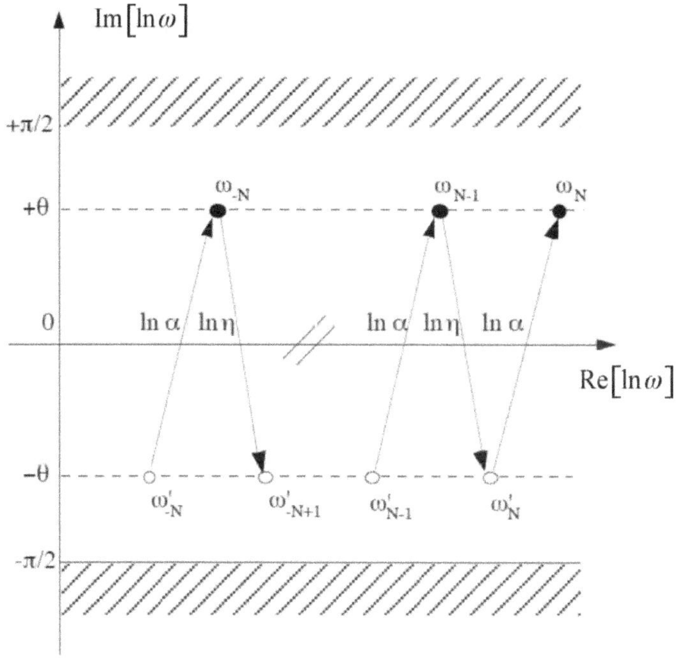

Figure 3.13. *Distribution of the synthesis zeros and poles corresponding to a complex non-integer differentiation order*

This condition on the pole real part becomes a condition on their argument, namely $\theta \in \left]-\pi/2, \pi/2\right[$, or given the expression of θ (relation [3.152]):

$$\frac{|b|}{2(2N+1)} \ln \frac{\omega_h}{\omega_b} < \frac{\pi}{2}. \qquad [3.157]$$

The stability condition of the synthesis transmittance $D_N(s)$ thus becomes a condition on its order $2N+1$. This order must be greater than a boundary defined by the inequality:

$$2N+1 > \frac{|b|}{\pi} \ln \frac{\omega_h}{\omega_b}. \qquad [3.158]$$

Such a condition is always verified when the imaginary order of differentiation b is nil, thus conveying the unconditional stability of the synthesis transmittance of a frequency bounded real non-integer differentiator (section 3.6.1).

3.6.4. *Distribution of synthesis zeros and poles: real and imaginary orders of differentiation*

Given [3.149], the natural logarithms of the synthesis zeros and poles are expressed, respectively, by:

$$\ln \omega_k^{'} = \ln \rho_k^{'} - i\theta \text{ and } \ln \omega_k = \ln \rho_k + i\theta. \qquad [3.159]$$

Thus, in the representation plane of the logarithm of the complex numbers, p_l, the zeros and poles are situated, respectively, on two straight lines parallel to the real axis, of ordinates $-\theta$ and $+\theta$ (Figure 3.13).

The change from the rank k zero to the same rank pole and then the change from this pole to the zero of the next rank are governed by the following relations obtained from [3.123]:

$$\ln \omega_k = \ln \alpha + \ln \omega_k^{'} \qquad [3.160]$$

and

$$\ln \omega_{k+1}^{'} = \ln \eta + \ln \omega_k. \qquad [3.161]$$

These relations show that the changes so defined are carried out by translations of affixes $\ln \alpha$ and $\ln \eta$ (Figure 3.13).

With a polar representation in the complex plane $\mathcal{C}_i$, the recursive factors α and η defined by [3.132] are written, respectively, as:

$$\alpha = |\alpha| e^{i2\theta} \text{ and } \eta = |\eta| e^{-i2\theta}, \qquad [3.162]$$

with

$$|\alpha| = \left(\frac{\omega_h}{\omega_b}\right)^{a/(2N+1)} \text{ and } |\eta| = \left(\frac{\omega_h}{\omega_b}\right)^{((1-a)/(2N+1))}, \qquad [3.163]$$

namely, in the plane P_l :

$$\ln \alpha = \ln|\alpha| + i2\theta \text{ and } \ln \eta = \ln|\eta| - i2\theta . \qquad [3.164]$$

When the differentiation imaginary order b is nil, relation [3.152] shows that the angle θ becomes nil. The two parallel zero and pole straight lines (of ordinates $-\theta$ and $+\theta$) become one with the real logarithmic axis.

Figure 3.13 illustrates in the plane P_l the distribution of the zeros and poles for a complex non-integer differentiation order. The shaded zones are the zones where the zeros and poles must be excluded for the stability of the synthesis transmittance $D_N(s)$.

Including the expressions of $\ln \alpha$ and $\ln \eta$ given by [3.164] into [3.126], generalized by α , η, $n \in \mathcal{C}_i$, provides the algebraic form of the complex non-integer differentiation order:

$$n = a + ib, \qquad [3.165]$$

where the differentiation real and imaginary orders, a and b , are given by:

$$a = \frac{\ln|\alpha|}{\ln|\alpha| + \ln|\eta|} \text{ and } b = \frac{2\theta}{\ln|\alpha| + \ln|\eta|} . \qquad [3.166]$$

Their quotient is expressed by the following relation:

$$\frac{b}{a} = \frac{2\theta}{\ln|\alpha|} . \qquad [3.167]$$

This result conveys that the ratio of the differentiation imaginary and real orders is equal to that of the imaginary and real parts of the natural logarithm of the recursive factor α (relation [3.164]).

3.6.5. *Determination of the number of synthesis zeros and poles*

In section 3.6.3, we demonstrated that the number $2N+1$ of synthesis zeros or poles must be greater than a limit defined by [3.158] to ensure the stability of the synthesis transmittance $D_N(s)$.

However, although this lower limit of $2N+1$ is sufficient for $D_N(s)$ to be stable, it is not at all sufficient for $D_N(s)$ to provide a good approximation of $D(s)$ and, in particular, of its frequency or time responses.

The minimum number of zeros or poles to ensure an impulse response of the synthesized differentiator with, as near as possible, the same energy as that of the differentiator to be synthesized, is determined below.

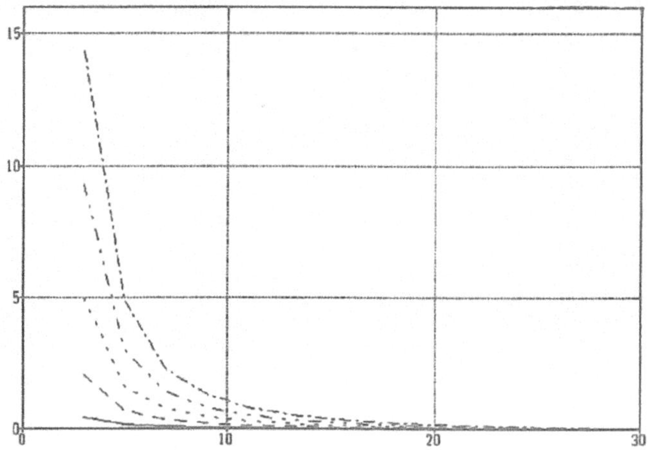

Figure 3.14. *Variations of* $E_{relative}$ *(%) versus* N *for* $n = 0.5 + i0.5$ *and various values of* μ *: —* $\mu = 10$ *; - - -* $\mu = 100$ *; - - - -* $\mu = 1{,}000$ *; - - - -* $\mu = 10{,}000$ *and - - -* $\mu = 100{,}000$

The energy of the impulse response of the synthesized differentiator with a given number of zeros or poles is calculated. Then the value of N is determined for the change from N to $N+1$ to correspond to an energy relative variation which is lower than a given, arbitrarily fixed, percentage. The graph of energy relative variation versus N reveals an elbow before which energy variation is highly sensitive to N and after which it is not. The value of N at the elbow can be considered as an optimal value which is not necessarily the value chosen by the designer.

3.6.5.1. *Impulse response energy*

As the impulse response of the synthesized differentiator $D_N(s)$ is a complex function of the form $s_r(t)+is_i(t)$, according to the Parseval theorem its energy admits an integral expression of the form:

$$E=\frac{1}{2\pi}\int_{-\infty}^{+\infty}\left|D_{N_r}(j\omega)\right|^2 d\omega+\frac{1}{2\pi}\int_{-\infty}^{+\infty}\left|D_{N_i}(j\omega)\right|^2 d\omega. \qquad [3.168]$$

Given that the degrees of the numerators and denominators of $D_{Nr}(j\omega)$ and $D_{Ni}(j\omega)$ are identical, the energy E is infinite. Thus, to limit the energy to a finite value, $D_{Nr}(j\omega)$ and $D_{Ni}(j\omega)$ must be replaced by the frequency responses $D'_{Nr}(j\omega)$ and $D'_{Ni}(j\omega)$ which are obtained by the cascading of a band-pass filter with $D_{Nr}(j\omega)$ and $D_{Ni}(j\omega)$, namely:

$$D'_{N_r}(j\omega)=\frac{\dfrac{j\omega}{\omega_b}}{\left(1+\dfrac{j\omega}{\omega_b}\right)\left(1+\dfrac{j\omega}{\omega_h}\right)}D_{N_r}(j\omega) \qquad [3.169]$$

and

$$D'_{N_i}(j\omega)=\frac{\dfrac{j\omega}{\omega_b}}{\left(1+\dfrac{j\omega}{\omega_b}\right)\left(1+\dfrac{j\omega}{\omega_h}\right)}D_{N_i}(j\omega). \qquad [3.170]$$

The bandwidth of the band-pass filter is limited by the low and high transitional frequencies of $D(s)$, ω_b and ω_h. Compared to a low-pass filter (of cutoff frequency ω_h, for example) which would have been sufficient to render the energy finite, the band-pass filter has the advantage of enhancing the effect of the zeros and poles ω'_k and ω_k.

Aström demonstrated that the expression [3.168] can be determined without having to calculate the integrals. It can be determined from the coefficients of the numerator and denominator of the transmittances $D'_{Nr}(s)$ and $D'_{Ni}(s)$.

Aström calculated an integral of the form:

$$I = \frac{1}{2\pi}\int_{-\infty}^{+\infty}\left|\frac{B(j\omega)}{A(j\omega)}\right|^2 d\omega, \qquad [3.171]$$

where $A(j\omega)$ and $B(j\omega)$ are $j\omega$ polynomials of integer degree with real coefficients, the degree of $B(j\omega)$ being less than that of $A(j\omega)$. The original procedure that he conceived [AST 70] enables the analytical expression of I from the elements of two tables made up of using the coefficients of $A(j\omega)$ and $B(j\omega)$ according to Routh's process.

In our study, the general form of integral [3.171] is compared to the integrals of [3.168] where $D_{Nr}(j\omega)$ and $D_{Ni}(j\omega)$ are replaced by $D'_{Nr}(j\omega)$ and $D'_{Ni}(j\omega)$ which are expressed as a ratio of two polynomials of different degrees, where the degree of the numerator is less than that of the denominator ([3.169], [3.170]). The energy that integral relation [3.168] defines is thus calculated according to Aström's method.

3.6.5.2. *Relative energy difference*

Some numerical results arising from the application of the Aström method show that a decade shift of the transitional frequencies of $D'_N(j\omega)$ toward high frequencies appears as a multiplication of the energy by a factor of 10. A decade shift toward low frequencies appears as a division by a factor of 10.

Furthermore, concerning the relative energy difference between two consecutive values of N, namely

$$E_{\text{relative}} = \frac{E(N) - E(N+1)}{E(N)}, \qquad [3.172]$$

where $E(N)$ and $E(N+1)$ denote the energies obtained for $2N+1$ and $2N+3$ couples of zeros and poles, other numerical results seem to show that this relative difference is, for a given ratio of the transitional frequencies ω_b and ω_h, independent of any frequency shift.

This property is used to provide standardized curves which represent the relative energy shift variations versus N for various values of the ratio $\mu = \omega_h / \omega_b$ (Figure 3.14). For $n = 0.5 + i0.5$ and $\mu = 1{,}000$, the optimal value of N at the elbow is four. The minimum value of N which gives an energy variation less than 0.1% equals nine.

3.6.6. *Validation in time domain*

In order to validate our synthesis method in time domain, it is convenient to compare the impulse responses of the differentiator to be synthesized and of the differentiator synthesized with various values of N.

3.6.6.1. *Impulse response of the differentiator to be synthesized*

The impulse response of the differentiator to be synthesized is defined as the original of the ideal transmittance:

$$D(s)=\left(\frac{\omega_u}{\omega_h}\right)^n\left(\frac{1+\dfrac{s}{\omega_b}}{1+\dfrac{s}{\omega_h}}\right)^n \quad \text{with } n\in \mathbb{C}_i . \qquad [3.173]$$

This transmittance can be rewritten as:

$$D(s)=\left(\frac{\omega_u}{\omega_b}\right)^n\left(1+\frac{\dfrac{\omega_b}{\omega_h}-1}{1+\dfrac{s}{\omega_h}}\right)^n , \qquad [3.174]$$

or, given the Newton expansion:

$$D(s)=\left(\frac{\omega_u}{\omega_b}\right)^n\sum_{k=0}^{\infty}\binom{n}{k}\left(\frac{\omega_b}{\omega_h}-1\right)^k\frac{1}{\left(1+\dfrac{s}{\omega_h}\right)^k} , \qquad [3.175]$$

from which we draw (for $\mathrm{Re}[n]>0$):

$$s(t)=\left(\frac{\omega_u}{\omega_b}\right)^n\left[\delta(t)+\sum_{k=1}^{\infty}\frac{\binom{n}{k}}{(k-1)!}(\omega_b-\omega_h)^k t^{k-1}e^{-t\omega_h}u(t)\right] . \qquad [3.176]$$

a)

b)

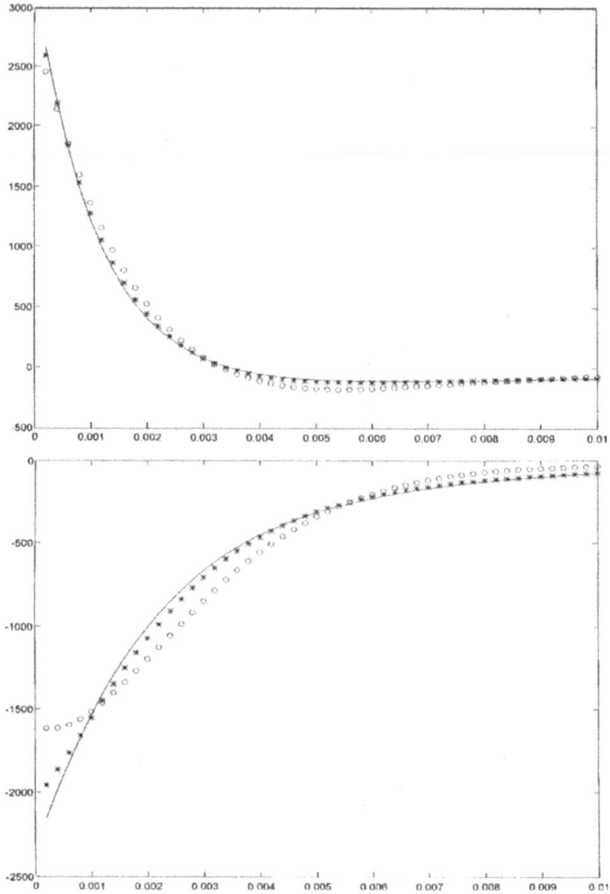

Figure 3.15. *a) Real and b) imaginary parts of the impulse responses of* $D(s)$ *and* $D_N(s)$ *for* $n = 0.5 + i0.5$, $\mu = 1{,}000$ *and various values of* N *:* — $D(s)$ *;* o $N = 1$ *;* ⋯⋯ * $N = 2$

3.6.6.2. *Impulse response of the synthesized differentiator*

The impulse response of the synthesized differentiator is defined as the original of the synthesis transmittance:

$$D_N(s) = \left(\frac{\omega_u}{\omega_h}\right)^n \prod_{k=-N}^{N} \frac{1+\dfrac{s}{\omega'_k}}{1+\dfrac{s}{\omega_k}} \text{ with } n \in \mathcal{C}_i. \qquad [3.177]$$

This transmittance can be rewritten under the form of relation [3.153] whose original is given by:

$$s(t) = K\delta(t) + \sum_{k=-N}^{N} A_k \omega_k e^{-\omega_k t} u(t), \qquad [3.178]$$

K and A_k being defined by [3.154] and [3.155].

3.6.6.3. *Comparative performances*

For $n = 0.5 + i0.5$, $\omega_b = 1\,\text{rd/s}$ and $\omega_h = 1{,}000\,\text{rd/s}$ $(\mu = 1{,}000)$, Figure 3.15 presents the time variations of the real and imaginary parts of the impulse responses of the transfers $D(s)$ and $D_N(s)$. The performances obtained well validate our approach. The figure indeed shows that the curves tend to be the same when N increases. For $N = 3$, it is already impossible to distinguish the difference between the curves.

3.6.6.4. *Achievement of the synthesis transmittance*

In order to achieve the real and imaginary parts of the synthesis transmittance $D_N(s)$, it is convenient to carry out (analytically or algorithmically) the expansion of [3.177] then the splitting into real and imaginary parts according to the ratio:

$$D_N(s) = \frac{P'(s) + iQ'(s)}{P(s) + iQ(s)}, \qquad [3.179]$$

where $P'(s)$, $Q'(s)$, $P(s)$ and $Q(s)$ are polynomials with real coefficients.

$D_N(s)$ can then be written in algebraic form of a complex quantity:

$$D_N(s) = D_{N_r}(s) + iD_{N_i}(s), \qquad [3.180]$$

where:

$$D_{N_r}(s) = \frac{P(s)P'(s) + Q(s)Q'(s)}{P^2(s) + Q^2(s)} \qquad [3.181]$$

and

$$D_{N_i}(s) = \frac{P(s)Q'(s) - P'(s)Q(s)}{P^2(s) + Q^2(s)}. \qquad [3.182]$$

These expressions do define the polynomial achievement of $D_N(s)$.

Its physical achievement through second-order active filters with operational amplifier results from the partial fraction expansion of $D_N(s)$. The corresponding circuit is patented by a French patent [OUS 90].

3.7. Bibliography

[AST 70] ASTRÖM K.J., *Introduction to Stochastic Control Theory*, Academic Press, New York, NY, pp. 116–138, 1970.

[ERD 62] ERDELYI A., *Operational Calculus and Generalized Functions*, Rinehart and Winston, Holt, 1962.

[GRU 67] GRÜNWALD A.K., "Ueberbegrenzte derivationen und dered anwendung", *Zeitschrift für Mathematik und Physik*, vol. 12, pp. 441–480, 1867.

[LET 68] LETNIKOV A.V., "Theory of differentiation of fractional order", *Matematiceskij Sbornik*, vol. 3, pp. 1–68, 1868.

[LIO 32] LIOUVILLE J., "Mémoire sur le calcul des différentielles à indices quelconques", *Journal Ecole Polytechnique*, vol. 13, no. 21, pp. 71–162, 1832.

[OLD 74] OLDHAM K.B., SPANIER J., *The Fractional Calculus*, Academic Press, New York, NY, 1974.

[OUS 91] OUSTALOUP A., *La commande CRONE*, Hermès, Paris, 1991.

[OUS 90] OUSTALOUP A., NOUILLANT M., Nouveau Système de Suspension, Patent No. 9004613, INPI, 1990.

[OUS 95] OUSTALOUP A., *La Dérivation Non Entière: Théorie, Synthèse et Applications*, Hermès, Paris, 1995.

[OUS 00] OUSTALOUP A., LEVRON F., MATHIEU B., *et al.*, "Frequency-band complex noninteger differentiator: characterization and synthesis", *IEEE Transactions on Circuits and Systems I: Fundamental Theory and Applications*, vol. 47, no. 1, pp. 25–39, January 2000.

[POS 30] POST E.L., "Generalized differentiation", *Transactions of the American Mathematical Society*, vol. 32, no. 4, pp. 732–781, 1930.

[TRI 11] TRIGEASSOU J.C., OUSTALOUP A., "Fractional integration: a comparative analysis of fractional integrators", *IEEE SSD'11*, Sousse, Tunisia, 2011.

[TRI 12] TRIGEASSOU J.C., MAAMRI N., SABATIER J., *et al.*, "State variables and transients of fractional order differential systems", *Computers and Mathematics with Applications*, vol. 64, no. 10, pp. 3117–3140, November 2012.

Chapter 4

On the CRONE Suspension

4.1. Introduction

This chapter deals with an application of non-integer differentiation in the (automotive) vehicle suspension area: the CRONE suspension. The term CRONE is a French abbreviation for *Commande (ou Comportement) Robuste d'Ordre Non Entier* (Non-integer order robust control (or behavior)). The various synthesis stages, from concept to practical achievement, are developed and performance tests revolving around prototypes validate the theoretical expectations.

The CRONE suspension presents two versions in conformity with the hydropneumatic and metallic technologies used for its achievement:

– a *hydropneumatic version* that results from the transposition, in vibratory insulation, of the porous dyke interpretation as defined in Chapter 1 and Appendix 2; to change from the porous dyke to the hydropneumatic version of the CRONE suspension, the dyke is rotated of a quarter of turn, air is replaced by nitrogen, water is replaced with oil and oil is set in motion with a piston linked to the wheel;

– a *metallic version* that is obtained from the usual (or traditional) suspension by replacing the order 1 traditional dashpot with a *non-integer order dashpot*. This dashpot is achieved by a *traditional dashpot with variable viscous friction coefficient* that is controlled (or piloted) so as to exert an effort identical to the one developed by the non-integer order dashpot.

As far as stability degree is concerned, robustness tests are carried out for different parametric states of the usual and the CRONE suspensions. These tests turn successively on the frequency and step responses of the two suspensions and also the roots of their characteristic equations. For the CRONE suspension, they reveal well the robustness of:

– the resonance ratio of the frequency response;

– the first *overshoot* of the step response;

– the *damping ratio* arising from the characteristic equation roots.

It is true that the resonance ratio in frequency domain and also the first overshoot and the damping ratio in time domain remain almost constant regardless of the load variations of the suspension.

4.2. From the porous dyke to the hydropneumatic version of the CRONE suspension

4.2.1. *Concept*

The idea of the CRONE suspension in hydropneumatic technology results from the transposition, in vibratory insulation, of the porous dyke interpretation as defined in Chapter 1 and Appendix 2.

To change from this interpretation to the hydropneumatic concept of the CRONE suspension (of a quarter vehicle), it suffices to:

– rotate the dyke representation of a quarter of turn (Figure 4.1(a));

– replace air with nitrogen (Figure 4.1(b));

– replace water with oil (Figure 4.1(b));

– set oil in motion with a piston linked to the wheel (Figure 4.1(c));

– transcribe technologically this new representation with a piston and spheres (Figure 4.2).

4.2.2. *From concept to achievement*

The recursivity of the stiffness is ensured by different calibration pressures of the spheres. The recursivity of the viscous friction coefficients is ensured by different laminating sections [OUS 90].

Thus, the principle of the *multisphere CRONE suspension* (Figure 4.2), representing its medium frequency behavior, is obtained.

a)
b)
c)

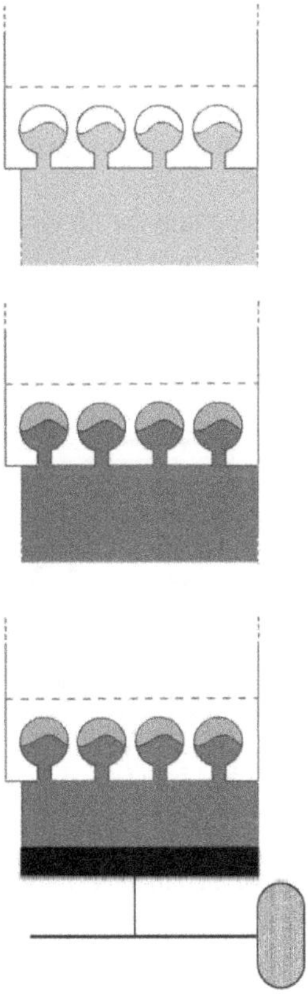

Figure 4.1. *From the porous dyke to the idea of the CRONE suspension hydropneumatic version*

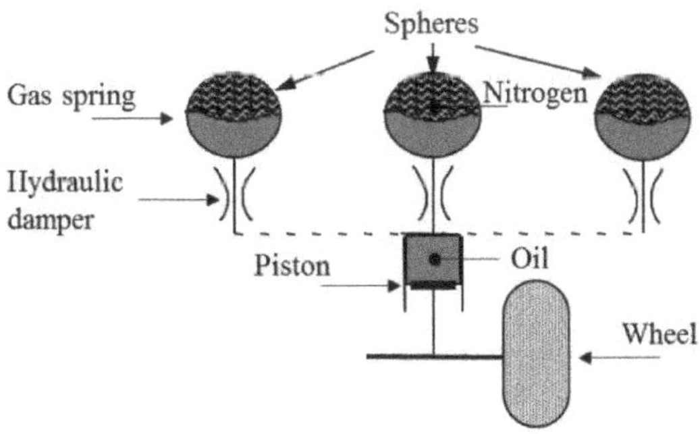

Figure 4.2. *Idea of the multisphere CRONE suspension*

For the *high frequencies*, a *silent-block* in elastomer (Figure 4.3), called an antinoise cupel, is introduced at the link between the suspension and the bodywork. This essentially elastic component is an additional stage which contributes to the high-frequency vibratory insulation. A *proportional type behavio*r is then obtained.

For the *low frequencies*, a *self-leveler* device (Figure 4.3), which achieves a position control ensuring, at rest, a constant altitude of the bodywork independent of its weight or, more generally, of any static effort on it, gives the suspension an *integrator type behavior*.

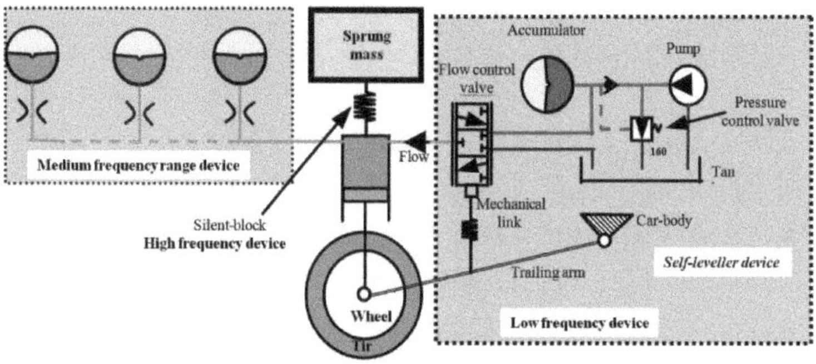

Figure 4.3. *From idea to its technological integration*

4.2.3. *Vehicle implementation*

Recursivity requires:

– a minimum of two ratios (to define their constancy);

– therefore a minimum of three cells;

– which is sufficient since the phase locking at $m\pi / 2$ (with $0 < m < 1$) must only be achieved between:

- the bodywork (sprung mass) natural frequency which is around 1 Hz;

- the wheel (unsprung mass) natural frequency which is around 18 Hz.

The minimal configuration so defined has been successfully used in the CRONE suspension as implemented on a Citroen BX (Figure 4.4) and awarded by the AFCET 1995 trophy distinguishing the best technological innovation in the university–industry relations (in this case, the CRONE team for the university component and PSA Peugeot-citroën for the industrial component).

a)

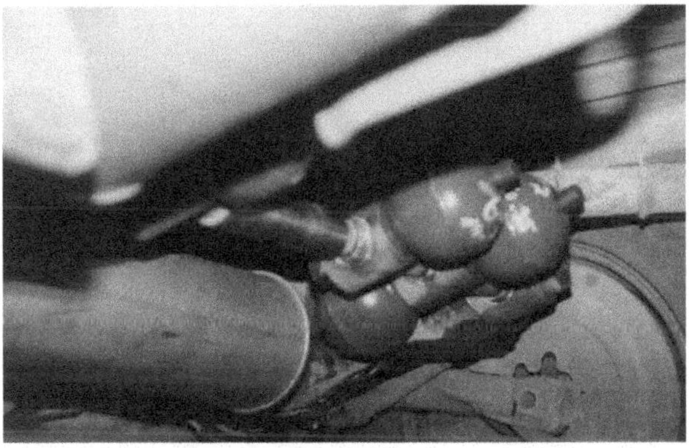

b)

Figure 4.4. *Photographs of the CRONE suspension implemented on a Citroen BX: a) in the front and b) at the rear*

Another minimal configuration, stemming from an approach by axle (front or rear), was exploited on PSA Peugeot-citroën prototype vehicles (e.g. Xantia and XM, among others).

Other technological solutions were exerted. Among them, let us note the *controlled CRONE suspension* (Figure 4.5), which is made up of a traditional spring (helicoidal in metallic version and with gas in hydropneumatic version) and a Continuously Variable Dashpot (AmCoVar) [DOR 95, ABA 97]. Such a dashpot (Figure 4.6) is equipped with a piloted valve whose role is to modify, through an electric control, the opening pressure of the check valves of a traditional dashpot. Here the principle of this dashpot is only defined. Its design and achievement are presented in the following section.

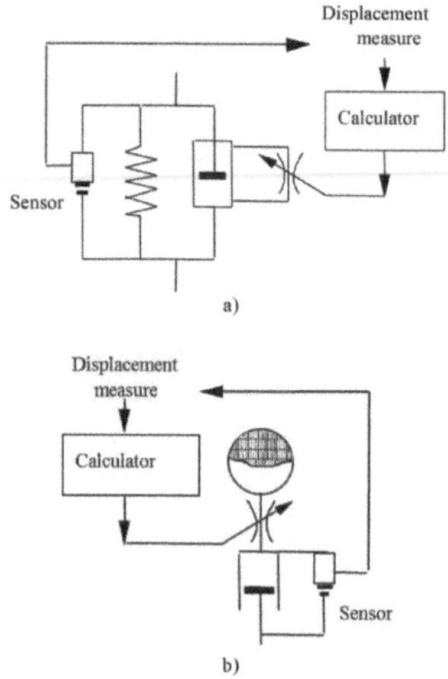

Figure 4.5. *Principle of the controlled CRONE suspension (equipped with an AmCoVar): a) metallic version and b) hydropneumatic version*

Figure 4.6. *Photographs of the AmCoVar: a) metallic version and b) hydropneumatic version*

4.3. Metallic version of the CRONE suspension

4.3.1. *A technological difference in terms of suspensions*

In statics, the vehicle bodywork weight is compensated by:

– the *self-leveler device*, in addition to the suspension, in a *hydropneumatic technology*;

– the *spring* of the suspension itself in a *metallic technology*, which implies that the CRONE suspension metallic version is, like the usual suspension:

- of type mass-spring-dashpot,

- apart from that the aforementioned dashpot is actually a damping device of another kind.

4.3.2. *Performance and robustness objective*

Just like the hydropneumatic version of the CRONE suspension, our aim is to ensure simultaneously to the vehicle bodywork:

– *a better vibration insulation* by a reduction of the vertical accelerations (*in terms of performance*);

– a better robustness of stability degree in relation to the carried load (in terms of robustness).

4.3.3. *Strategy*

Contrary to the effective strategies that consist of *reparameterizing* a usual suspension through a modification (linked or unlinked) of the spring *stiffness* and the dashpot viscous damping (or viscous friction coefficient), our strategy consists of *restructuring* a usual suspension through a modification of the *order* of the dashpot (defined by a force-displacement transfer), which comes to replace:

– the usual dashpot (of order 1);

– by a dashpot of non-integer order.

4.3.4. *Contract collaboration*

The industrial partner is the PSA Peugeot-Citroën company. The contribution of the collaboration is the conception and the achievement of the non-integer order dashpot (according to the patents 90 046 13 and 95 050 84 registered in 1990 and 1995, respectively), and also its implementation on a 406 Peugeot in 1998 with the help of the Coverplant company (Figure 4.7).

a)

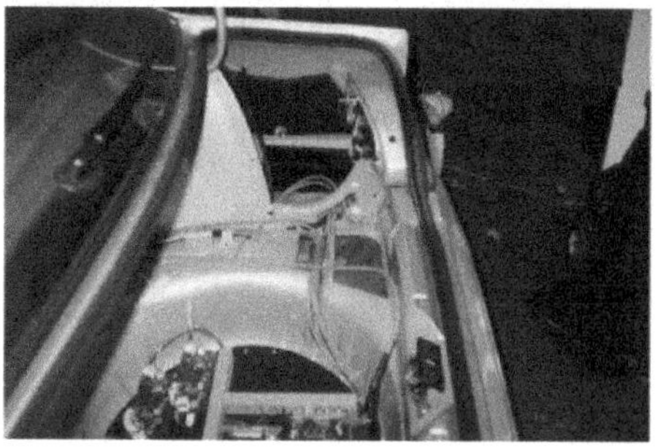

b)

Figure 4.7. *406 Peugeot implementation: a) in the front and b) at the rear*

4.3.5. *Principle of the CRONE suspension*

Let a suspension be of mass-spring-dashpot type (Figure 4.8) in which:

– M denotes the *bodywork mass* supported by each wheel;

– k denotes the *spring stiffness*;

– C denotes the *dashpot viscous damping;*

– $x(t)$ and $y(t)$ denote the vertical displacements of the wheel and the bodywork, respectively.

The *usual suspension* uses a dashpot which exerts a force proportional to its relative (or differential) speed, therefore to the *first derivative of its relative (or differential) displacement.*

The principle of the *CRONE suspension* consists of replacing the order 1 dashpot so defined by a non-integer order dashpot. This dashpot develops a force proportional to the *non-integer derivative of its relative displacement*, namely:

$$F(t) = C\left(\frac{d}{dt}\right)^m [x(t) - y(t)], \text{with } 0 < m < 1. \qquad [4.1]$$

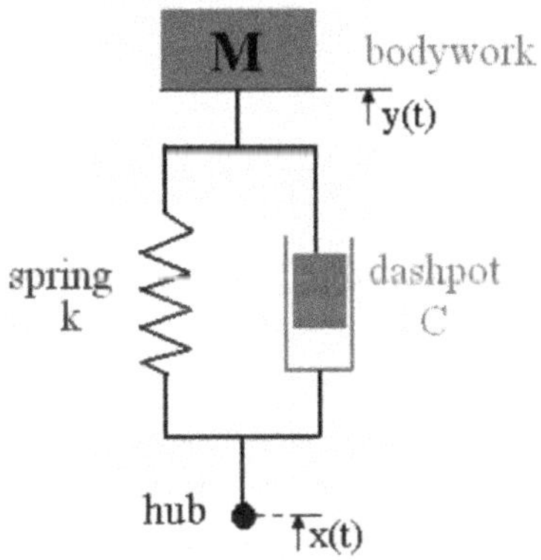

Figure 4.8. *Suspension scheme*

4.3.6. *Transfers of the usual and CRONE suspensions*

Applying *Newton's second principle* to the usual and CRONE suspensions makes it possible to formulate the differential equations:

$$M\frac{d^2y(t)}{dt^2} = -k[y(t) - x(t)] - C\left(\frac{d}{dt}\right)[y(t) - x(t)] \qquad [4.2]$$

and

$$M\frac{d^2y(t)}{dt^2} = -k\left[y(t) - x(t)\right] - C\left(\frac{d}{dt}\right)^m \left[y(t) - x(t)\right], \quad [4.3]$$

namely:

$$M\frac{d^2y(t)}{dt^2} + C\frac{dy(t)}{dt} + ky(t) = C\frac{dx(t)}{dt} + kx(t) \quad [4.4]$$

and

$$M\frac{d^2y(t)}{dt^2} + C\left(\frac{d}{dt}\right)^m y(t) + ky(t) = C\left(\frac{d}{dt}\right)^m x(t) + kx(t), \quad [4.5]$$

from which we draw the transmittances:

$$H(s) = \frac{Y(s)}{X(s)} = \frac{k + Cs}{k + Cs + Ms^2} \quad [4.6]$$

for the *usual suspension* ($m = 1$), and

$$H(s) = \frac{Y(s)}{X(s)} = \frac{k + Cs^m}{k + Cs^m + Ms^2} \quad [4.7]$$

for the *CRONE suspension* ($0 < m < 1$).

4.3.7. *Initial behavior: no initial acceleration for the CRONE suspension*

The deformation (by crushing) of a tire while climbing a pavement, which constitutes an extreme test case in vibratory insulation by impact absorption, means that *the wheel hub describes a profile that looks more like a ramp than a step.*

So, the *ramp function* turns out to be a *model of elementary deterministic solicitation sufficiently representative of reality*, thus enabling us to express the vertical displacement of the wheel hub in the form:

$$x(t) = t\,u(t), \quad [4.8]$$

where $u(t)$ denotes the unit step function.

As regards the *response* $y(t)$ *to this ramp try*, it is possible to write, from the initial value theorem

$$\lim_{t\to 0} \ddot{y}(t) = \lim_{s\to\infty} s[s^2 H(s)X(s)], \qquad [4.9]$$

particularized by

$$\lim_{t\to 0} \ddot{y}(t) = \lim_{s\to\infty} sH(s) \quad \text{for } X(s) = \frac{1}{s^2}: \qquad [4.10]$$

$$\ddot{y}(0^+) = \lim_{s\to\infty} \frac{ks + Cs^2}{k + Cs + Ms^2} = \frac{C}{M} \qquad [4.11]$$

for the *usual suspension* ($m = 1$), and

$$\ddot{y}(0^+) = \lim_{s\to\infty} \frac{ks + Cs^{1+m}}{k + Cs^m + Ms^2} = 0 \qquad [4.12]$$

for the *CRONE suspension* $(0 < m < 1)$.

The comparison of these relations shows that *the initial acceleration of the bodywork is finite for the usual suspension and nil for the CRONE suspension.*

Beyond the robustness of the CRONE suspension that we are going to show, this property is already remarkable. It indeed ensures *a better comfort for the passengers.*

4.3.8. *Stability degree robustness*

4.3.8.1. *Robustness tests*

For different parametric states of the usual and CRONE suspensions, namely $k = 17055\ N/m$ and $C = 5500\ Ns/m$ for $m = 1$ and also $k = 4264\ N/m$ and $C = 9000\ Ns^{0.8}/m$ for $m = 0.8$, these states ensuring approximately the same dynamics to the two suspensions for a mass of 300 kg, the tests on the degree stability robustness in relation to the carried load are completed for different values of the mass M, namely 150, 300, 600 and 900 kg.

4.3.8.2. *Stability degree measurement*

Stability degree is quantified (or measured) by:

– the *resonance ratio*, defined from the *frequency response* (Figure 4.9);

– the *first overshoot*, defined from the *step response* (Figure 4.10), or;

– the *damping ratio*, defined from the *characteristic equation roots* (Figure 4.11).

This leads us to say that the robustness tests as defined turn successively on:

– the frequency and step responses of the two suspensions;

– the roots of their characteristic equation.

4.3.8.3. *Frequency and step responses: robustness of the resonance ratio and the first overshoot for the CRONE suspension*

Figures 4.9 and 4.10 present the frequency and step responses for both usual and CRONE suspensions.

These responses, given for the different values of the vehicle load, reveal the robustness of stability degree for the CRONE suspension through the constancy of:

– the resonance ratio in frequency domain;

– the first overshoot in time domain.

4.3.8.4. *Characteristic equation roots: robustness of the damping ratio for the CRONE suspension*

The algorithm, that the module "*Generalized characteristic equation*" of the CRONE software uses, is applied to find the roots of the characteristic equation of:

– the *usual suspension*, namely ($m = 1$)

$$Ms^2 + Cs + k = 0; \qquad [4.13]$$

– the *CRONE suspension*, namely ($m = 0.8$)

$$Ms^2 + Cs^{0.8} + k = 0. \qquad [4.14]$$

For the parametric states of the usual and CRONE suspensions as previously defined, Figure 4.11 illustrates *the image of the roots (or the poles)* so obtained for the different mass values defined by the test conditions. Knowing that the damping ratio is given by the cosine of the half-center angle that the conjugate complex pole pair forms, *the angular quasi-constancy of the CRONE suspension poles is representative of its robustness*.

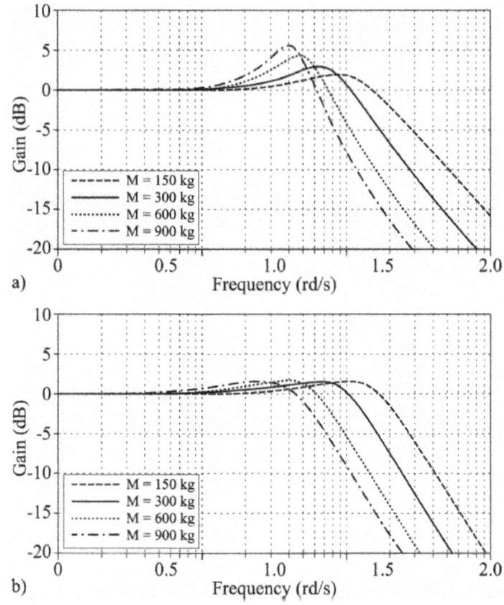

Figure 4.9. *Gain diagrams of a) the usual suspension and b)the CRONE suspension:* $--M = 150\ kg$ *;*$— M = 300\ kg$*;* **.....** $M = 600\ kg$ *; and* $-\cdot M = 900\ kg$

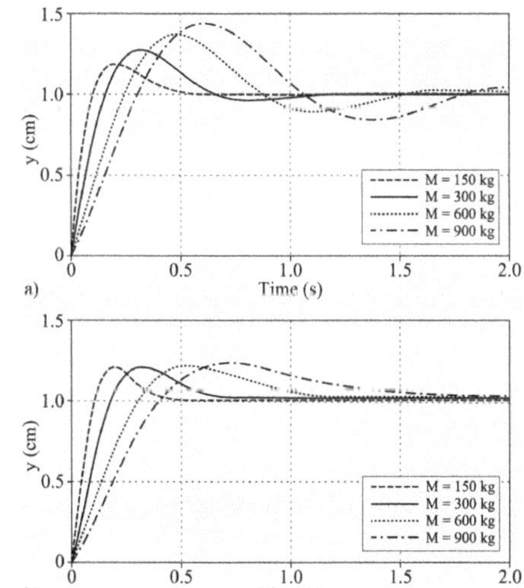

Figure 4.10. *Step responses of a) the usual suspension and b) the CRONE suspension:* $--M = 150\ kg$ *;*$— M = 300\ kg$ *;* $\cdots$ $M = 600\ kg$ *; and* $-\cdot$ $M = 900\ kg$

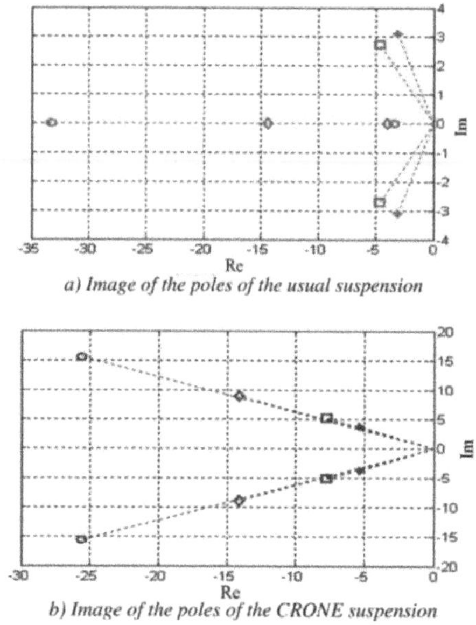

Figure 4.11. *Image of the poles of a) the usual suspension and b) the CRONE suspension for different loads:* ○ $M = 150$ *kg* ; ◊ $M = 300$ *kg* ; □ $M = 600$ *kg* ; *and* * $M = 900$ *kg*

4.3.9. *Idea of the synthesis of a non-integer order dashpot*

The synthesis of a non-integer order dashpot is based on the *non-stationarity of a usual dashpot* through the time variation of the viscous damping (or viscous friction coefficient). Indeed, the force developed by a non-integer order dashpot, namely

$$F(t) = C\left(\frac{d}{dt}\right)^m [x(t) - y(t)], \qquad [4.15]$$

can be interpreted as resulting from a *usual dashpot with variable viscous damping*, namely

$$F(t) = C(t)\left(\frac{d}{dt}\right)[x(t) - y(t)], \qquad [4.16]$$

from where we draw, by identifying these two equations:

$$C(t) = C\frac{\left(\frac{d}{dt}\right)^m [x(t) - y(t)]}{\left(\frac{d}{dt}\right)[x(t) - y(t)]}. \qquad [4.17]$$

Thus, the synthesis process consists of:

– first, computing (in conformity with this relation) the *instantaneous viscous damping* $C(t)$ from the *relative displacement recorded by a position sensor*;

– second, computing the *instantaneous section* $s(t)$ *of the dashpot hole* from the value of $C(t)$ so obtained.

4.3.10. *Active character of the CRONE suspension*

The time variation of the viscous damping $C(t)$ for a mass of 300 kg and for step and harmonic responses as shown in Figure 4.12 presents positive and negative values that convey the existence of dissipative and active phases, thus expressing that the damping device which is at stake is in fact:

– a (piloted) genuine dashpot for the dissipative phases;

– relayed by a *hydraulic actuator controlled in force* for the *active phases* (which makes the CRONE suspension *active*).

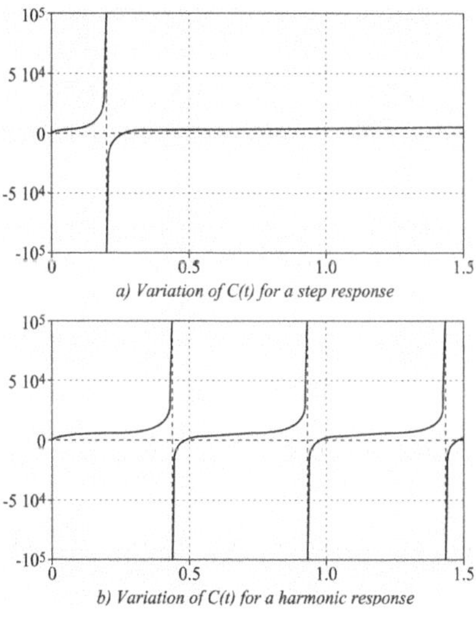

Figure 4.12. *Variation of* $C(t)$ *for a) step response and b) harmonic response*

4.3.11. *Piloted passive CRONE suspension*

Given that the energies at stake during the active phases are negligible compared to those corresponding to the dissipative phases (quantified study in sinusoidal state), which confers a *quasi-dissipative character* to the dashpot of non-integer order (thus justifying its dashpot denomination), it is convenient to adopt the following process:

– *a priori, cancel* $C(t)$ *when it becomes negative;*

– *a posteriori, verify that the cancellation of* $C(t)$ *so defined does not practically affect the performances* (with an eventual reparameterization of the suspension).

The modification of $C(t)$ in conformity with a new viscous damping, denoted by $C^*(t)$, such as

$$C^*(t) = \begin{cases} 0 \quad \text{when } C(t) \le 0 \\ C \dfrac{\left(\dfrac{d}{dt}\right)^m [x(t) - y(t)]}{\left(\dfrac{d}{dt}\right)[x(t) - y(t)]} \quad \text{when } C(t) > 0, \end{cases} \qquad [4.18]$$

defines the CRONE suspension said to be *modified* or, in other words, the *piloted passive CRONE suspension* that constitutes the most economical solution as it needs no exterior energy supply and that approximately keeps the same performances (Figure 4.13) with the following reparameterization: $k = 10{,}000\ N/m$, $C = 9{,}000\ Ns^{0.7}/m$ and $m = 0.7$.

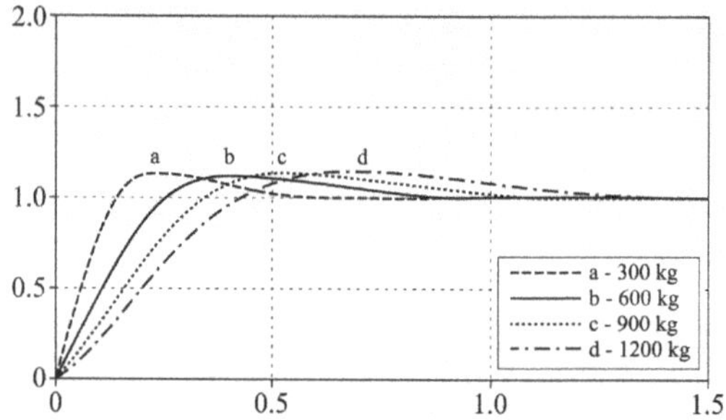

Figure 4.13. *Step responses of the modified CRONE suspension: (a) 300 kg; (b) 600 kg; (c) 900 kg; and (d) 1,200 kg*

4.4. Bibliography

[ABA 97] ABADIE V., MOREAU X., NOUILLANT M., *et al.*, Système d'amortissement réglable en continu, Patent no. 9705089, INPI, 1997.

[DOR 95] DORE P., MOREAU X., NOUILLANT M., *et al.*, Amortisseur piloté, Patent no. 9505084, INPI, 1995. [Extension to Japan and USA in 1996]

[MOR 04] MOREAU X., ALTET O., OUSTALOUP A., "The CRONE suspension: management of dilemma comfort-road holding", *Journal of Nonlinear Dynamics*, vol. 38, pp. 461–484, 2004.

[OUS 90] OUSTALOUP A., NOUILLANT M., Nouveau système de suspension, Patent no. 9004613, INPI, 1990.

[OUS 91a] OUSTALOUP A., *La commande CRONE*, Hermès, Paris, 1991.

[OUS 91b] OUSTALOUP A., NOUILLANT M., BALLOUK A., "Non integer derivation in mechanics through the CRONE suspension", *13th IMACS World Congress on Computation and Applied Mathematics*, Dublin, Ireland, 22–26 July 1991.

[OUS 93] OUSTALOUP A., MOREAU X., NOUILLANT M., "From the second generation CRONE control to the CRONE suspension", *IEEE International Conference on Systems, Man and Cybernetics, Systems Engineering in the Service of Humans, Conference Proceedings*, Le Touquet, France, vol. 2, pp. 143–148, 17–20 October 1993.

[OUS 95] OUSTALOUP A., *La dérivation non entière: théorie, synthèse et applications*, Hermès, Paris, 1995.

[OUS 96] OUSTALOUP A., MOREAU X., NOUILLANT M., "The CRONE suspension", *Control Engineering Pratice*, vol. 4, no. 8, pp. 1101–1108, 1996.

[OUS 10] OUSTALOUP A., MOREAU X., "Mechanical version of the CRONE suspension", *Advances in the Theory of Control, Signals and Systems with Physical Modeling*, LNCIS 407, Springer-Verlag, Berlin Heidelberg, pp. 99–112, 2010.

[SER 07] SERRIER P., MOREAU X., OUSTALOUP A., "Limited-bandwidth fractional differentiator: synthesis and application in vibration isolation", *Advances in Fractional Calculus*, Springer Netherlands, pp. 287–302, 2007.

Chapter 5

On the CRONE Control

5.1. Introduction

Robustness is a very wide notion that always conveys the same idea, that is to say *insensitivity* (or *quasi-insensitivity*). So, in the same domain, there are as many robustness types as there are insensitive (or quasi-insensitive) magnitudes and magnitudes to which these ones are insensitive (or quasi-insensitive).

In automatic control, it is frequent to consider *stability robustness* whose aim is to maintain stability (*robustness in stability*). But in the non-integer approach that the CRONE control uses, the considered robustness is *stability degree robustness* whose aim is to maintain or *desensitize* frequential or temporal dynamic performances which measure stability degree, notably the damping ratio which is a temporal dynamic performance (*robustness in performance*).

In the case of the relaxation of water on a porous dyke, we have shown that the robustness of the damping ratio is illustrated by two isodamping half-straight lines in the operational plane, and by a (vertical) frequency template in the Nichols plane whose form and vertical sliding ensure the invariance of the phase margin.

The transposition of this robustness in automatic control is at the origin of the initial approach of the CRONE control. Indeed, its first principles arise from the interpretation of the model of water relaxation on a porous dyke in conformity with two legitimate interpretations which, respectively, define:

– the first generation CRONE control through a *controller phase locking* (at a non-integer multiple of $\pi / 2$);

– the second generation CRONE control through an *open-loop phase locking* (at a non-integer multiple of $-\pi/2$ (between $-\pi/2$ and $-\pi$)).

The change from second to third generation CRONE control results from the generalization of the *vertical template* in conformity with two generalization levels:

– in the first generalization level, the vertical template is replaced by a straight line segment of any direction, called a *generalized template*, the direction of which is given by that of the main axis of the uncertainty domain (of the plant or the open loop) calculated at

- the open-loop unit gain frequency, ω_u

- or the closed-loop resonance frequency in tracking, ω_r

– in the second generalization level, the generalized template is replaced by a set of generalized templates, called *multi-template*, which leads to a *curvilinear template* generalizing the rectilinear template formed by the generalized template; the tangent at each point of the curvilinear template is the main axis of the corresponding uncertainty domain.

5.2. From the porous dyke to the CRONE control of first and second generations

To change from the porous dyke to the CRONE control of first and second generations, it is convenient to interpret the model of the water relaxation on a porous dyke according to two possible interpretations that, respectively, define:

– the first generation CRONE control (or strategy) through a *controller phase locking* (around the open-loop unit gain frequency ω_u);

– the second generation CRONE control (or strategy) through an *open-loop phase locking* (around the frequency ω_u and for the nominal parametric state and parametric variations of the plant).

To that effect, let us recall here the differential equations that model the water relaxation. It is about the differential equation that represents the model of the water-dyke interface, namely

$$Q(t)=\frac{1}{\omega_0^{\,n-1}}\left(\frac{d}{dt}\right)^{n-1}P(t) \text{ with } 1<n<2, \qquad [5.1]$$

and the differential equation arising from the application of the dynamics fundamental law to the motion water mass, M, namely

$$\frac{M}{S^2}\frac{dQ(t)}{dt}+P(t)=0\,. \qquad [5.2]$$

5.2.1. *First interpretation of the relaxation model: first generation CRONE strategy*

5.2.1.1. *Dynamic behavior of the water mass*

Introducing the water displacement $X(t)$ with respect to the resting position, thus enabling a new formulation of the flow according to

$$Q(t)=S\frac{dX(t)}{dt}, \qquad [5.3]$$

relation [5.2] becomes:

$$\frac{M}{S^2}\frac{d^2X(t)}{dt^2}+P(t)=0, \qquad [5.4]$$

a differential equation from which we can draw a symbolic equation of the form:

$$X(s)=-\frac{S}{Ms^2}P(s)\,. \qquad [5.5]$$

5.2.1.2. *Dynamic behavior of the water-dyke interface*

The combination of relations [5.1] and [5.3] determines the differential equation:

$$\frac{1}{\omega_0^{\,n-1}}\left(\frac{d}{dt}\right)^{n-1}P(t)=S\frac{dX(t)}{dt}, \qquad [5.6]$$

from which we draw the symbolic equation:

$$\frac{1}{\omega_0^{\,n-1}}s^{n-1}P(s)=SsX(s), \qquad [5.7]$$

which can be written as:

$$P(s)=S\omega_0^{\,n-1}s^{2-n}X(s), \qquad [5.8]$$

namely, by putting $n = 2 - m$ with $0 < m < 1$:

$$P(s) = S\omega_0^{1-m} s^m X(s). \qquad [5.9]$$

This symbolic equation conveys that *the pressure* $P(t)$ *at the water-dyke interface is proportional to the non-integer derivative (of order* m*) of the water displacement* $X(t)$.

Knowing that the physics of the dyke limits the order m differentiation to a medium frequency range, it is convenient to replace the operator, s, with the *frequency bounded operator,*

$$\omega_b \frac{1 + s/\omega_b}{1 + s/\omega_h} \text{ with } \omega_b < \omega_h, \qquad [5.10]$$

where ω_b and ω_h denote the low and high transitional frequencies that define the frequency truncation of the operator s.

Equation [5.9] can thus be restructured into a more realistic form:

$$P(s) = C_0 \left(\frac{1 + s/\omega_b}{1 + s/\omega_h} \right)^m X(s), \qquad [5.11]$$

by putting

$$C_0 = S\omega_0^{1-m} \omega_b^m. \qquad [5.12]$$

5.2.1.3. *Functional representation highlighting a constant phase CRONE controller*

Let us take again two symbolic equations [5.5] and [5.11], namely:

$$X(s) = -\frac{S}{Ms^2} P(s) \qquad [5.13]$$

and

$$P(s) = C_0 \left(\frac{1 + s/\omega_b}{1 + s/\omega_h} \right)^m X(s). \qquad [5.14]$$

This equation system can be translated by the functional diagram of Figure 5.1, which recalls a free control loop (nil reference input $E(s)$).

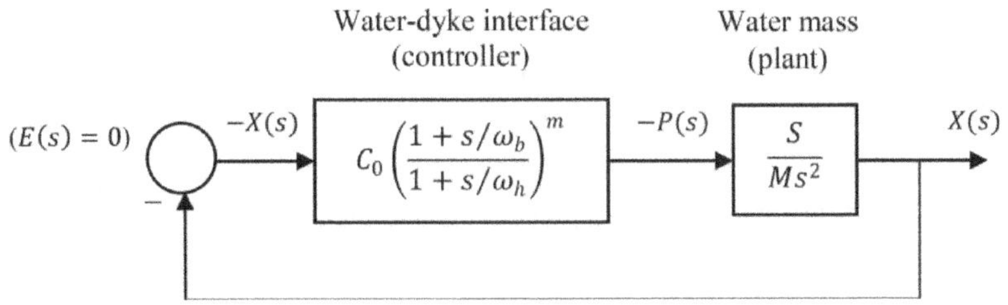

Figure 5.1. *Functional diagram governing relaxation*

The action chain structure successively reveals that:

– the transmittance that describes the motion of the water mass M, namely

$$\frac{X(s)}{-P(s)} = \frac{s}{Ms^2}, \qquad [5.15]$$

can be interpreted as the transmittance of a *plant* $G(s)$ of input $-P(s)$ and output $X(s)$;

– the transmittance of the water-dyke interface, namely

$$\frac{-P(s)}{-X(s)} = C_0\left(\frac{1+s/\omega_b}{1+s/\omega_h}\right)^m, \qquad [5.16]$$

can be interpreted as the transmittance of a *controller* $C_m(s)$ of input $-X(s)$ and output $-P(s)$.

The structure of the controller $C_m(s)$ so defined is nothing but the *ideal structure* of a *constant phase CRONE controller* whose *real structure* is obtained in section 3.6.1 through the synthesis of a frequency bounded real non-integer differentiator.

5.2.1.4. *Idea of the first generation CRONE strategy*

Located in a medium frequency range, the open-loop unit gain frequency, ω_u, belongs to the *order m asymptotic behavior* of the controller. The phase locking at $m\pi/2$ that characterizes such a behavior then ensures the invariance of the controller phase around the frequency ω_u.

Thus, the controller presents at least the merit not to participate in the variations of phase margin; notably, it does not reinforce them in any case as a variable phase controller would do around the frequency ω_u, not to mention the PID controller. Actually, on account of its transparency to the phase variations, a simple constant phase CRONE controller generally ensures a better robustness than a PID controller.

In automatic control, seeking the synthesis of such a controller defines the approach (said to be non-integer) that the *first generation CRONE control* uses.

5.2.2. *Second interpretation of the relaxation model: second generation CRONE strategy*

5.2.2.1. *Non-integer order differential equation as a relaxation model*

Putting the expression of $Q(t)$ given by [5.1] into relation [5.2] enables us to establish a linear differential equation of non-integer order n between 1 and 2, namely:

$$\frac{M}{S^2}\frac{1}{\omega_0^{\,n-1}}\left(\frac{d}{dt}\right)^n P(t)+P(t)=0\,, \qquad [5.17]$$

or, in a canonical form:

$$\tau^n\left(\frac{d}{dt}\right)^n P(t)+P(t)=0\,, \qquad [5.18]$$

by putting

$$\tau=\left(\frac{M}{S^2}\frac{1}{\omega_0^{\,n-1}}\right)^{\frac{1}{n}}. \qquad [5.19]$$

5.2.2.2. *Functional representation leading to an open-loop frequency template*

The Laplace transform of equation [5.18], namely

$$\tau^n s^n P(s)+P(s)=0\,, \qquad [5.20]$$

enables us to write the following relation:

$$P(s)=-\left(\frac{1}{\tau s}\right)^n P(s). \qquad [5.21]$$

This symbolic equation is translated by the functional diagram of Figure 5.2, which recalls a free control loop (nil reference input $E(s)$).

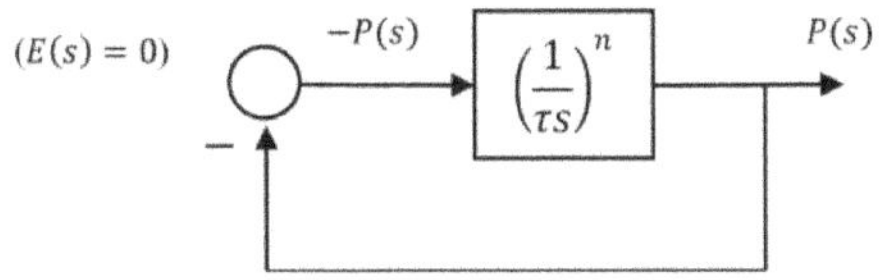

Figure 5.2. *Functional diagram enabling us to define an open-loop transfer*

Because of a unit feedback of the control loop, the action (or direct) chain determines an open-loop transmittance of the below form:

$$\beta(s) = \left(\frac{1}{\tau s}\right)^n = \left(\frac{\omega_u}{s}\right)^n, \qquad [5.22]$$

which is nothing but the transmittance of a *non-integer integrator* of unit gain (or transition) frequency $\omega_u = 1/\tau$.

Knowing that $\arg \beta(j\omega) = -n\pi/2$ and $1 < n < 2$, the Nichols locus of $\beta(j\omega)$ is a *vertical straight line* of abscissa between $-\pi/2$ and $-\pi$.

For the same physical reason as in the first interpretation case, the order n integration is limited to a medium frequency range. Thus, the vertical straight line of abscissa $-n\pi/2$ is reduced to a *vertical straight line segment* stretching around the unit gain frequency ω_u (Figure 5.3.). This segment is called an *open-loop frequency template* (or more simply *template*).

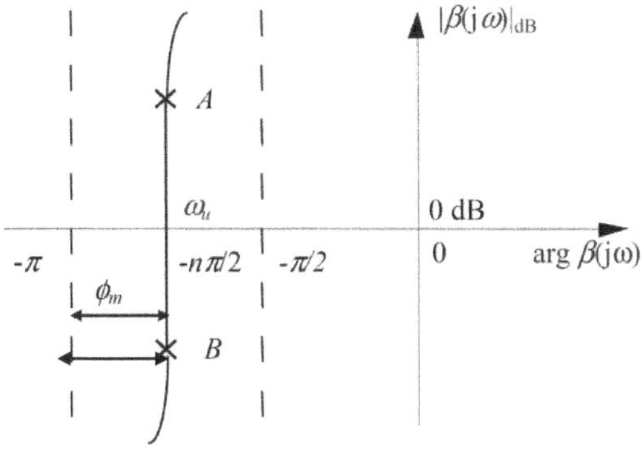

Figure 5.3. *A vertical straight line segment defines the template in the Nichols plane*

5.2.2.3. *Idea of the second generation CRONE strategy*

When the water mass M changes, the frequency ω_u is modified in conformity with the relation

$$\omega_u = \frac{1}{\tau} = \left(\omega_0^{n-1} \frac{S^2}{M} \right)^{1/n} . \qquad [5.23]$$

This leads us to say that the template so defined slides on itself at the time of a water mass variation. Given its form, such a template vertical displacement ensures the constancy of the phase margin Φ_m (Figure 5.3.) and, consequently, the constancy of the "distance to the critical point", which is then directly measured by Φ_m, thus conveying the stability degree robustness. It is easy to conceive that robustness improves as the length AB of the template increases, the template displacement being linked to the modification of the parameter M.

In automatic control, a purpose is to obtain a similar behavior, that is to say (Figure 5.3.):

– *an open-loop Nichols locus which forms the vertical template so defined for the nominal parametric state of the plant*;

– *a sliding of the template on itself at the time of a reparameterization of the plant* (this condition is verified when the reparameterization only leads to gain variations around ω_u).

The form and the vertical sliding of the template ensure not only the constancy of the phase margin Φ_m, but also (Figure 5.4.):

– *the constancy of the reduced first overshoot* $D_1^{'}$ *of the step response in tracking or in regulation*, through the tangency of the template to a same *iso-overshoot contour* (of graduation $D_1^{'}$), a performance contour which is a Nichols amplitude contour that we have *validated* as an iso-overshoot contour by using complex non-integer integration [BAL 92];

– *the constancy of the damping ratio* ζ *in tracking and in regulation*, through the tangency of the template to a same *isodamping contour* (of graduation ζ), a performance contour which is nothing but a Oustaloup isodamping contour that we have *built* due to an ingenious use of the complex non-integer integration and the covering techniques [OUS 95a, OUS 95b, OUS 99b].

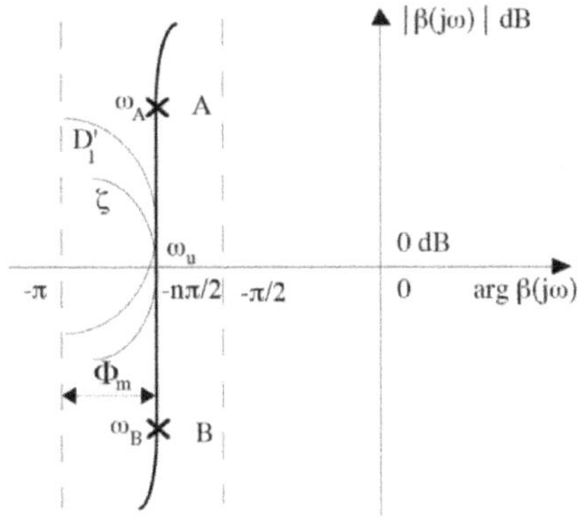

Figure 5.4. *The form and the vertical sliding of the template AB simultaneously ensure the robustness of the phase margin Φ_m, the reduced first overshoot $D_1^{'}$ and the damping ratio ζ*

Seeking the synthesis of such a template defines the approach (said to be non-integer) that the *second generation CRONE control* uses.

5.3. Second generation CRONE control and uncertainty domains

5.3.1. *Uncertainty domains*

Case of the plant: an uncertainty domain of the plant is defined as the domain of the Nyquist or Nichols plane, which groups together the image points of the frequency responses of the plant obtained, at a given frequency, for its various parametric states (Figure 5.5).

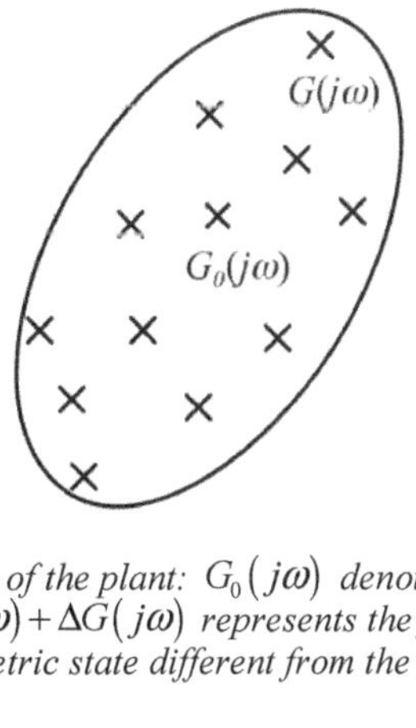

Figure 5.5. *Uncertainty domain of the plant: $G_0(j\omega)$ denotes the frequency response of the nominal plant; $G(j\omega) = G_0(j\omega) + \Delta G(j\omega)$ represents the frequency response of the plant for a parametric state different from the nominal one*

Case of the open loop: an uncertainty domain of the open loop is defined as the domain of the Nyquist or Nichols plane, which regroups the image points of the open-loop frequency responses obtained, at a given frequency, for the various parametric states of the plant (Figure 5.6).

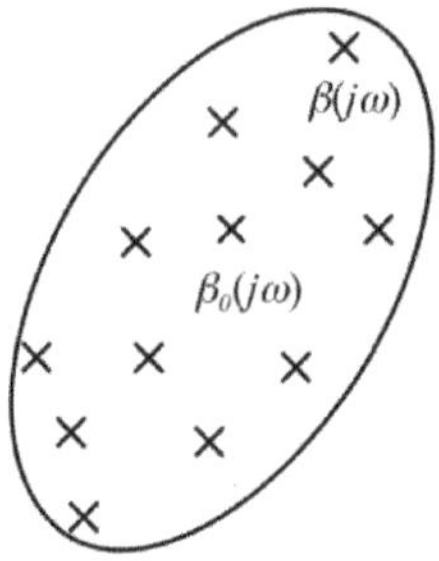

Figure 5.6. *Uncertainty domain of the open loop: $\beta_0(j\omega)=C(j\omega)G_0(j\omega)$ denotes the nominal open-loop frequency response; $\beta(j\omega)=\beta_0(j\omega)+\Delta\beta(j\omega)$ with $\Delta\beta(j\omega)=C(j\omega)\Delta G(j\omega)$ represents the open-loop frequency response for a plant parametric state different from the nominal one*

Given that (Figure 5.7)

$$\beta(j\omega)=C(j\omega)G(j\omega), \qquad [5.24]$$

it is possible to write :

$$|\beta(j\omega)|dB=|C(j\omega)|dB+|G(j\omega)|dB \qquad [5.25]$$

and

$$\arg\beta(j\omega)=\arg C(j\omega)+\arg G(j\omega). \qquad [5.26]$$

Such relations express that the change from an image point of $G(j\omega)$ to an image point of $\beta(j\omega)$ is carried out, for each point, through the same translation operations $|C(j\omega)|$ dB and $\arg C(j\omega)$, the controller $C(j\omega)$ providing the same gain and the same phase at a given frequency ω.

This leads us to say that the uncertainty domains of the open loop are identical to those of the plant, except for their positions which are different in conformity with the relation $\beta_0(j\omega)=C(j\omega)G_0(j\omega)$.

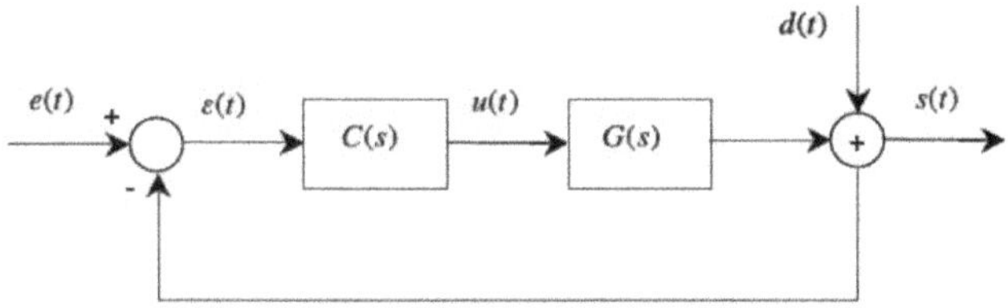

Figure 5.7. *Elementary control loop: $e(t)$, reference input; $d(t)$, disturbance; $s(t)$, output; $\varepsilon(t)$, error signal; and $u(t)$, plant input*

5.3.2. *Particular open-loop uncertainty domains*

When the relaxation water mass M changes, the frequency ω_u is modified according to relation [5.23], namely:

$$\omega_u = \frac{1}{\tau} = \left(\omega_0^{n-1} \frac{S^2}{M} \right)^{\frac{1}{n}}. \qquad [5.27]$$

If the phase in open loop remains constant, namely $\arg \beta(j\omega) = -n\pi/2$, the open loop gain varies in conformity with the extremal values:

$$|\beta(j\omega)|_{\min} = \left(\frac{\omega_{u_{\min}}}{\omega} \right)^n \qquad [5.28]$$

and

$$|\beta(j\omega)|_{\max} = \left(\frac{\omega_{u_{\max}}}{\omega} \right)^n, \qquad [5.29]$$

from which we draw:

$$\frac{|\beta(j\omega)|_{\max}}{|\beta(j\omega)|_{\min}} = \left(\frac{\omega_{u_{\max}}}{\omega_{u_{\min}}} \right)^n, \qquad [5.30]$$

namely in decibels:

$$|\beta(j\omega)|_{\max} dB - |\beta(j\omega)|_{\min} dB = 20n \log \left(\frac{\omega_{u_{\max}}}{\omega_{u_{\min}}} \right). \qquad [5.31]$$

In the Nichols plane, the uncertainty domains are thus vertical straight line segments of constant length, $20n\log\left(\omega_{u_{max}}/\omega_{u_{min}}\right)$ (Figure 5.8).

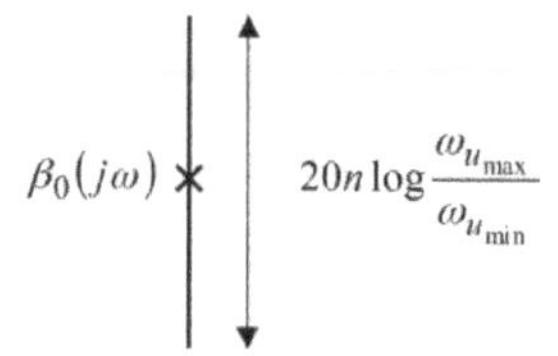

Figure 5.8. *Type of uncertainty domains corresponding to water mass variations*

5.3.3. *Adequacy of the second generation CRONE control template to the particular uncertainty domains*

The frequency configuration defined by the vertical template and the uncertainty domains so particularized express the part played by the template in the positioning of the uncertainty domains.

Figure 5.9 shows that the template ensures an optimal positioning of the uncertainty domains such that they do not overlap the low stability degree areas. The low stability degree area illustrated as an example is the area which is delimited by a Nichols amplitude contour.

It turns out that the optimal positioning of the uncertainties ensured by the template results from the fact that *the direction of the template is the same as that of the uncertainty domains*. This property is at the origin of the generalization of the template that less particular uncertainty domains require.

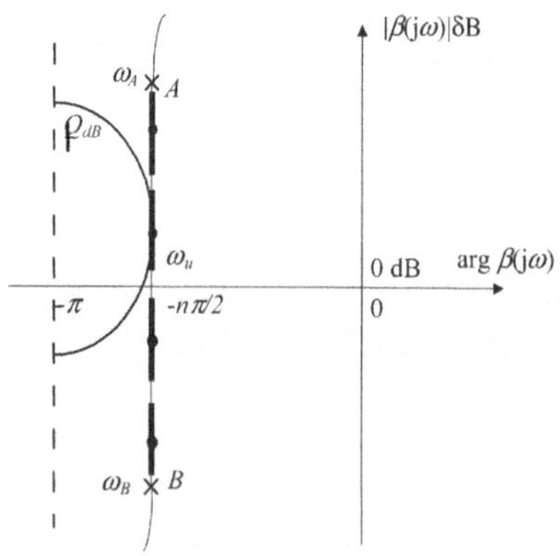

Figure 5.9. *The vertical template positions the uncertainty domains so that they do not overlap the amplitude contour* Q_{dB}

5.4. Generalization of the vertical template through the third generation CRONE control

5.4.1. *First level of generalization*

5.4.1.1. *Generalized template*

When the uncertainty domains are not reduced to vertical straight line segments, it is convenient to generalize the vertical template, particularly by a template:

– *always defined as a straight line segment for the nominal parametric state of the plant*;

– *but which no longer systematically determines a phase locking around the frequency* ω_u.

The *vertical straight line segment* defining the template in the Nichols plane until now is thus replaced by a *straight line segment of any direction* called a *generalized template* (Figure 5.10).

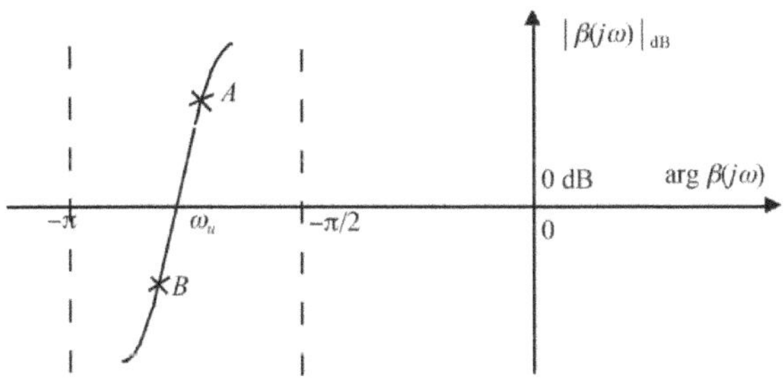

Figure 5.10. *Representation of the generalized template in the Nichols plane*

In order that the generalized template ensures an optimal positioning of the uncertainty domains such that they overlap the low stability degree areas as little as possible, the direction of the generalized template can be given by the mean direction of the main axes of the uncertainty domains calculated around the frequency ω_u (Figure 5.11).

Given the continuity of the direction of the main axes of the uncertainty domains, a good approximation consists of giving to the generalized template the direction of the main axis of the uncertainty domain calculated at the frequency ω_u (Figure 5.12(a)).

The direction of the generalized template can also be that of the main axis of the uncertainty domain calculated at the resonance frequency in tracking, ω_r; this choice appears preferential since the generalized template is tangent to the amplitude contour Q_{dB} at this frequency (Figure 5.12(b)).

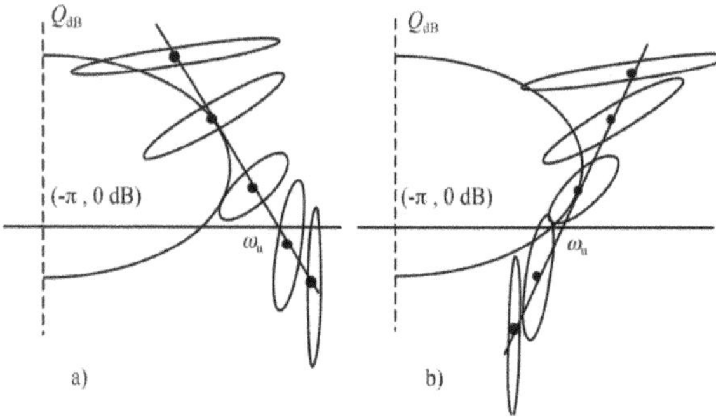

Figure 5.11. *The generalized template positions the uncertainty domains so that they overlap the amplitude contour Q_{dB} as little as possible: a) any generalized template and b) optimal generalized template*

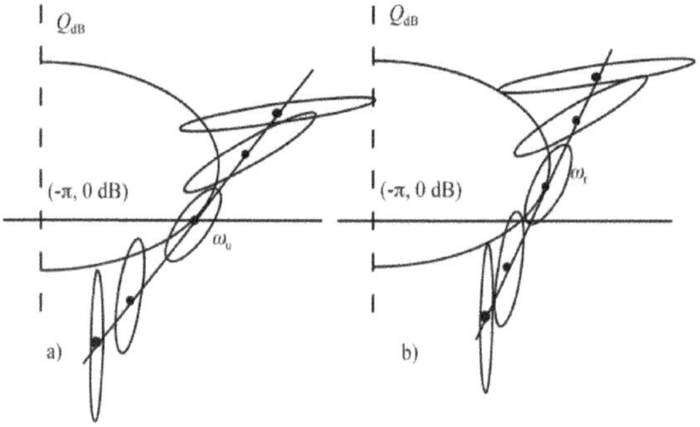

Figure 5.12. *Optimal generalized template whose direction is given by that of the main axis of the uncertainty domain corresponding to a) the frequency ω_u and b) the frequency ω_r*

5.4.1.2. *Generalized template and complex non-integer integration*

The vertical template is characterized by a real non-integer integration order, n, which determines its phase placement at $-n\pi/2$. It is indeed described by the transmittance of a real non-integer integrator, namely:

$$\mathrm{B}(s) = \left(\frac{\omega_u}{s}\right)^n \quad \text{with } n \in \mathbb{R}. \qquad [5.32]$$

In the study which consists of extending the description of the vertical template [OUS 91], we have shown that the generalized template can be characterized by a complex non-integer integration order, n, whose real part, a, determines its phase placement at the frequency ω_u, namely $-a\pi/2$, and whose imaginary part, b, determines its incline to the vertical (Figure 5.13.).

The generalized template is indeed described by a transmittance based on the complex non-integer integration, namely:

$$B(s) = \left(ch\left(b\frac{\pi}{2}\right)\right)^{sign(b)} \left(\frac{\omega_u}{s}\right)^a \left(\left[\left(\frac{\omega_u}{s}\right)^{ib}\right]_{\mathcal{C}_j}\right)^{-sign(b)}, \qquad [5.33]$$

or even, developing the part reduced to the operational plane $\mathcal{C}_j$:

$$B(s) = \left(ch\left(b\frac{\pi}{2}\right)\right)^{sign(b)} \left(\frac{\omega_u}{s}\right)^a \left(\cos\left(b\ln\left(\frac{\omega_u}{s}\right)\right)\right)^{-sign(b)}. \qquad [5.34]$$

In the case when the description transmittance of the generalized template is the subject of a frequency truncation at low and high frequencies, it is convenient to replace relation [5.33] by a more general expression:

$$B(s) = C^{sign(b)} \left(C_0 \frac{1+\dfrac{s}{\omega_h}}{1+\dfrac{s}{\omega_b}}\right)^a \left(\left[\left(C_0 \frac{1+\dfrac{s}{\omega_h}}{1+\dfrac{s}{\omega_b}}\right)^{ib}\right]_{\mathcal{C}_j}\right)^{-sign(b)}, \qquad [5.35]$$

in which:

$$C = ch\left[b\left(arctg\left(\frac{\omega_u}{\omega_b}\right) - arctg\left(\frac{\omega_u}{\omega_h}\right)\right)\right] \qquad [5.36]$$

and

$$C_0 = \left(\frac{1+\left(\dfrac{\omega_u}{\omega_b}\right)^2}{1+\left(\dfrac{\omega_u}{\omega_h}\right)^2}\right)^{\frac{1}{2}}, \qquad [5.37]$$

the expression of C_0 being reduced to $C_0 = \omega_u / \omega_b$ in the particular case when the low and high transitional frequencies, ω_b and ω_h, are geometrically distributed in relation to ω_u, namely $(\omega_b \omega_h)^{1/2} = \omega_u$.

To ensure the closed-loop stability, it has been shown that the condition must be verified. This condition turns on the integration imaginary order, b, and the frequency range between ω_b and ω_h [LAN 10, LAN 12].

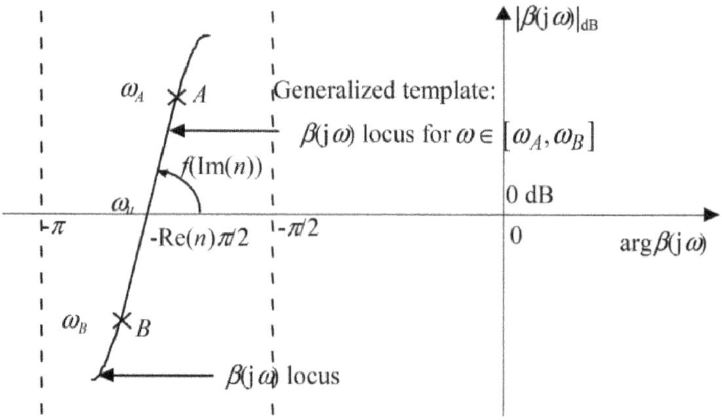

Figure 5.13. *The integration real order, $a = \mathrm{Re}(n)$, determines the phase placement at the frequency ω_u of the generalized template, namely $-a\pi/2$; the integration imaginary order, $b = \mathrm{Im}(n)$, determines its angle to the vertical*

5.4.2. *Second level of generalization*

5.4.2.1. *Curvilinear template*

A rectilinear template whose direction is that of the main axis of the uncertainty domain calculated at the frequency ω_r only ensures an optimal positioning of this uncertainty domain; the other domains can overlap the amplitude contour Q_{dB} too much.

In order to process each uncertainty domain as that calculated at the frequency ω_r (whose generalized template takes the direction), it is convenient to replace the generalized template by a *curvilinear template* whose tangent at each frequency point is the main axis of the corresponding uncertainty domain (Figure 5.14). This strategy indeed uses at best the *fractal robustness property* through which *the direction of the template is the same as that of the uncertainty domains.*

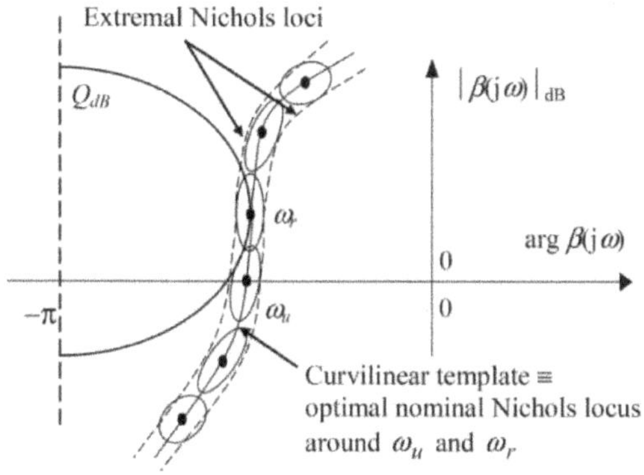

Figure 5.14. *The area delimited by the extremal Nichols loci defines, around ω_u or ω_r, an uncertain Nichols locus whose "thickness" determines its uncertainty*

5.4.2.2. *Operational description of the curvilinear template*

To obtain a description of the curvilinear template by a non-integer transmittance, it is convenient to interpret the asymptotic locus of the curvilinear template as a *multi-template* defined by a set of generalized templates (Figure 5.15).

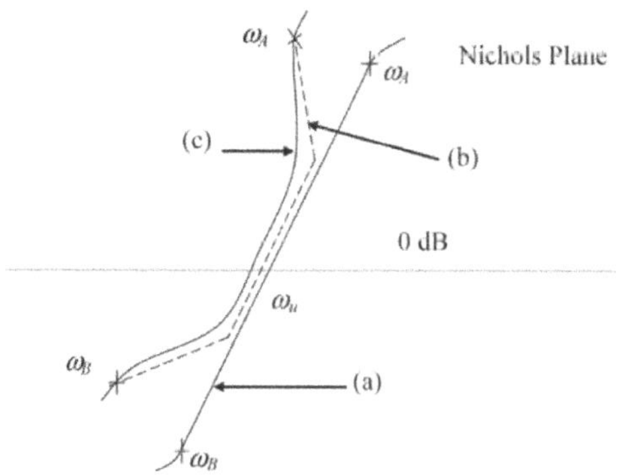

Figure 5.15. *From the generalized template a) to the multi-template b); from the rectilinear template a) to the curvilinear template c)*

The interpretation as defined through the multi-template makes it possible to describe the curvilinear template by a transmittance resulting from the product of frequency bounded transmittances of the same form as expression [5.35], namely:

$$B(s) = \prod_{k=-N_-}^{N^+} B_k(s), \qquad [5.38]$$

with (for $k \neq 0$):

$$\mathrm{B}_k(s) = C_k^{sign(b_k)} \left(\frac{\omega_k^{'}}{\omega_k} \frac{1+\frac{s}{\omega_{k+1}}}{1+\frac{s}{\omega_k}} \right)^{a_k} \left(\left[\left(\frac{\omega_k^{'}}{\omega_k} \frac{1+\frac{s}{\omega_{k+1}}}{1+\frac{s}{\omega_k}} \right)^{ib_k} \right]_{C_j} \right)^{-sign(b_k)} \quad ; \qquad [5.39]$$

for $k=0$, $\mathrm{B}_k(s)$ is defined by relation [5.35] in which ω_b and ω_h are in this case replaced by ω_0 and ω_1.

ω_k and ω_{k+1} denote the low and high transitional frequencies of the generalized template of rank k. $\omega_k^{'}$ represents its "central" frequency when k is different from zero, that is to say:

$$\omega_k^{'} = (\omega_k \omega_{k+1})^{1/2} \text{ for } k \neq 0 . \qquad [5.40]$$

The rank 0 generalized template, said to be *median*, is that which is defined in relation to $\omega_0^{'} = \omega_u$. The frequency $\omega_0^{'}$ does not systematically represent its central frequency in order to ensure an additional degree of freedom.

The gain C_k makes it possible to ensure a unit gain to $B_k(s)$ at the frequency ω_u fixed by the conceptor.

N_- denotes the number of generalized templates defined in relation to the frequencies $\omega_k^{'}$ lower than ω_u. N^+ denotes the number of generalized templates defined in relation to the frequencies $\omega_k^{'}$ higher than ω_u. In the particular case when $N_- = N^+ = 0$, $\mathrm{B}(s)$ is reduced to the frequency bounded transmittance defined by [5.35].

5.4.3. *Open-loop transfer integrating the curvilinear template*

The aim is to describe analytically the open-loop behavior (for the nominal plant) which takes into account at once:

– the accuracy specifications at the low frequencies;

– the curvilinear template close to the frequency ω_u;

– the behavior of the plant at the high frequencies and the specifications on the input sensitivity at these frequencies.

For phase minimum stable plants, it turns out that the behavior so defined can be described by a *transmittance based on frequency bounded complex non-integer integration* [OUS 99a, OUS 99b], namely:

$$\beta(s) = \beta_b(s)\mathrm{B}(s)\beta_h(s). \qquad [5.41]$$

– The transmittance $B(s)$, based on complex non-integer integration, is nothing but the description transmittance of the curvilinear template defined by relations [5.38] and [5.39]. Its extremal transitional frequencies are ω_{-N^-} in low frequency and $\omega_{N^+ +1}$ in high frequency.

– The transmittance $\beta_b(s)$ is that of an order n_b proportional-integrator, whose transitional frequency is equal to the low extremal transitional frequency of $B(s)$, in order that connecting $\beta_b(s)$ with $B(s)$ does not lead to introducing additional parameters. $\beta_b(s)$ is then defined by:

$$\beta_b(s) = C_b\left(\frac{\omega_{-N^-}}{s}+1\right)^{n_b}; \qquad [5.42]$$

if n_{pb} represents the asymptotic behavior order of the plant in low frequency ($\omega < \omega_{-N^-}$), the order n_b is given by $n_b \geq 1$ if $n_{pb} = 0$ and $n_b \geq n_{pb}$ if $n_{pb} \geq 1$, knowing that $n_b = 1$ cancels the position error and $n_b = 2$ cancels the velocity error; finally, C_b permits us to ensure a unit gain to $\beta_b(s)$ at the frequency ω_u.

– The transmittance $\beta_h(s)$ is that of an order n_h low-pass filter, whose transitional frequency is equal to the high extremal transitional frequency of $B(s)$, in order that connecting $\beta_h(s)$ with $B(s)$ does not introduce additional parameters. $\beta_h(s)$ is then defined by:

$$\beta_h(s) = \frac{C_h}{\left(1+\dfrac{s}{\omega_{N^+ +1}}\right)^{n_h}}; \qquad [5.43]$$

if n_{ph} denotes the asymptotic behavior order of the plant in high frequency ($\omega > \omega_{N^+ +1}$), the order n_h is given by $n_h \geq n_{ph}$, knowing that (for $\omega > \omega_{N^+ +1}$)

$n_h = n_{ph}$ ensures the constancy versus frequency of the input sensitivity and $n_h > n_{ph}$ ensures its decreasing versus frequency; finally, C_h permits us to ensure a unit gain to $\beta_h(s)$ at the frequency ω_u.

5.4.4. *Optimization of the open-loop behavior*

In order to simplify the developments of this section, the curvilinear template is reduced to the generalized template by taking $N_- = N^+ = 0$. B(s) is then defined by expression [5.35], the extremal transitional frequencies ω_{-N^-} and $\omega_{N^+ +1}$ being thus reduced to ω_b and ω_h.

5.4.4.1. *Optimal template*

The optimization of the generalized template as for:

– its *position* (along the axis 0dB),

– its *direction* (or angle to the vertical),

– and its *length*,

consists of determining the optimal values of the following four parameters:

– the *integration real order*, a, which determines the position of the template along the axis 0dB ;

– the *integration imaginary order*, b, which determines its incline with respect to the vertical;

– the *low and high transitional frequencies*, ω_b and ω_h, which determine its length through their action on its extremal frequencies ω_A and ω_B.

These parameters indeed "preparameterize" the open-loop transmittance $\beta(s)$. In the particular case when $n_b = 3$ and $n_h = 4$, Figure 5.16 shows how they act in the search for the optimal form of $\beta(s)$.

Finally, given that the structure of the open-loop transmittance is known and the frequency ω_u is fixed by the conceptor, the four optimal parameters, a_{opt}, b_{opt}, ω_{bopt} and ω_{hopt}, completely determine the optimal form of $\beta(s)$. This one is indeed obtained by putting the optimal parameters so defined into expression [5.41], in which B(s) is given by [5.35] and $\beta_b(s)$ and $\beta_h(s)$ by [5.42] and [5.43], the parameters C_b and C_h being deduced from the others.

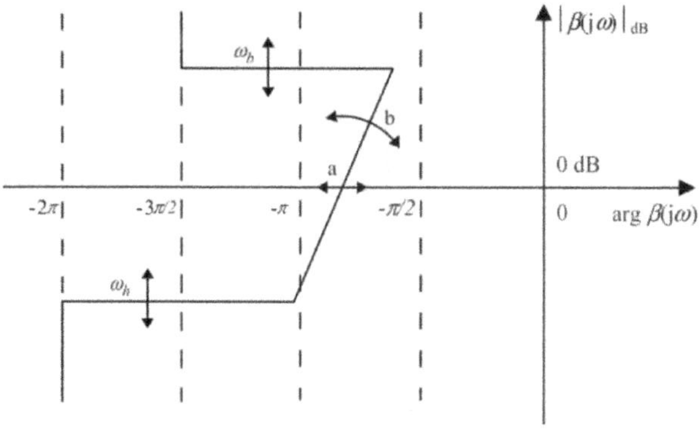

Figure 5.16. *Illustration of the effects of the parameters a, b, ω_b and ω_h on the Nichols asymptotic locus of $\beta(j\omega)$*

5.4.4.2. *Criterion to be minimized by the optimal template*

The search for the optimal template is a matter for optimization under constraints. The optimal template must minimize a quadratic criterion of the form

$$J = (D_{1_{max}} - D_{1_d})^2 + (D_{1_{min}} - D_{1_d})^2, \qquad [5.44]$$

namely, in frequency domain [OUS 95a, OUS 99a],

$$J = (Q_{a_{max}} - Q_{a_d})^2 + (Q_{a_{min}} - Q_{a_d})^2, \qquad [5.45]$$

while respecting the frequency constraints on the entirety of the sensitivity functions [OUS 99b] of the control (Figure 5.17).

D_1 represents the first overshoot of the unit step response in tracking or regulation. Q_a denotes the resonance ratio in tracking. D_{1_d} and Q_{a_d} are the desired values of D_1 and Q_a for the nominal parametric state of the plant. $D_{1_{min}}$, $D_{1_{max}}$ and $Q_{a_{min}}$, $Q_{a_{max}}$ are the minimal and maximal values of D_1 and Q_a obtained for its various parametric states.

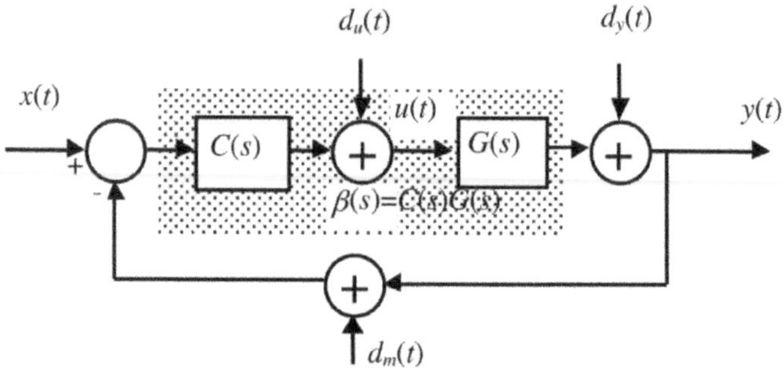

Figure 5.17. *Control scheme: $x(t)$, reference input; $u(t)$, plant input; $y(t)$, plant or control output; $d_u(t)$, input disturbance; $d_y(t)$, output disturbance; and $d_m(t)$, measurement disturbance*

5.4.5. *Structure and parametric estimation of the controller*

The controller in cascade with the plant is defined by a *non-integer transmittance*, $C_{ne}(s)$, deduced from the ratio:

$$C_{ne}(s) = \frac{\beta(s)}{G_0(s)}, \qquad [5.46]$$

where $\beta(s)$ is given by [5.41] and $G_0(s)$ denotes the nominal transmittance of the plant.

The *integer transmittance* of the controller, $C_{en}(s)$, results from the identification of $C_{ne}(s)$ through a *low-dimension integer rational model*, namely:

$$C_{en}(s) = \frac{R(s)}{S(s)}, \qquad [5.47]$$

where $R(s)$ and $S(s)$ are polynomials of specified degrees n and d:

$$R(s) = \sum_{k=0}^{n} r_k s^k \text{ and } S(s) = \sum_{k=0}^{d} s_k s^k . \qquad [5.48]$$

Two parametric estimation techniques are used. The first technique is based on the elementary symmetrical functions of the roots of Viète [LAN 94, OUS 02]. The second technique is based on solving a linear programming problem [MAT 94].

5.4.6. *Application*

The study plant considered as an example is described by the transfer function:

$$G(s) = G_0 \frac{\left(1 + \frac{s}{\omega_1'}\right)}{\left(1 + \frac{s}{\omega_1}\right)\left(1 + 2\varsigma\frac{s}{\omega_n} + \frac{s^2}{\omega_n^2}\right)}. \qquad [5.49]$$

Its nominal parametric state is defined by $G_{0_{nom}} = 50$, $\omega'_{1_{nom}} = 0.01$, $\omega_{1_{nom}} = 0.02$, $\omega_{n_{nom}} = 1$ and $\zeta_{nom} = 0.8$. The parameters G_0, ω_1', ω_1 and ω_n are uncertain and can take any value within the following intervals:

$$G_{0_{nom}}/2 \le G_0 \le 2G_{0_{nom}}, \quad \omega'_{1_{nom}}/2 \le \omega_1' \le 2\omega'_{1_{nom}},$$

$$\omega_{1_{nom}}/2 \le \omega_1 \le 2\omega_{1_{nom}} \text{ and } \omega_{n_{nom}}/1.5 \le \omega_n \le 1.5\omega_{n_{nom}}.$$

The aim of the control law to be synthesized is to ensure:

– a tracking resonance ratio of $2.3\,\mathrm{dB}$;

– an open-loop unit gain frequency of $20rd/s$.

The controller which responds to these objectives is given by relation [5.46] in which the optimal open-loop transmittance $\beta(s)$ is defined by relation [5.41] with $N^- = N^+ = 0$, $n_b = 1$, $n_h = 2$, $a = 1.44$, $b = -0.209$, $\omega_b = 0.12rd/s$ and $\omega_h = 955.44\,rd/s$.

As illustrated in Figure 5.18, this optimal open-loop behavior positions the uncertainty domains so as to ensure the robustness of the closed-loop stability degree.

In the time domain, the robustness of stability degree is illustrated by the desensitization to the parametric state of the plant, of the first overshoot of the response to a step output disturbance (Figure 5.19).

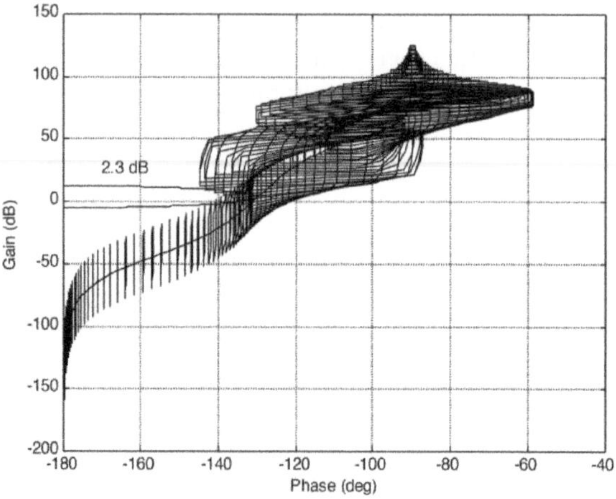

Figure 5.18. *Optimal open-loop Nichols locus and associated uncertainty domains*

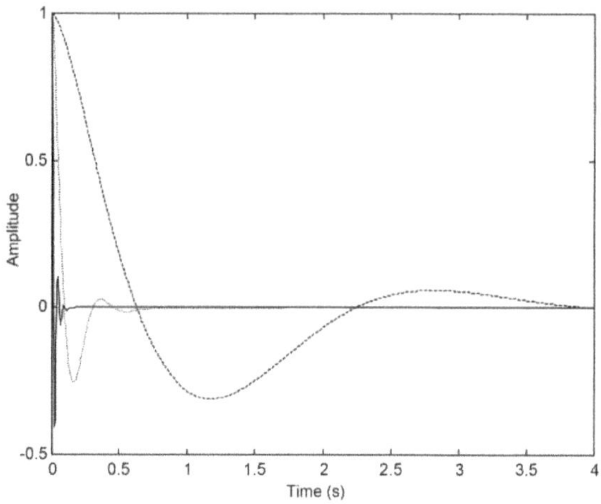

Figure 5.19. *Closed-loop responses to a step output disturbance for the nominal parametric state of the plant and two extremal parametric states*

5.5. Bibliography

[BAL 92] BALLOUK A., La dérivation non entière réelle et complexe: synthèse et applications dans les sciences de l'ingénieur, PhD Thesis, University of Bordeaux 1, 1992.

[LAN 94] LANUSSE P., De la commande CRONE de première génération à la commande CRONE de troisième génération, PhD Thesis, University of Bordeaux 1, 1994.

[LAN 10] LANUSSE P., OUSTALOUP A., Stability of closed-loop systems based on fractional complex order Integration – 4th IFAC Workshop on "Fractional Differentiation and its Applications" (FDA'10). Badajoz, Spain, October 18–20, 2010

[LAN 12] LANUSSE P., OUSTALOUP A., POMMIER-BUDINGER V., "Stability of closed loop fractional – order systems and definition of damping contours for the design of controllers", *International Journal of Bifurcation and Chaos (JCR IF 0.92)*, SCOPUS SJR 0.54 SNIP 0.765), vol. 22, no. 04, April 2012.

[MAT 94] MATHIEU B., OUSTALOUP A., LEVRON F., Une solution globale à l'identification paramétrique dans le domaine fréquentiel, Pôle SARTA du GdR Automatique CNRS, Paris, 7–8 November 1994.

[OUS 75] OUSTALOUP A., Etude et réalisation d'un système d'asservissement d'ordre 3/2 de la fréquence d'un laser à colorant continu, PhD Thesis, University of Bordeaux 1, France, 1975.

[OUS 78] OUSTALOUP A., "Oscillateur sinusoïdal d'ordre demi-entier", Patent No. 7835728, INPI, 1978.

[OUS 79] OUSTALOUP A., "Systèmes de modulation de fréquence BF à grande linéarité", *L'onde Electrique*, vol. 59, no. 8–9, pp. 102–107, 1979.

[OUS 81a] OUSTALOUP A., "Fractional order sinusoidal oscillators: optimization and their use in highly linear FM modulation", *IEEE Transactions on Circuits and Systems*, vol. 28, no. 10, pp. 1007–1009, 1981.

[OUS 81b] OUSTALOUP A., "Linear feedback control systems of fractional order between 1 and 2", *IEEE International Symposium on Circuits and Systems*, Chicago, IL, 27–29 April 1981.

[OUS 81c] OUSTALOUP A., Systèmes asservis linéaires d'ordre fractionnaire, PhD Thesis, University of Bordeaux 1, 1981.

[OUS 83] OUSTALOUP A., *Fractional Order Control*, North-Holland Publishing Company, 1983.

[OUS 89] OUSTALOUP A., "From fractality to noninteger derivation through recursivity, a property common to these two concepts: a fundamental idea for a new process control strategy", *Computing and Computers for Control Systems*, pp. 421–426, December 1989.

[OUS 91] OUSTALOUP A., *La commande CRONE*, Hermès, Paris, 1991.

[OUS 92a] OUSTALOUP A., "The CRONE control: template and real non integer derivation", *IMACS-WORLD CONGRESS, Computation and applied mathematics*, pp. 409–422, Elsevier, 1992.

[OUS 92b] OUSTALOUP A., LANUSSE P., "New trends in CRONE control: generalized template and complex non integer derivation", *IMACS-WORLD CONGRESS, Computation and applied mathematics*, pp. 395–408, Elsevier, 1992.

[OUS 95a] OUSTALOUP A., *La dérivation non entière: théorie, synthèse et applications*, Hermès, Paris, 1995.

[OUS 95b] OUSTALOUP A., MATHIEU B., LANUSSE P., "Intégration non entière complexe et contours d'isoamortissement", *RAIRO-APII*, vol. 29, no. 2, pp. 177–203, 1995.

[OUS 95c] OUSTALOUP A., MATHIEU B., LANUSSE P., "The CRONE control of resonant plants: application to a flexible transmission", *European Journal of Control*, vol. 1, no. 2, pp. 113–121, 1995.

[OUS 99a] OUSTALOUP A., SABATIER J., LANUSSE P., "From fractal robustness to the CRONE control", *Fractional Calculus and Applied Analysis (FCAA): An International Journal for Theory and Applications*, vol. 2, no. 1, pp. 1–30, January 1999.

[OUS 99b] OUSTALOUP A., MATHIEU B., *La commande CRONE: du scalaire au multivariable*, Hermès, Paris, 1999.

[OUS 02] OUSTALOUP A., LANUSSE P., LEVRON F., "Frequency-domain synthesis of a filter using Viète root functions", *IEEE Transactions on Automatic Control*, vol. 47, no. 5, pp. 837–841, May 2002.

Chapter 6

Recursivity and Non-Integer Differentiation

6.1. Introduction

Defined as a *recurrence relation independent of rank*, "recursivity" is a property common to fractality and non-integer differentiation or integration or, more precisely, to the construction of a fractal figure and of a non-integer differentiator or integrator. Indeed, the construction of a geometrical figure according to a recursive process leads to a figure of *non-integer (*or *fractal) dimension*, as the recursive construction of a physical system (e.g. electric network) leads us to a system of *non-integer order*. This leads us to say that recursivity is fundamental in the non-integer differentiation synthesis.

More generally, the *recursivity/non-integer differentiation interdependence* is fundamental in the context of the non-integer derivative considered as a synthesizable operator and as a modeling tool. Such an interdependence is indeed at the origin of the idea of the non-integer derivative synthesis and also of the idea of using the non-integer derivative as a modeling tool of the dynamic behavior of physical phenomena whose essence is of a fractal nature.

Section 6.2 presents an exhaustive study of an indefinite recursive parallel arrangement of series RC cells. As an indefinite sum of the admittances of each of the cells, the arrangement admittance is expressed by an unlimited series. The interest of the arrangement study is due to calculations directly led on this series in order to avoid a reduction to a common denominator, then explosive given the number of cells. After showing that the poles are geometrically distributed on $\mathbb{R}^-$, the zeros are determined as for their localization, their alternating with the poles, their geometric progression and their position with respect to the poles. The zeros and poles are then transposed to the frequency domain, and the impulse and step behaviors are the subject of significant developments. Still, if the arrangement indefinite character leads to an unlimited series that simplifies the mathematical developments, at the practical level, it will be convenient to limit the results so obtained to those corresponding to a medium frequency range, this remark also being valid for the arrangements of section 6.4.

Section 6.3 is devoted to the study of a recursive arborescent arrangement of gamma RLC cells, the arborescence being obtained according to a *bifurcation iterative process* in conformity with the lung intern structure seen under the respiratory angle. Led in such a way, this study presents the advantage of reducing the arrangement so obtained to a *recursive cascade arrangement of gamma RLC cells* whose study, founded on a *continued fraction expansion,* theoretically enriches the link between recursivity and non-integer differentiation.

Section 6.4 puts forward a unified study of indefinite recursive parallel arrangements of RL, RC and RLC cells. The originality of this study results from methods said to be heuristic as they are founded on *heuristic assumptions* that, in this case, enable us to quickly obtain exact results with remarkable simplicity. Notably, the non-integer order and the "frequency periodicity" of the admittance and also the non-integer power and the "time periodicity" of the impulse response inscribe their determination in these heuristic approaches.

Section 6.5 deals with eight recursive arrangements of RC and RL cells obtained from four possible combinations of a resistance and a capacitance and from four possible combinations of a resistance and an inductance. It concerns the two arrangements stemming from a recursive parallel arrangement of series RC or RL cells, the two arrangements stemming from a recursive series arrangement of parallel RC or RL cells, the two arrangements stemming from a recursive cascade arrangement of gamma RC or RL cells, and also the two arrangements stemming from a recursive cascade arrangement of gamma CR or LR cells. The transitional frequency distribution, common to these eight arrangements, expresses the same behavior of these arrangements on the condition of considering the admittance for RC cell arrangements and the impedance for RL cell arrangements. Considering the admittance for all the arrangements, it appears that RC cell arrangements approximate a non-integer differentiator, whereas RL cell arrangements approximate a non-integer integrator. The same recursivity of the components is studied in conformity with identical recursive factors, first greater than unit, then equal to unit: in both cases, a *differentiation half-integer order* is obtained by considering the admittance for RC cell arrangements and the impedance for RL cell arrangements.

Section 6.6 shows how to reparameterize the cells to define a unit gain frequency of the indefinite recursive parallel arrangements of series RC and RL cells.

Section 6.7 deals with the energy stored first by a non-integer differentiator and then by a non-integer integrator. Two dual energetic approaches are led through an indefinite recursive parallel arrangement of series RC cells for the non-integer differentiator and an indefinite recursive series arrangement of parallel RL cells for the non-integer integrator. The energy stored by the capacitors for the first arrangement or by the coils for the second one turns out to be infinite through an *infinite capacitance* or an *infinite inductance*. In both cases, the stored energy release during the relaxation phase is carried out through infinitely fast discharges and also infinitely slow discharges that are at the origin of the *long memory* (in this case, *infinite*) which is well known on the subject.

6.2. Indefinite recursive parallel arrangement of series RC cells

The considered arrangement is a *ladder network* (Figure 6.1):

– whose *indefinite* character is due to an infinity of series RC cells;

– whose *recursive* character is due to a geometric progression of resistances and capacitances such as

$$R_i = \frac{R}{\alpha^i} \text{ and } C_i = \frac{C}{\eta^i}, \qquad [6.1]$$

where $\alpha > 1$ and $\eta > 1$, called *recursive factors*, the rank (or subscript) i of the cell R_iC_i belonging to the set of integers, namely $i \in \mathbb{Z}$, the rank 0 cell being the median (or central) cell RC.

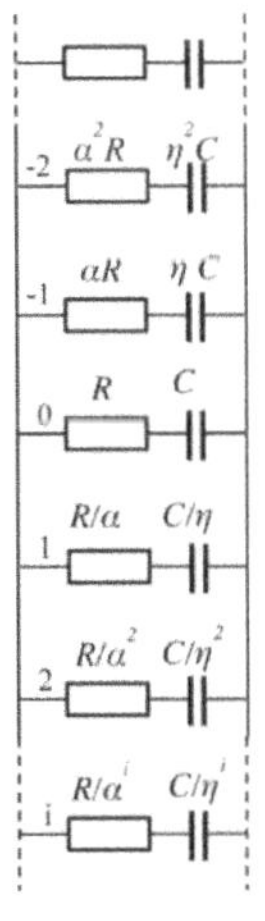

Figure 6.1. *Indefinite recursive parallel arrangement of series RC cells*

If $Y_i(s)$ represents the admittance (in s) of the series RC cell of rank i, namely

$$Y_i(s) = \frac{1}{R_i + \frac{1}{C_i s}} = \frac{1}{\frac{R}{\alpha^i} + \frac{\eta^i}{Cs}} = \frac{\alpha^i Cs}{RCs + \lambda^i} \text{ with } \lambda = \alpha\eta > 1, \qquad [6.2]$$

the network admittance is expressed by the indefinite sum:

$$Y(s) = \sum_{i=-\infty}^{+\infty} Y_i(s) = \sum_{i=-\infty}^{+\infty} \frac{\alpha^i Cs}{RCs + \lambda^i}, \qquad [6.3]$$

namely:

$$Y(s) = Cs \sum_{i=-\infty}^{+\infty} \frac{\alpha^i}{RCs + \lambda^i}, \qquad [6.4]$$

or, by denoting the unlimited series by $S(s)$, namely

$$S(s) = \sum_{i=-\infty}^{+\infty} \frac{\alpha^i}{RCs + \lambda^i}: \qquad [6.5]$$

$$Y(s) = CsS(s). \qquad [6.6]$$

As far as the zeros and poles of $Y(s)$ are concerned, the last two expressions successively show that:

– the pole of rank i, solution of $RCs + \lambda^i = 0$, is $p_i = -\lambda^i / RC$, the set of $Y(s)$ poles (in geometric progression) then being

$$\mathcal{P} = \left\{ p_i = -\frac{\lambda^i}{RC}; i \in \mathbb{Z} \right\}; \qquad [6.7]$$

– an obvious zero is $s = 0$;

– the other zeros are such that $S(s) = 0$, hence the following developments to determine them as for their localization, their alternating with the poles, their geometric progression and their position in relation to the poles.

6.2.1. *Localization of zeros*

The series $S(s)$ is decomposable into two parts:

$$S(s)=\sum_{i=-\infty}^{i=-1}\frac{\alpha^i}{RCs+\lambda^i}+\sum_{i=0}^{+\infty}\frac{\alpha^i}{RCs+\lambda^i}. \quad [6.8]$$

Now, if i becomes $-i$, the first part becomes:

$$\sum_{-i=-\infty}^{-i=-1}\frac{\alpha^{-i}}{RCs+\lambda^{-i}}=\sum_{i=1}^{+\infty}\frac{\alpha^{-i}}{RCs+\lambda^{-i}}, \quad [6.9]$$

or, by dividing the numerator and the denominator by α^{-i}:

$$\sum_{i=1}^{+\infty}\frac{1}{RCs\alpha^i+\dfrac{1}{\eta^i}}, \quad [6.10]$$

or even, by multiplying the numerator and the denominator by η^i :

$$\sum_{i=1}^{+\infty}\frac{\eta^i}{RCs\lambda^i+1}. \quad [6.11]$$

Then $S(s)$ can be written as:

$$S(s)=\sum_{i=1}^{+\infty}\frac{\eta^i}{RCs\lambda^i+1}+\sum_{i=0}^{+\infty}\frac{\alpha^i}{RCs+\lambda^i}, \quad [6.12]$$

or, by reorganizing the two parts of $S(s)$ in such a way that the first term of the series corresponds to the subscript 0:

$$S(s)=\sum_{i=0}^{+\infty}\frac{\alpha^i}{RCs+\lambda^i}+\sum_{i=1}^{+\infty}\frac{\eta^i}{RCs\lambda^i+1}, \quad [6.13]$$

namely:

$$S(s)=S_1(s)+S_2(s). \quad [6.14]$$

Let us put $s = a + jb$, its conjugate is then $\overline{s} = a - jb$.

The series $S_1(s)$ becomes:

$$S_1 = \sum_{i=0}^{+\infty} \frac{\alpha^i \left(RC\overline{s} + \lambda^i \right)}{\left(RCs + \lambda^i \right)\left(RC\overline{s} + \lambda^i \right)}, \qquad [6.15]$$

or, given the general formula $Z\overline{Z} = |Z|^2$, it becomes:

$$S_1 = \sum_{i=0}^{+\infty} \frac{\alpha^i \left(RC\overline{s} + \lambda^i \right)}{\left| RCs + \lambda^i \right|^2}, \qquad [6.16]$$

or even:

$$S_1 = \sum_{i=0}^{+\infty} \frac{\alpha^i \left(RCa + \lambda^i \right) - j\alpha^i RCb}{\left| RCs + \lambda^i \right|^2}, \qquad [6.17]$$

namely:

$$S_1 = \mathrm{Re}(S_1) + \mathrm{Im}(S_1), \qquad [6.18]$$

with:

$$\mathrm{Re}(S_1) = \sum_{i=0}^{+\infty} \frac{\alpha^i \left(RCa + \lambda^i \right)}{\left| RCs + \lambda^i \right|^2} \qquad [6.19]$$

and

$$\mathrm{Im}(S_1) = -RCb \sum_{i=0}^{+\infty} \frac{\alpha^i}{\left| RCs + \lambda^i \right|^2}. \qquad [6.20]$$

As for the series $S_2(s)$, it becomes:

$$S_2 = \sum_{i=1}^{+\infty} \frac{\eta^i \left(RC\overline{s}\lambda^i + 1 \right)}{\left(RCs\lambda^i + 1 \right)\left(RC\overline{s}\lambda^i + 1 \right)}, \qquad [6.21]$$

namely:

$$S_2 = \mathrm{Re}(S_2) + \mathrm{Im}(S_2), \tag{6.22}$$

with:

$$\mathrm{Re}(S_2) = \sum_{i=1}^{+\infty} \frac{\eta^i \left(RCa\lambda^i + 1\right)}{\left|RCs\lambda^i + 1\right|^2} \tag{6.23}$$

and

$$\mathrm{Im}(S_2) = -RCb \sum_{i=1}^{+\infty} \frac{(\eta\lambda)^i}{\left|RCs\lambda^i + 1\right|^2}. \tag{6.24}$$

The real and imaginary parts of S, namely

$$\mathrm{Re}(S) = \mathrm{Re}(S_1) + \mathrm{Re}(S_2) \tag{6.25}$$

and

$$\mathrm{Im}(S) = \mathrm{Im}(S_1) + \mathrm{Im}(S_2), \tag{6.26}$$

then admit expressions of the form:

$$\mathrm{Re}(S_2) = \sum_{i=1}^{+\infty} \frac{\alpha^i \left(RCa + \lambda^i\right)}{\left|RCs + \lambda^i\right|^2} + \sum_{i=1}^{+\infty} \frac{\eta^i \left(RCa\lambda^i + 1\right)}{\left|RCs\lambda^i + 1\right|^2} \tag{6.27}$$

and

$$\mathrm{Im}(S_2) = -RCb \left[\sum_{i=1}^{+\infty} \frac{\alpha^i}{\left|RCs + \lambda^i\right|^2} + \sum_{i=1}^{+\infty} \frac{(\eta\lambda)^i}{\left|RCs\lambda^i + 1\right|^2} \right]. \tag{6.28}$$

CONCLUSION.–

If $b > 0$, then $\mathrm{Im}(S) < 0$, and therefore $S \neq 0$.

If $b < 0$, then $\mathrm{Im}(S) > 0$, and therefore $S \neq 0$.

Thus, if $S=0$, then $b=0$, and therefore $s=a\in\mathbb{R}$,

which expresses that *the zeros of* S *are real.*

If $a\geq 0$, $\alpha^i\left(RCa+\lambda^i\right)>0$ and $\eta^i\left(RCa\lambda^i+1\right)>0$,

then $\mathrm{Re}(S)>0$, and therefore $S\neq 0$.

Thus, if $S=0$, then $a<0$, and therefore $s=a\in\mathbb{R}^-$,

which expresses that *the zeros of* S *are real and strictly negative.*

6.2.2. *Zero and pole alternating*

As S vanishes for values of the negative real variable s, let us denote by x this variable ($x<0$) and let us consider the function $S(x)$ (which is real as $b=0$) and also the derivative function $S'(x)$.

$S(x)$ and $S'(x)$ are, respectively, written as:

$$S(x)=S_1(x)+S_2(x) \qquad [6.29]$$

and

$$S'(x)=S_1'(x)+S_2'(x), \qquad [6.30]$$

$S_1(x)$ and $S_2(x)$ arising from relations [6.13] and [6.14]:

$$S_1(x)=\sum_{i=0}^{+\infty}\frac{\alpha^i}{RCx+\lambda^i} \qquad [6.31]$$

and

$$S_2(x)=\sum_{i=1}^{+\infty}\frac{\eta^i}{RCx\lambda^i+1}, \qquad [6.32]$$

the derivatives $S_1^{'}(x)$ and $S_2^{'}(x)$ being computed from these expressions in conformity with:

$$S_1^{'}(x) = -RC\sum_{i=0}^{+\infty}\frac{\alpha^i}{\left(RCx+\lambda^i\right)^2} \qquad [6.33]$$

and

$$S_2^{'}(x) = -RC\sum_{i=1}^{+\infty}\frac{(\eta\lambda)^i}{\left(RCx\lambda^i+1\right)^2}, \qquad [6.34]$$

$S_1^{'}(x)$ and $S_2^{'}(x)$ being negative outside the set of poles $\mathcal{P} = \left\{p_i = -\dfrac{\lambda^i}{RC}; i \in \mathbb{Z}\right\}$.

To determine the number of zeros between two consecutive poles, it is convenient to study the variation of $S(x)$ for $x \in \left]p_{i+1}, p_i\right[$, the limits of the interval, p_{i+1} and p_i, being such that $p_{i+1} = \lambda p_i$ with $\lambda > 1$ and $p_i = -\dfrac{\lambda^i}{RC} < 0$.

As $S_1^{'}(x)$ and $S_2^{'}(x)$ are negative outside the poles, so it is for the derivative function $S^{'}(x)$, which implies that the function $S(x)$ is constantly decreasing between p_{i+1} and p_i (Figure 6.2).

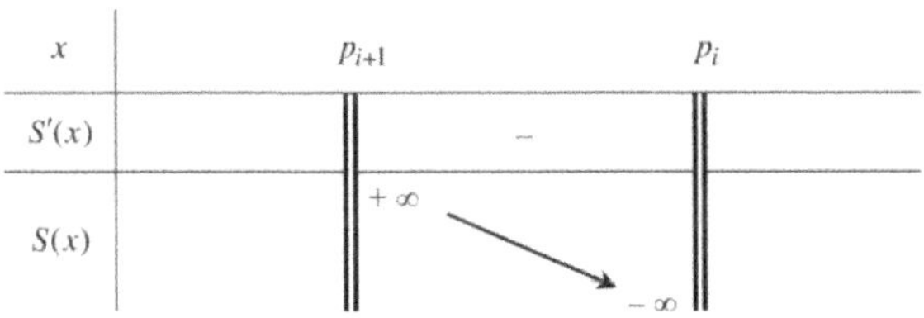

Figure 6.2. *Variation of S(x) between two consecutive poles*

CONCLUSION.–

Given that this decreasing is carried out between +∞ and −∞, $S(x)$ only vanishes for a value of x between p_{i+1} and p_i, which proves:

– the existence of a single zero z_{i+1} between the consecutive poles p_{i+1} and p_i;

– that is to say, the alternating of the zeros and poles of $Y(s)$.

6.2.3. *Geometric progression of zeros*

The aim of this section is to show that the zeros of $Y(s)$ are also in geometric progression.

To that effect, let us take again the expression of $Y(s)$, namely

$$Y(s) = Cs\sum_{i=-\infty}^{+\infty} \frac{\alpha^i}{RCs + \lambda^i}, \qquad [6.35]$$

then let us express $Y(\lambda s)$. It happens:

$$Y(\lambda s) = C\lambda s\sum_{i=-\infty}^{+\infty} \frac{\alpha^i}{RC\lambda s + \lambda^i}, \qquad [6.36]$$

namely:

$$Y(\lambda s) = Cs\sum_{i=-\infty}^{+\infty} \frac{\alpha^i \lambda}{RC\lambda s + \lambda^i}, \qquad [6.37]$$

or, by dividing the numerator and denominator by λ:

$$Y(\lambda s) = Cs\sum_{i=-\infty}^{+\infty} \frac{\alpha^i}{RCs + \lambda^{i-1}}, \qquad [6.38]$$

or even, by multiplying and dividing by α:

$$Y(\lambda s) = \alpha Cs\sum_{i=-\infty}^{+\infty} \frac{\alpha^{i-1}}{RCs + \lambda^{i-1}}, \qquad [6.39]$$

or else, by putting $j = i - 1$:

$$Y(\lambda s) = \alpha Cs\sum_{i=-\infty-1}^{+\infty-1} \frac{\alpha^j}{RCs + \lambda^j}, \qquad [6.40]$$

namely (as $-\infty - 1 = -\infty$ and $+\infty - 1 = +\infty$):

$$Y(\lambda s) = \alpha Y(s) \quad \forall s \in \mathbb{C} - \mathcal{P}. \qquad [6.41]$$

CONCLUSION.–

If $s \in \mathcal{Z} = \{zeros\ of\ Y(s)\}$, then $Y(s)=0$, and therefore $Y(\lambda s)=0$, which expresses that λs is also a zero of $Y(s)$. Thus, if $s \in \mathcal{Z}$, then $\lambda s \in \mathcal{Z}$, and therefore (as there exists a zero between each pair of consecutive poles):

$$\lambda^i s \in \mathcal{Z} \quad \forall i \in \mathbb{Z}. \qquad [6.42]$$

Thus, *the geometric progression of the zeros of* $Y(s)$ is proved.

6.2.4. *Position of the zeros relatively to the poles*

A zero being between two consecutive poles, the aim of this section is to determine the position of the zero in relation to the poles by determining the ratio between the zero and one of the poles.

To that effect, let us take again the expression of $S(s)$, namely

$$S(s) = \sum_{i=-\infty}^{+\infty} \frac{\alpha^i}{RCs + \lambda^i}, \qquad [6.43]$$

then let us express $S\left(-\frac{\eta}{RC}\right)$. We obtain:

$$S\left(-\frac{\eta}{RC}\right) = \sum_{i=-\infty}^{+\infty} \frac{\alpha^i}{RC\left(-\frac{\eta}{RC}\right)+\lambda^i} = \sum_{i=-\infty}^{+\infty} \frac{\alpha^i}{\lambda^i - \eta}. \qquad [6.44]$$

The positivity of $\lambda^i - \eta$ for $i \geq 1$ leads us to decompose this series into two parts:

$$S\left(-\frac{\eta}{RC}\right) = \sum_{i=-\infty}^{i=0} \frac{\alpha^i}{\lambda^i - \eta} + \sum_{i=1}^{+\infty} \frac{\alpha^i}{\lambda^i - \eta}. \qquad [6.45]$$

The second part can be written as:

$$\sum_{i=1}^{+\infty} \frac{\alpha^i}{\lambda^i - \eta} = \sum_{i=1}^{+\infty} \frac{\alpha^i}{\lambda^i\left(1-\frac{\eta}{\lambda^i}\right)} = \sum_{i=1}^{+\infty} \frac{1}{\eta^i} \frac{1}{1-\frac{\eta}{\lambda^i}}, \qquad [6.46]$$

namely, given that

$$0<\frac{\eta}{\lambda^i}<1 \text{ and } \frac{1}{1-u}=\sum_{k=0}^{+\infty}u^k \text{ for } |u|<1: \tag{6.47}$$

$$\sum_{i=1}^{+\infty}\frac{\alpha^i}{\lambda^i-\eta}=\sum_{i=1}^{+\infty}\frac{1}{\eta^i}\sum_{k=0}^{+\infty}\left(\frac{\eta}{\lambda^i}\right)^k, \tag{6.48}$$

or, knowing that everything is positive:

$$\sum_{i=1}^{+\infty}\frac{\alpha^i}{\lambda^i-\eta}=\sum_{k=0}^{+\infty}\eta^k\sum_{i=1}^{+\infty}\frac{1}{\left(\eta\lambda^k\right)^i}, \tag{6.49}$$

or even, willing to bring out a sum from $i=0$ to $i=+\infty$:

$$\sum_{i=1}^{+\infty}\frac{\alpha^i}{\lambda^i-\eta}=\sum_{k=0}^{+\infty}\eta^k\left[\sum_{i=0}^{+\infty}\frac{1}{\left(\eta\lambda^k\right)^i}-\frac{1}{\left(\eta\lambda^k\right)^0}\right], \tag{6.50}$$

or else, given that

$$0<\frac{1}{\eta\lambda^k}<1 \text{ and } \sum_{i=0}^{+\infty}u^i=\frac{1}{1-u} \text{ for } |u|<1: \tag{6.51}$$

$$\begin{aligned}\sum_{i=1}^{+\infty}\frac{\alpha^i}{\lambda^i-\eta} &= \sum_{k=0}^{+\infty}\eta^k\left[\frac{1}{1-\frac{1}{\eta\lambda^k}}-1\right]\\ &= \sum_{k=0}^{+\infty}\eta^k\frac{\frac{1}{\eta\lambda^k}}{1-\frac{1}{\eta\lambda^k}}=\sum_{k=0}^{+\infty}\frac{\eta^k}{\eta\lambda^k-1}.\end{aligned} \tag{6.52}$$

As for the first part of the series, it can be written, by putting $i=-k$, as:

$$\begin{aligned}\sum_{i=-\infty}^{i=0}\frac{\alpha^i}{\lambda^i-\eta} &= \sum_{k=0}^{+\infty}\frac{\alpha^{-k}}{\lambda^{-k}-\eta}=\sum_{k=0}^{+\infty}\frac{1}{\alpha^k\left(\frac{1}{\lambda^k}-\eta\right)}\\ &= \sum_{k=0}^{+\infty}\frac{\lambda^k}{\alpha^k\left(1-\eta\lambda^k\right)}=-\sum_{k=0}^{+\infty}\frac{\eta^k}{\eta\lambda^k-1}.\end{aligned} \tag{6.53}$$

CONCLUSION.–

As the two parts of the series are opposite, the series is nil, namely:

$$S\left(-\frac{\eta}{RC}\right)=0, \quad [6.54]$$

which expresses that $-\eta / RC$ is a zero of $Y(s)$ and the set of the zeros of $Y(s)$ is then, because of their geometric progression:

$$\mathcal{Z}=\left\{z_i=-\frac{\eta}{RC}\lambda^{i-1};i\in\mathbb{Z}\right\}. \quad [6.55]$$

This set, which defines the position of the zeros, enables us to place them in relation to the poles whose position has already been defined by the set

$$\mathcal{P}=\left\{p_i=-\frac{\lambda^i}{RC};i\in\mathbb{Z}\right\}, \quad [6.56]$$

notably the positioning of a zero in relation to the preceding rank pole being defined by the ratio:

$$\frac{z_{i+1}}{p_i}=\eta. \quad [6.57]$$

6.2.5. *Zero and pole transposition to frequency domain*

The zeros and poles of the admittance $Y(s)$ now being perfectly defined, it is convenient to transpose them to the frequency domain through transitional frequencies that they engender.

Given the structure imposed on the admittance $Y(s)$ by the zero and pole distribution (said recursive) as defined, $Y(s)$ can be put under the form of an indefinite product of first-order factors resulting, for each of them, from one zero and one pole, namely:

$$Y(s) = K \prod_{i=-\infty}^{+\infty} \frac{1+\frac{s}{\omega_i^{'}}}{1+\frac{s}{\omega_i}}, \qquad [6.58]$$

the transitional frequencies $\omega_i^{'}$ and ω_i being distributed in a recursive manner in conformity with the zero and pole recursivity, namely:

$$\omega_i^{'} = -z_i = \frac{\eta}{RC}\lambda^{i-1} \qquad [6.59]$$

and

$$\omega_i = -p_i = \frac{\lambda^i}{RC}. \qquad [6.60]$$

These relations enable us to establish the various ratios between transitional frequencies (Figure 6.3):

$$\frac{\omega_{i+1}^{'}}{\omega_i^{'}} = \frac{z_{i+1}}{z_i} = \lambda = \alpha\eta ; \qquad [6.61]$$

$$\frac{\omega_{i+1}}{\omega_i} = \frac{p_{i+1}}{p_i} = \lambda = \alpha\eta ; \qquad [6.62]$$

$$\frac{\omega_i}{\omega_i^{'}} = \frac{p_i}{z_i} = \frac{\lambda^i}{RC}\frac{RC}{\eta\lambda^{i-1}} = \frac{\lambda}{\eta} = \alpha ; \qquad [6.63]$$

$$\frac{\omega_{i+1}^{'}}{\omega_i} = \frac{z_{i+1}}{p_i} = \frac{\eta}{RC}\lambda^i \frac{RC}{\lambda^i} = \eta . \qquad [6.64]$$

6.2.6. *Impulse behavior*

The aim of this section is to study the impulse behavior of the ladder network:

– when the recursive factors α and η tend toward 1 in such a way that the non-integer order $0 < n < 1$ remains constant;

– in other words, when α and η are replaced, for instance, by $\alpha^{1/N}$ and $\eta^{1/N}$, N tending toward infinity;

– the invariance of n being then well ensured in conformity with:

$$n=\frac{\ln\left(\alpha^{\frac{1}{N}}\right)}{\ln\left(\alpha^{\frac{1}{N}}\eta^{\frac{1}{N}}\right)}=\frac{\frac{1}{N}\ln(\alpha)}{\frac{1}{N}\left(\ln(\alpha)+\ln(\eta)\right)}=\frac{\ln(\alpha)}{\ln(\alpha\eta)}. \qquad [6.65]$$

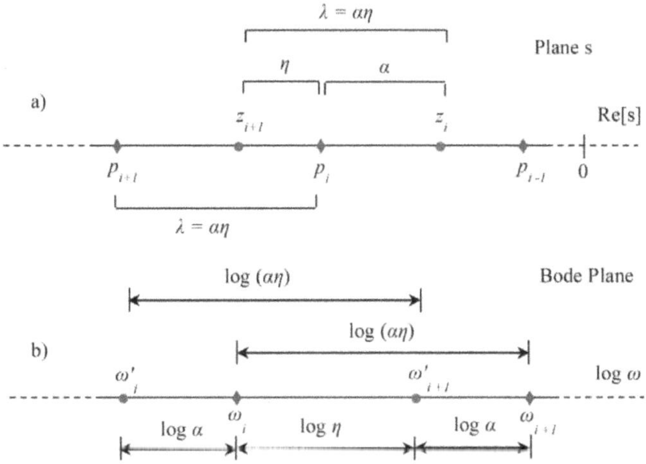

Figure 6.3. *Representation of the zeros and poles in the operational plane a) and of the transitional frequencies in the Bode plane b)*

The impulse behavior study is here carried out through:

– the admittance $Y(s)$ in the operational domain (section 6.2.6.1);

– the impulse response $y(t)=\mathcal{L}^{-1}\left[Y(s)\right]$ in the time domain (section 6.2.6.2).

6.2.6.1. *Operational approach*

Let us take again the ladder network admittance defined by relation [6.6], namely

$$Y(s)=CsS(s), \qquad [6.66]$$

where

$$S(s)=\sum_{i=-\infty}^{+\infty}\frac{\alpha^i}{RCs+\lambda^i} \text{ with } \lambda=\alpha\eta>1 \ (\alpha \text{ and } \eta>1), \qquad [6.67]$$

then let us put $q = (RC)s$, the unlimited series $S(s)$ then being reduced to:

$$S = \sum_{i=-\infty}^{+\infty} \frac{\alpha^i}{q+\lambda^i}. \qquad [6.68]$$

Let $u_i = \lambda^i$, then $\Delta u_i = u_{i+1} - u_i = \lambda^{i+1} - \lambda^i = \lambda^i(\lambda - 1) = u_i(\lambda - 1)$, therefore, $u_i = \dfrac{\Delta u_i}{\lambda - 1}$.

Moreover, $n = \dfrac{\ln(\alpha)}{\ln(\alpha\eta)} = \dfrac{\ln(\alpha)}{\ln(\lambda)}$; hence, $\ln(\alpha) = n\ln(\lambda) = \ln(\lambda^n)$, namely $\alpha = \lambda^n$, and therefore $\alpha^i = (\lambda^n)^i = (\lambda^i)^n = u_i^n = u_i^{n-1}u_i = u_i^{n-1}\dfrac{\Delta u_i}{\lambda - 1}$.

Thus, the term in i of the series S can be written as:

$$\frac{\alpha^i}{q+\lambda^i} = \frac{1}{\lambda-1}\frac{u_i^{n-1}}{q+u_i}\Delta u_i, \qquad [6.69]$$

the series S then admitting the following expression:

$$S = \frac{1}{\lambda-1}\sum_{i=-\infty}^{+\infty} \frac{u_i^{n-1}}{q+u_i}\Delta u_i, \qquad [6.70]$$

an expression which shows well that when λ tends toward 1, the series S comes closer to the integral

$$I = \frac{1}{\lambda-1}\int_0^{+\infty} \frac{u^{n-1}}{q+u}du, \qquad [6.71]$$

its integration bounds arising from the fact that u varies from zero to infinity since from $u_i = \lambda^i$ we draw $u_{-\infty} = \lambda^{-\infty} = 1/\lambda^{\infty} = 0$ and $u_{+\infty} = \lambda^{+\infty} = \infty$.

Now, this integral is well known, indeed being given by

$$I = \frac{1}{\lambda-1}\frac{\pi}{\sin(\pi n)}q^{n-1}, \; 0 < n < 1, \qquad [6.72]$$

namely, given that $q = (RC)s$:

$$I(s) = \frac{1}{\lambda - 1} \frac{\pi}{\sin(\pi n)} (RC)^{n-1} s^{n-1}. \quad [6.73]$$

Finally, knowing that

$$Y(s) = CsS(s), \quad [6.74]$$

it is possible to write, for λ tending toward 1 with n constant:

$$Y(s) = CsI(s), \quad [6.75]$$

namely, given [6.73]:

$$Y(s) = \frac{C(RC)^{n-1}}{\lambda - 1} \frac{\pi}{\sin(\pi n)} s^n. \quad [6.76]$$

The admittance so obtained well describes the impulse behavior of the network in the operational domain, as it represents the Laplace transform of the impulse response defined as by the original of [6.76], namely for $t > 0$:

$$y(t) = \frac{C(RC)^{n-1}}{\lambda - 1} \frac{\pi}{\sin(\pi n)} \frac{t^{-n-1}}{\Gamma(-n)}. \quad [6.77]$$

6.2.6.2. *Time approach*

Let us take again the expression of $Y(s)$ (relation [6.4]), namely

$$Y(s) = Cs \sum_{i=-\infty}^{+\infty} \frac{\alpha^i}{RCs + \lambda^i}, \quad [6.78]$$

then let us rewrite it as:

$$Y(s) = Cs \sum_{i=-\infty}^{+\infty} \frac{\alpha^i / RC}{s + \lambda^i / RC} = \frac{1}{R} s \sum_{i=-\infty}^{+\infty} \frac{\alpha^i}{s + \lambda^i / RC}, \quad [6.79]$$

from which we draw:

$$\frac{Y(s)}{s} = \frac{1}{R} \sum_{i=-\infty}^{+\infty} \frac{\alpha^i}{s + \lambda^i / RC}, \quad [6.80]$$

the corresponding original (or time) equation being written as:

$$\mathcal{L}^{-1}\left[\frac{Y(s)}{s}\right]=\frac{1}{R}\sum_{i=-\infty}^{+\infty}\alpha^{i}\exp(-\lambda^{i}\frac{t}{RC}), \qquad [6.81]$$

namely:

$$\mathcal{L}^{-1}\left[\frac{Y(s)}{s}\right]=\frac{1}{R}S(t), \qquad [6.82]$$

where

$$S(t)=\sum_{i=-\infty}^{+\infty}\alpha^{i}\exp(-\lambda^{i}\frac{t}{RC}), \qquad [6.83]$$

the unlimited series $S(t)$ being reduced to:

$$S=\sum_{i=-\infty}^{+\infty}\alpha^{i}\exp(-\lambda^{i}\tau) \text{ with } \tau=\frac{t}{RC}. \qquad [6.84]$$

Let $u_i=\lambda^i$, then $\Delta u_i=u_{i+1}-u_i=\lambda^{i+1}-\lambda^i=\lambda^i(\lambda-1)=u_i(\lambda-1)$, and therefore $u_i=\frac{\Delta u_i}{\lambda-1}$.

Moreover, $n=\frac{\ln(\alpha)}{\ln(\alpha\eta)}=\frac{\ln(\alpha)}{\ln(\lambda)}$, hence, $\ln(\alpha)=n\ln(\lambda)=\ln(\lambda^n)$, namely $\alpha=\lambda^n$, and therefore $\alpha^i=\left(\lambda^n\right)^i=\left(\lambda^i\right)^n=u_i^n=u_i^{n-1}u_i=u_i^{n-1}\frac{\Delta u_i}{\lambda-1}$.

Thus, the i th term of the series S can be written as:

$$\alpha^{i}\exp(-\lambda^{i}\tau)=\frac{1}{\lambda-1}u_i^{n-1}\exp(-u_i\tau)\Delta u_i, \qquad [6.85]$$

the series S then admitting the following expression:

$$S=\frac{1}{\lambda-1}\sum_{i=-\infty}^{+\infty}u_i^{n-1}\exp(-u_i\tau)\Delta u_i, \qquad [6.86]$$

an expression that shows well that when λ tends toward one, the series S comes closer to the integral

$$I = \frac{1}{\lambda - 1} \int_0^\infty u^{n-1} e^{-u\tau} \, \mathrm{d}u \, , \qquad [6.87]$$

its integration bounds arising (as previously) from the fact that u varies from zero to infinity when i varies from $-\infty$ to $+\infty$.

Now, this integral is well known, indeed being given by

$$I = \frac{1}{\lambda - 1} \Gamma(n) \tau^{-n} \, , \qquad [6.88]$$

namely, given that $\tau = \dfrac{t}{RC}$:

$$I(t) = \frac{1}{\lambda - 1} \Gamma(n) \left(\frac{t}{RC} \right)^{-n} . \qquad [6.89]$$

Finally, knowing that

$$\mathcal{L}^{-1} \left[\frac{Y(s)}{s} \right] = \frac{1}{R} S(t) \qquad [6.90]$$

it is possible to write, for λ tending toward 1 with n constant:

$$\mathcal{L}^{-1} \left[\frac{Y(s)}{s} \right] = \frac{1}{R} I(t) \, , \qquad [6.91]$$

or, given [6.89]:

$$\mathcal{L}^{-1} \left[\frac{Y(s)}{s} \right] = \frac{1}{R} \frac{1}{\lambda - 1} \Gamma(n) \left(\frac{t}{RC} \right)^{n} , \qquad [6.92]$$

namely, by expressing the change from the original to its Laplace transform:

$$\mathcal{L} \left[\frac{1}{R} \frac{(RC)^n}{\lambda - 1} \Gamma(n) t^{-n} \right] = \frac{Y(s)}{s} \, , \qquad [6.93]$$

which enables us to write:

$$\mathcal{L}\left[\frac{\mathrm{d}}{\mathrm{d}t}\left[\frac{1}{R}\frac{(RC)^n}{\lambda-1}\Gamma(n)t^{-n}\right]\right]=Y(s), \qquad [6.94]$$

therefore, knowing that $y(t)=\mathcal{L}^{-1}\left[Y(s)\right]$:

$$y(t)=\frac{1}{\lambda-1}\frac{(RC)^n}{R}\Gamma(n)\frac{\mathrm{d}t^{-n}}{\mathrm{d}t},\ t>0, \qquad [6.95]$$

or:

$$y(t)=-\frac{1}{\lambda-1}\frac{(RC)^{n-1}RC}{R}\Gamma(n)nt^{-n-1}=-\frac{C(RC)^{n-1}}{\lambda-1}n\Gamma(n)t^{-n-1}, \qquad [6.96]$$

or even, given that $n\Gamma(n)=\Gamma(n+1)$:

$$y(t)=-\frac{C(RC)^{n-1}}{\lambda-1}\Gamma(n+1)t^{-n-1}. \qquad [6.97]$$

Now, Euler's reflection formula

$$\Gamma(z)\,\Gamma(1-z)=\frac{\pi}{\sin(\pi z)} \text{ with } z \text{ non-integer}, \qquad [6.98]$$

becomes for $z=-n$:

$$\Gamma(-n)\Gamma(1+n)=-\frac{\pi}{\sin(\pi n)}, \qquad [6.99]$$

from which we draw:

$$\Gamma(1+n)=-\frac{\pi}{\sin(\pi n)}\frac{1}{\Gamma(-n)}, \qquad [6.100]$$

a formula that put into [6.97] leads to:

$$y(t)=\frac{C(RC)^{n-1}}{\lambda-1}\frac{\pi}{\sin(\pi n)}\frac{t^{-n-1}}{\Gamma(-n)}, \qquad [6.101]$$

an expression of the impulse response that well verifies the one arising from the operational approach (relation [6.77]).

6.2.7. *Step behavior*

The aim of this section is to determine:

– the step response of the ladder network;

– its gap in relation to the one obtained when the recursive factors α and η tend toward 1 in such a way that the non-integer order $0 < n < 1$ remains constant.

6.2.7.1. *Step response of the ladder network*

Let us again take the ladder network admittance defined by relation [6.4], namely

$$Y(s) = Cs \sum_{i=-\infty}^{+\infty} \frac{\alpha^i}{RCs + \lambda^i}, \qquad [6.102]$$

then let us determine the ratio $Y(s)/s$, that is to say the product $Y(s).\frac{1}{s}$ which represents the Laplace transform of the network unit step response, $f(t)$. We obtain:

$$\frac{Y(s)}{s} = \frac{C}{RC} \sum_{i=-\infty}^{+\infty} \frac{\alpha^i}{s + \frac{\lambda^i}{RC}}, \qquad [6.103]$$

namely:

$$\frac{Y(s)}{s} = \frac{1}{R} \sum_{i=-\infty}^{+\infty} \frac{\alpha^i}{s + \frac{\lambda^i}{RC}}, \qquad [6.104]$$

the unit step response $f(t)$, indeed given by

$$f(t) = \mathcal{L}^{-1}\left[\frac{Y(s)}{s}\right], \qquad [6.105]$$

then being expressed by:

$$f(t) = \frac{u(t)}{R} \sum_{i=-\infty}^{+\infty} \alpha^i e^{-\frac{\lambda^i}{RC}t}, \qquad [6.106]$$

or, for $t > 0$:

$$f(t) = \frac{1}{R} \sum_{i=-\infty}^{+\infty} \alpha^i e^{-\frac{\lambda^i}{RC}t}, \qquad [6.107]$$

or even, by putting $u_i = \dfrac{\lambda^i}{RC}$ and knowing that $\alpha = \lambda^n$:

$$f(t) = \frac{1}{R} \sum_{i=-\infty}^{+\infty} \left(\lambda^n\right)^i e^{-u_i t}, \qquad [6.108]$$

or else:

$$\begin{aligned} f(t) &= \frac{1}{R} \sum_{i=-\infty}^{+\infty} \left(\lambda^i\right)^n e^{-u_i t} \\ &= \frac{1}{R} \sum_{i=-\infty}^{+\infty} \left(RCu_i\right)^n e^{-u_i t} \qquad [6.109] \\ &= \frac{\left(RC\right)^n}{R} \sum_{i=-\infty}^{+\infty} u_i^n e^{-u_i t}. \end{aligned}$$

6.2.7.2. *Framing of the step response* $f(t)$

Let $\varphi(u) = u^{n-1} e^{-ut}$, then $\varphi'(u) = u^{n-2} e^{-ut}\left(n-1-tu\right) < 0$, and therefore the function $\varphi(u)$ is decreasing.

If $u_i \le u \le u_{i+1}$, then $\varphi(u_{i+1}) \le \varphi(u) \le \varphi(u_i)$.

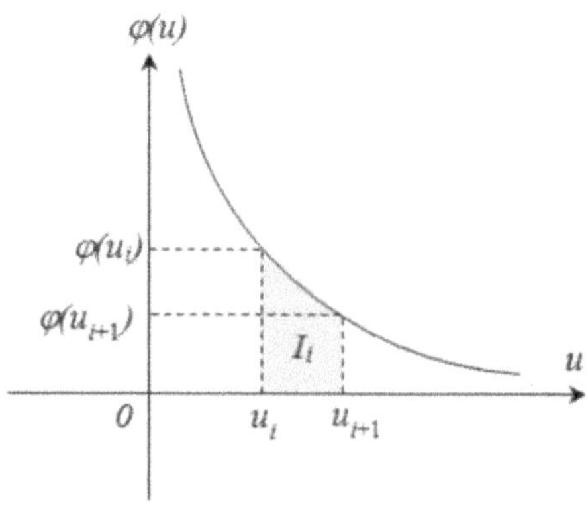

Figure 6.4. *Area under the curve between u_i and u_{i+1}*

Let $I_i = \int_{u_i}^{u_{i+1}} \varphi(u)\mathrm{d}u$ (Figure 6.4).

The area under the curve between u_i and u_{i+1}, I_i, is such that:

$$(u_{i+1} - u_i)\varphi(u_{i+1}) \le I_i \le (u_{i+1} - u_i)\varphi(u_i), \tag{6.110}$$

or, given that $u_{i+1} - u_i = (\lambda - 1)u_i$:

$$(\lambda - 1)u_i\varphi(u_{i+1}) \le I_i \le (\lambda - 1)u_i\varphi(u_i), \tag{6.111}$$

or even, knowing that $u_i = u_{i+1} / \lambda$:

$$\frac{1}{\lambda}(\lambda - 1)u_{i+1}\varphi(u_{i+1}) \le I_i \le (\lambda - 1)u_i\varphi(u_i), \tag{6.112}$$

a double inequality that enables us to write a new one:

$$\frac{1}{\lambda}(\lambda - 1)\sum_{i=-\infty}^{+\infty} u_{i+1}\varphi(u_{i+1}) \le \sum_{i=-\infty}^{+\infty} I_i \le (\lambda - 1)\sum_{i=-\infty}^{+\infty} u_i\varphi(u_i), \tag{6.113}$$

whose various sums are given by the following relations:

$$\sum_{i=-\infty}^{+\infty} I_i = \int_0^\infty \varphi(u)\,du = \int_0^\infty u^{n-1}e^{-ut}\,du, \tag{6.114}$$

a well known integral, equal to $t^{-n}\Gamma(n)$, namely

$$\sum_{i=-\infty}^{+\infty} I_i = t^{-n}\Gamma(n); \tag{6.115}$$

$$\begin{aligned} \sum_{i=-\infty}^{+\infty} u_{i+1}\varphi(u_{i+1}) &= \sum_{i=-\infty}^{+\infty} u_i\varphi(u_i) = \sum_{i=-\infty}^{+\infty} u_i u_i^{n-1} e^{-u_i t} \\ &= \sum_{i=-\infty}^{+\infty} u_i^n e^{-u_i t} = \frac{R}{(RC)^n} f(t). \end{aligned} \tag{6.116}$$

Double inequality [6.113] thus becomes:

$$\frac{1}{\lambda}(\lambda - 1)\frac{R}{(RC)^n} f(t) \le t^{-n}\Gamma(n) \le (\lambda - 1)\frac{R}{(RC)^n} f(t). \tag{6.117}$$

A framing of the step response $f(t)$ can then be determined in conformity with:

$$\frac{(RC)^n}{(\lambda-1)R}t^{-n}\Gamma(n) \le f(t) \le \lambda\frac{(RC)^n}{(\lambda-1)R}t^{-n}\Gamma(n). \qquad [6.118]$$

6.2.7.3. *Asymptotic behavior of the response* $f(t)$

If λ tends toward 1, the lower and upper bounds of $f(t)$ so obtained get closer to merge in the same function to the limit:

$$g(t) = \frac{(RC)^n}{(\lambda-1)R}t^{-n}\Gamma(n). \qquad [6.119]$$

Double inequality [6.118] can then be written as:

$$g(t) \le f(t) \le \lambda g(t). \qquad [6.120]$$

6.2.7.4. *Property of the response* $f(t)$

Let us again take the expression of $f(t)$ as given by relation [6.109], namely

$$f(t) = \frac{(RC)^n}{R}\sum_{i=-\infty}^{+\infty} u_i^n e^{-u_i t}, \qquad [6.121]$$

then let us compute $f(\lambda t)$. It happens:

$$f(\lambda t) = \frac{(RC)^n}{R}\sum_{i=-\infty}^{+\infty} u_i^n e^{-\lambda u_i t}, \qquad [6.122]$$

or, knowing that

$$\lambda u_i = u_{i+1} \text{ and } u_i = \frac{u_{i+1}}{\lambda}: \qquad [6.123]$$

$$f(\lambda t) = \frac{(RC)^n}{R}\sum_{i=-\infty}^{+\infty} \frac{1}{\lambda^n} u_{i+1}^n e^{-u_{i+1} t}, \qquad [6.124]$$

or even, as $\lambda^n = \alpha$:

$$\begin{aligned} f(\lambda t) &= \frac{1}{\alpha}\frac{(RC)^n}{R}\sum_{i=-\infty}^{+\infty} u_{i+1}^n e^{-u_{i+1} t} \\ &= \frac{1}{\alpha}\frac{(RC)^n}{R}\sum_{i=-\infty}^{+\infty} u_i^n e^{-u_i t} = \frac{1}{\alpha} f(t), \end{aligned} \qquad [6.125]$$

from which we draw the general relation that $f(t)$ verifies:

$$\alpha f(\lambda t) = f(t) \text{ for all } t > 0 . \qquad [6.126]$$

6.2.7.5. *Property of the asymptotic response* $g(t)$

Let us again take the expression of $g(t)$ given by relation [6.119], namely

$$g(t) = \frac{(RC)^n}{(\lambda - 1)R} t^{-n} \Gamma(n) , \qquad [6.127]$$

then let us compute $g(\lambda t)$. It happens:

$$g(\lambda t) = \frac{(RC)^n}{(\lambda - 1)R} \lambda^{-n} t^{-n} \Gamma(n) , \qquad [6.128]$$

or, as $\lambda^n = \alpha$:

$$g(\lambda t) = \frac{1}{\alpha} \frac{(RC)^n}{(\lambda - 1)R} t^{-n} \Gamma(n) = \frac{1}{\alpha} g(t) , \qquad [6.129]$$

from which we draw the general relation that $g(t)$ verifies:

$$\alpha g(\lambda t) = g(t) \text{ for all } t > 0 . \qquad [6.130]$$

6.2.7.6. *Properties of the quotient* $h(t) = f(t) / g(t)$

Knowing that the relative error between the response $f(t)$ and the asymptotic response $g(t)$ is expressed by

$$e_r(t) = \frac{f(t) - g(t)}{g(t)} = h(t) - 1 \text{ with } h(t) = \frac{f(t)}{g(t)} , \qquad [6.131]$$

it is convenient to determine the ratio $h(t)$ from the expressions of $f(t)$ and $g(t)$, respectively, given by [6.109] and [6.119], namely:

$$h(t) = \frac{(RC)^n}{R} \frac{(\lambda - 1)R}{(RC)^n t^{-n} \Gamma(n)} \sum_{i=-\infty}^{+\infty} u_i^n e^{-u_i t} , \qquad [6.132]$$

or else:

$$h(t)=(\lambda-1)\frac{t^n}{\Gamma(n)}\sum_{i=-\infty}^{+\infty}u_i^n e^{-u_i t}\,. \qquad [6.133]$$

By again taking double inequality [6.120] and by dividing it by $g(t)$, we obtain:

$$1\le\frac{f(t)}{g(t)}\le\lambda\,, \qquad [6.134]$$

hence, a first property (of boundedness) of $h(t)$ through which $h(t)$ responds to the double inequality:

$$1\le h(t)\le\lambda \text{ for all } t>0\,. \qquad [6.135]$$

Furthermore, $h(\lambda t)$ can be written as:

$$h(\lambda t)=\frac{f(\lambda t)}{g(\lambda t)}, \qquad [6.136]$$

namely, given that

$$\alpha f(\lambda t)=f(t) \text{ and } \alpha g(\lambda t)=g(t): \qquad [6.137]$$

$$h(\lambda t)=\frac{f(t)}{g(t)}, \qquad [6.138]$$

hence, a second property (of periodicity) of $h(t)$ according to which $h(t)$ satisfies the general relation:

$$h(\lambda t)=h(t) \text{ for all } t>0\,. \qquad [6.139]$$

6.2.7.7. *Properties of the function* $k(u)=h(\lambda^u)$

Let us take the function

$$k(u)=h(\lambda^u) \text{ with } u \text{ real}, \qquad [6.140]$$

which, given the expression of $h(t)$, is written as:

$$\begin{aligned} k(u) &= (\lambda-1)\frac{\left(\lambda^u\right)^n}{\Gamma(n)}\sum_{i=-\infty}^{+\infty}u_i^n e^{-u_i\lambda^u} \\ &= \frac{\lambda-1}{\Gamma(n)}\sum_{i=-\infty}^{+\infty}\left(u_i\lambda^u\right)^n e^{-u_i\lambda^u}, \end{aligned} \qquad [6.141]$$

or, given that

$$u_i\lambda^u = \frac{\lambda^i}{RC}\lambda^u = \frac{\lambda^{i+u}}{RC} = v_i(u): \qquad [6.142]$$

$$k(u) = \frac{\lambda-1}{\Gamma(n)}\sum_{i=-\infty}^{+\infty}\left(v_i(u)\right)^n e^{-v_i(u)} \text{ with } v_i(u) = \frac{\lambda^{i+u}}{RC}. \qquad [6.143]$$

As $1 \le h(t) \le \lambda$, $k(u) = h(\lambda^u)$ is bounded in conformity with:

$$1 \le k(u) \le \lambda. \qquad [6.144]$$

Moreover, as $h(\lambda t) = h(t)$, $k(u+1)$ can be written as:

$$k(u+1) = h(\lambda^{u+1}) = h(\lambda\lambda^u) = h(\lambda^u), \qquad [6.145]$$

namely:

$$k(u+1) = k(u), \qquad [6.146]$$

a relation that expresses the periodicity of $k(u)$.

Finally, $k(u)$ is a bounded and periodic function of period 1, defined for all real u. It is therefore expandable in Fourier series.

6.2.7.8. *Expansion of* $k(u)$ *in Fourier series*

Through its boundedness and its periodicity of period 1, $k(u)$ admits the Fourier series expansion:

$$k(u) = a_0 + \sum_{k-1}^{\infty}\left(a_k\cos(2k\pi u) + b_k\sin(2k\pi u)\right), \qquad [6.147]$$

in which:

$$a_0 = \int_0^1 k(u)\,du, \qquad [6.148]$$

$$a_k = 2\int_0^1 k(u)\cos(2k\pi u)\,du \qquad [6.149]$$

and

$$b_k = 2\int_0^1 k(u)\sin(2k\pi u)\,du\,. \qquad [6.150]$$

6.2.7.8.1. Calculation of the series first coefficient

Because of the expression of $k(u)$ given by [6.143], a_0 can be written as:

$$\begin{aligned} a_0 &= \int_0^1 \left[\frac{\lambda-1}{\Gamma(n)}\sum_{i=-\infty}^{+\infty}\left(v_i(u)\right)^n e^{-v_i(u)}\right]du \\ &= \frac{\lambda-1}{\Gamma(n)}\sum_{i=-\infty}^{+\infty}\int_0^1 \left(v_i(u)\right)^n e^{-v_i(u)}\,du \\ &= \frac{\lambda-1}{\Gamma(n)}\sum_{i=-\infty}^{+\infty} J_i, \quad \text{where} \quad J_i = \int_0^1 \left(v_i(u)\right)^n e^{-v_i(u)}\,du. \end{aligned} \qquad [6.151]$$

Given the form of J_i, let

$$x = v_i(u) = \frac{\lambda^{i+u}}{RC} = \frac{\lambda^i}{RC}\lambda^u = \frac{\lambda^i}{RC}e^{u\ln(\lambda)}, \qquad [6.152]$$

then

$$dx = \frac{\lambda^i}{RC}e^{u\ln(\lambda)}\ln(\lambda)du = x\ln(\lambda)du\,, \qquad [6.153]$$

hence

$$du = \frac{dx}{x\ln(\lambda)}, \qquad [6.154]$$

therefore:

$$J_i = \int_{v_i(0)}^{v_i(1)} x^n e^{-x}\frac{dx}{x\ln(\lambda)} = \frac{1}{\ln(\lambda)}\int_{u_i}^{u_{i+1}} x^{n-1}e^{-x}\,dx\,. \qquad [6.155]$$

a_0 then becomes:

$$\begin{aligned} a_0 &= \frac{\lambda-1}{\Gamma(n)}\sum_{i=-\infty}^{+\infty}\left[\frac{1}{\ln(\lambda)}\int_{u_i}^{u_{i+1}} x^{n-1}e^{-x}\,dx\right] \\ &= \frac{\lambda-1}{\Gamma(n)\ln(\lambda)}\sum_{i=-\infty}^{+\infty}\int_{u_i}^{u_{i+1}} x^{n-1}e^{-x}\,dx \\ &= \frac{\lambda-1}{\Gamma(n)\ln(\lambda)}\int_{u_{-\infty}}^{u_{+\infty}} x^{n-1}e^{-x}\,dx \\ &= \frac{\lambda-1}{\Gamma(n)\ln(\lambda)}\int_{0}^{\infty} x^{n-1}e^{-x}\,dx = \frac{\lambda-1}{\Gamma(n)\ln(\lambda)}\Gamma(n), \end{aligned} \quad [6.156]$$

namely:

$$a_0 = \frac{\lambda-1}{\ln(\lambda)}. \quad [6.157]$$

6.2.7.8.2. Calculation of the series other coefficients

By putting relation [6.143] into the general expression of a_k ($k \geq 1$), a_k can be written as:

$$\begin{aligned} a_k &= 2\int_0^1\left[\frac{\lambda-1}{\Gamma(n)}\sum_{i=-\infty}^{+\infty}(v_i(u))^n e^{-v_i(u)}\right]\cos(2k\pi u)\,du \\ &= 2\frac{\lambda-1}{\Gamma(n)}\sum_{i=-\infty}^{+\infty}\int_0^1 (v_i(u))^n e^{-v_i(u)}\cos(2k\pi u)\,du \\ &= 2\frac{\lambda-1}{\Gamma(n)}\sum_{i=-\infty}^{+\infty} J_i, \text{ where } J_i = \int_0^1 (v_i(u))^n e^{-v_i(u)}\cos(2k\pi u)\,du. \end{aligned} \quad [6.158]$$

Given the form of J_i, let us again take

$$x = v_i(u) = \frac{\lambda^{i+u}}{RC}, \quad [6.159]$$

from which we again draw (relation [6.154])

$$du = \frac{dx}{x\ln(\lambda)}, \quad [6.160]$$

but also

$$\begin{aligned} \ln(x) &= -\ln(RC)+(i+u)\ln(\lambda) \\ &= -\ln(RC)+i\ln(\lambda)+u\ln(\lambda), \end{aligned} \qquad [6.161]$$

hence

$$u=\frac{\ln(RCx)}{\ln(\lambda)}-i\,, \qquad [6.162]$$

therefore, knowing that

$$\begin{aligned} \cos(2k\pi u) &= \cos\left(2k\pi\frac{\ln(RCx)}{\ln(\lambda)}-2k\pi i\right) \\ &= \cos\left(2k\pi\frac{\ln(RCx)}{\ln(\lambda)}\right): \end{aligned} \qquad [6.163]$$

$$\begin{aligned} J_i &= \int_{v_i(0)}^{v_i(1)} x^n e^{-x}\cos\left(2k\pi\frac{\ln(RCx)}{\ln(\lambda)}\right)\frac{dx}{x\ln(\lambda)} \\ &= \frac{1}{\ln(\lambda)}\int_{u_i}^{u_{i+1}} x^{n-1}e^{-x}\cos\left(2k\pi\frac{\ln(RCx)}{\ln(\lambda)}\right)dx. \end{aligned} \qquad [6.164]$$

a_k then becomes:

$$a_k=2\frac{\lambda-1}{\Gamma(n)}\sum_{i=-\infty}^{+\infty}\left[\frac{1}{\ln(\lambda)}\int_{u_i}^{u_{i+1}} x^{n-1}e^{-x}\cos\left(2k\pi\frac{\ln(RCx)}{\ln(\lambda)}\right)dx\right], \qquad [6.165]$$

namely:

$$a_k=2\frac{\lambda-1}{\Gamma(n)\ln(\lambda)}\int_0^{\infty} x^{n-1}e^{-x}\cos\left(2k\pi\frac{\ln(RCx)}{\ln(\lambda)}\right)dx\,. \qquad [6.166]$$

The calculation of b_k is more identical than similar. We find indeed:

$$b_k=\frac{2(\lambda-1)}{\Gamma(n)\ln(\lambda)}\int_0^{\infty} x^{n-1}e^{-x}\sin\left(2k\pi\frac{\ln(RCx)}{\ln(\lambda)}\right)dx\,. \qquad [6.167]$$

Let us form the complex quantity z_k, such as

$$z_k = a_k + jb_k . \qquad [6.168]$$

It happens:

$$z_k = \frac{2(\lambda-1)}{\Gamma(n)\ln(\lambda)} \int_0^{\infty} x^{n-1} e^{-x} e^{j2k\pi\frac{\ln(RCx)}{\ln(\lambda)}} dx , \qquad [6.169]$$

namely, given that

$$\begin{aligned} e^{j2k\pi\frac{\ln(RCx)}{\ln(\lambda)}} &= e^{j\frac{2k\pi}{\ln(\lambda)}\ln(RC)} e^{j\frac{2k\pi}{\ln(\lambda)}\ln(x)} \\ &= e^{\ln(RC)^{j\frac{2k\pi}{\ln(\lambda)}}} e^{\ln(x)^{j\frac{2k\pi}{\ln(\lambda)}}} \\ &= (RC)^{j\frac{2k\pi}{\ln(\lambda)}} x^{j\frac{2k\pi}{\ln(\lambda)}} : \end{aligned} \qquad [6.170]$$

$$z_k = \frac{2(\lambda-1)}{\Gamma(n)\ln(\lambda)} \int_0^{\infty} x^{n-1} e^{-x} (RC)^{j\frac{2k\pi}{\ln(\lambda)}} x^{j\frac{2k\pi}{\ln(\lambda)}} dx , \qquad [6.171]$$

or:

$$z_k = \frac{2(\lambda-1)}{\Gamma(n)\ln(\lambda)} (RC)^{j\frac{2k\pi}{\ln(\lambda)}} \int_0^{\infty} x^{n-1+j\frac{2k\pi}{\ln(\lambda)}} e^{-x} dx , \qquad [6.172]$$

or even:

$$z_k = \frac{2(\lambda-1)}{\ln(\lambda)} (RC)^{j\frac{2k\pi}{\ln(\lambda)}} \frac{\Gamma\left(n + j\frac{2k\pi}{\ln(\lambda)}\right)}{\Gamma(n)} . \qquad [6.173]$$

Thus, the coefficients a_k and b_k can easily be determined in conformity with:

$$a_k = \mathrm{Re}[z_k] \text{ and } b_k = \mathrm{Im}[z_k] \text{, for all } k \geq 1 . \qquad [6.174]$$

6.2.7.8.3. Behavior of the series coefficients

The effective calculation of the coefficients so obtained shows a very fast decreasing of the Fourier series coefficients. In practice, the error is indistinguishable if we approximate $k(u)$ with the series limited to the first harmonic:

$$k_1(u) = a_0 + a_1 \cos(2\pi u) + b_1 \sin(2\pi u), \qquad [6.175]$$

namely:

$$k_1(u) = a_0 + A_1 \cos(2\pi u - \phi_1), \qquad [6.176]$$

where:

$$A_1 = \sqrt{a_1^2 + b_1^2} \qquad [6.177]$$

and

$$\phi_1 = arctg \frac{b_1}{a_1} = \arg z_1, \qquad [6.178]$$

the expressions of A_1 and ϕ_1 such as formulated being determined as follows:

$$\begin{aligned} A_1 \cos(2\pi u - \phi_1) &= a_1 \cos(2\pi u) + b_1 \sin(2\pi u) \\ &= a_1 \cos(2\pi u) + b_1 \cos(2\pi u - \frac{\pi}{2}); \end{aligned} \qquad [6.179]$$

$$A_1 e^{j(2\pi u - \phi_1)} = a_1 e^{j2\pi u} + b_1 e^{j(2\pi u - \frac{\pi}{2})}; \qquad [6.180]$$

$$A_1 e^{-j\phi_1} = a_1 + b_1 e^{-j\frac{\pi}{2}} = a_1 - jb_1; \qquad [6.181]$$

$$A_1 = |a_1 - jb_1| = \sqrt{a_1^2 + b_1^2} = |z_1|; \qquad [6.182]$$

$$-\phi_1 = \arg(a_1 - jb_1) = -arctg \frac{b_1}{a_1} = -\arg z_1. \qquad [6.183]$$

The approximation of $k(u)$ defined as such is justified by rewriting z_k as:

$$z_k = \frac{2(\lambda - 1)}{\ln(\lambda)} (RC)^{jx} \frac{\Gamma(n + jx)}{\Gamma(n)} \text{ where } x = \frac{2k\pi}{\ln(\lambda)} > 0, \qquad [6.184]$$

namely:

$$z_k = \frac{2(\lambda - 1)}{\ln(\lambda)} e^{jx\ln(RC)} \frac{\Gamma(n + jx)}{\Gamma(n)} . \qquad [6.185]$$

Indeed, the modulus of z_k, $|z_k| = A_k$, namely

$$|z_k| = \frac{2(\lambda - 1)}{\ln(\lambda)} \Psi(x), \text{ where } \Psi(x) = \frac{|\Gamma(n + jx)|}{\Gamma(n)}, \qquad [6.186]$$

presents, through the function $\Psi(x)$, a decrease and more precisely a very fast tendency toward zero when x tends toward infinity.

Thus, by putting $a = 2\pi/\ln(\lambda) > 0$, the series (u_k), such that $u_k = \Psi(ka)$, decreases and converges very quickly toward zero, its convergence being studied due to Stirling's asymptotic formula, namely

$$\Gamma(z) \approx \sqrt{2\pi} z^{z-1} e^{-z} \text{ for } z \text{ tending toward infinity.} \qquad [6.187]$$

If $z = \rho e^{j\theta}$ and if $z = n + jb$, then (if ρ tends toward infinity) this formula becomes:

$$\begin{aligned} \Gamma(z) &\approx \sqrt{2\pi} (\rho e^{j\theta})^{n-1+jb} e^{-n-jb} \\ &\approx \sqrt{2\pi} \rho^{n-1} \rho^{jb} e^{j\theta(n-1)} e^{-\theta b} e^{-n} e^{-jb} \\ &\approx \sqrt{2\pi} \rho^{n-1} e^{jb\ln(\rho)} e^{j\theta(n-1)} e^{-\theta b} e^{-n} e^{-jb}, \end{aligned} \qquad [6.188]$$

therefore:

$$|\Gamma(z)| \approx \sqrt{2\pi} \rho^{n-1} e^{-(n+\theta b)}, \qquad [6.189]$$

or:

$$|\Gamma(n + jb)| \approx \sqrt{2\pi} \rho^{n-1} e^{-(n+\theta b)}, \qquad [6.190]$$

or even, for $b = ka$, $z = n + jb$ then becoming $z_k = n + jka$ or $z_k = \rho_k e^{j\theta_k}$:

$$|\Gamma(n + jka)| \approx \sqrt{2\pi} \rho_k^{\,n-1} e^{-(n+\theta_k ka)} . \qquad [6.191]$$

Thus, for ρ_k tending toward infinity, $u_k = \Psi(ka)$ can be particularized in conformity with:

$$u_k \approx \frac{\sqrt{2\pi}}{\Gamma(n)} \rho_k^{\ n-1} e^{-(n+\theta_k ka)} , \qquad [6.192]$$

or, given that $\rho_k \approx ka$ and $\theta_k \approx \pi/2$ (as $z_k = n + jka$ tends toward jka when k tends toward infinity in conformity with the tendency toward infinity of ρ_k):

$$\begin{aligned} u_k &\approx \frac{\sqrt{2\pi}}{\Gamma(n)} (ka)^{n-1} e^{-(n+\frac{\pi}{2}ka)} \\ &\approx \frac{\sqrt{2\pi}}{\Gamma(n)} a^{n-1} e^{-n} (k^{n-1} e^{-\frac{\pi}{2}ka}), \end{aligned} \qquad [6.193]$$

u_{k+1} then being written as:

$$\begin{aligned} u_{k+1} &\approx \frac{\sqrt{2\pi}}{\Gamma(n)} a^{n-1} e^{-n} \left((k+1)^{n-1} e^{-\frac{\pi}{2}(k+1)a} \right) \\ &\approx \frac{\sqrt{2\pi}}{\Gamma(n)} a^{n-1} e^{-n} k^{n-1} e^{-\frac{\pi}{2}ka} e^{-\frac{\pi}{2}a} \\ &\approx u_k e^{-\frac{\pi}{2}a} , \end{aligned} \qquad [6.194]$$

from which we draw:

$$\frac{u_{k+1}}{u_k} \approx e^{-\frac{\pi}{2}a} , \qquad [6.195]$$

namely, finally, knowing that $a = 2\pi/\ln(\lambda)$:

$$\frac{u_{k+1}}{u_k} \approx e^{-\frac{\pi^2}{\ln(\lambda)}} . \qquad [6.196]$$

Thus, the closer the parameter λ is to 1, the faster the convergence of the series (u_k) toward zero.

6.3. Recursive arborescent network as a lung respiratory model

6.3.1. *Modeling by an equivalent electric network*

From the respiratory point of view, a lung can be modeled by a *bifurcation arborescent structure* where each branch presents (in electrical equivalence):

– a *dissipative effect* which can be characterized by a *resistance*, due to viscosity and turbulence energy losses;

– an *inductive effect* which can be characterized by an *inductance*, due to inertial effects of motion air;

– a *capacitive effect* which can be characterized by a *capacitance*, due to the conjoint elasticity of the air and the branch which makes (by a *tank effect*) the ingoing flow different from the outgoing one.

This leads us to say that the electric model equivalent to a branch is a gamma cell (Figure 6.5) whose horizontal branch is constituted of a resistance in series with an inductance and whose vertical branch is constituted of a capacitance.

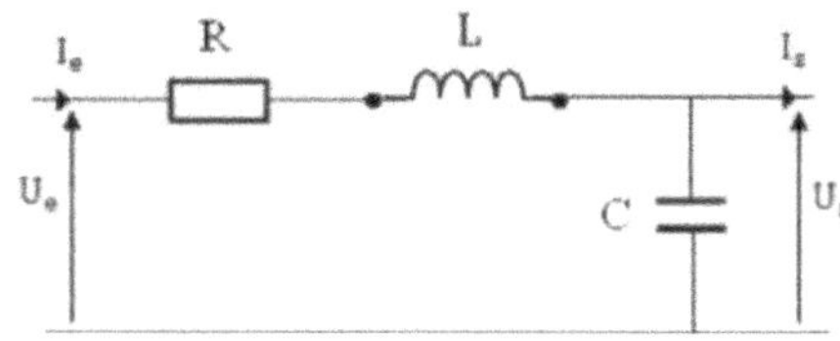

Figure 6.5. *Gamma RLC cell equivalent to a branch: I_s is different from I_e through the capacitance C*

In a bifurcation arborescence (Figure 6.6), each branch splits into two branches, where each branch splits itself into two other branches, and so on for the following steps.

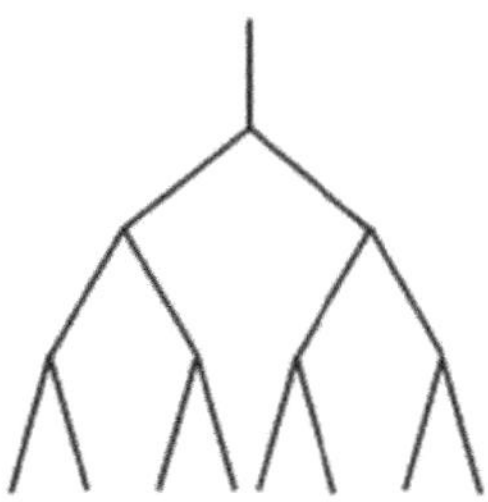

Figure 6.6. *Bifurcation arborescence*

Given that each branch admits, as an electrical equivalence, a *gamma RLC cell*, the branch set that achieves the bifurcation arborescence admits an equivalent electric network under the form of an *arborescent arrangement of gamma RLC cells* (Figure 6.7).

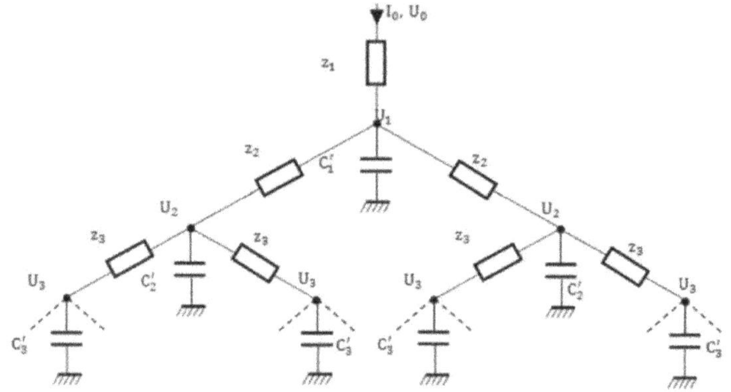

Figure 6.7. *Arborescent arrangement of gamma RLC cells:* $z_i(j\omega) = R_i' + jL_i'\omega$

By gathering in one single node the node set of the same potential, and by repeating this operation for each network potential, the arborescent arrangement of gamma RLC cells comes down to a *cascade arrangement of gamma RLC cells* (Figure 6.8) in which:

$$Z_i(j\omega) = \frac{z_i(j\omega)}{2^{i-1}} = \frac{R_i' + jL_i'\omega}{2^{i-1}}, \qquad [6.197]$$

namely:

$$Z_i(j\omega) = R_i + jL_i\omega, \qquad [6.198]$$

by putting:

$$R_i = \frac{R_i'}{2^{i-1}} \text{ and } L_i = \frac{L_i'}{2^{i-1}}, \qquad [6.199]$$

expressions with which it is convenient to associate:

$$C_i = 2^{i-1} C_i'. \qquad [6.200]$$

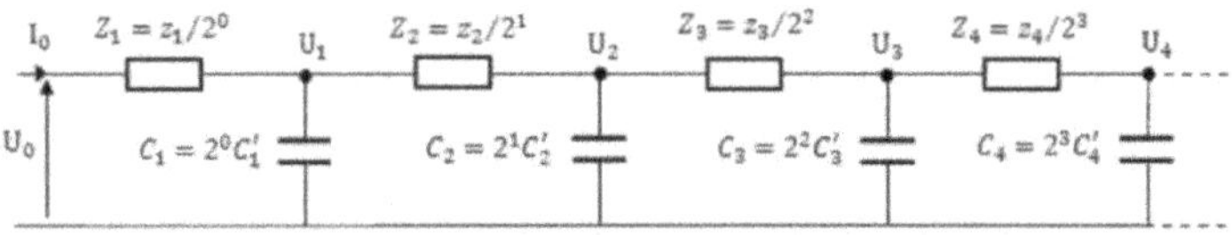

Figure 6.8. *Cascade arrangement of gamma RLC cells*

Under the *assumption of the resistance, inductance and capacitance recursivity* of the arborescent network of Figure 6.7, namely

$$R'_{i+1} = \alpha' R'_i,\ \ L'_{i+1} = \lambda' L'_i, \text{ and } C'_{i+1} = \eta' C'_i, \qquad [6.201]$$

it is possible to write the following equations for the resistances, inductances and capacitances of the ladder network of Figure 6.8:

$$R_{i+1} = \frac{R'_{i+1}}{2^i} = \frac{\alpha' R'_i}{2^i} = \frac{\alpha'}{2}\frac{R'_i}{2^{i-1}} = \frac{\alpha'}{2} R_i, \qquad [6.202]$$

$$L_{i+1} = \frac{L'_{i+1}}{2^i} = \frac{\lambda' L'_i}{2^i} = \frac{\lambda'}{2}\frac{L'_i}{2^{i-1}} = \frac{\lambda'}{2} L_i \qquad [6.203]$$

and

$$C_{i+1} = 2^i C'_{i+1} = 2^i \eta' C'_i = 2\eta' 2^{i-1} C'_i = 2\eta' C_i, \qquad [6.204]$$

namely:

$$R_{i+1} = \alpha R_i \text{ with } \alpha = \frac{\alpha'}{2} > 1 \qquad [6.205]$$

$$L_{i+1} = \lambda L_i \text{ with } \lambda = \frac{\lambda'}{2} \leq \alpha \qquad [6.206]$$

and

$$C_{i+1} = \eta C_i \text{ with } \eta = 2\eta' > 1, \qquad [6.207]$$

the later study conditions imposing indeed a constraint between λ and α.

By again taking the ladder network of Figure 6.8 and *by adopting for the vertical branches a representation in accordance with the horizontal branches*, we obtain, for N gamma cells, the *ladder network* of Figure 6.9 in which:

$$Z_i(j\omega) = R_i + jL_i\omega \tag{6.208}$$

and

$$Z_i^{'}(j\omega) = \frac{1}{jC_i\omega}, \tag{6.209}$$

with

$$\frac{R_{i+1}}{R_i} = \alpha > 1, \ \frac{L_{i+1}}{L_i} = \lambda \leq \alpha \ \text{and} \ \frac{C_{i+1}}{C_i} = \eta > 1, \tag{6.210}$$

a network for which the general representation under the N gamma ZZ' cell form has the merit to enable generic calculations applicable to any expression of $Z_i(j\omega)$ or $Z_i'(j\omega)$.

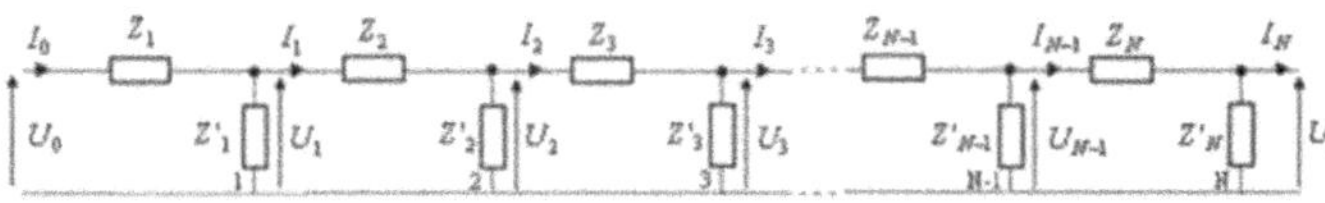

Figure 6.9. *Ladder network resulting from the cascade setting of N gamma ZZ' cells:* $X_0 = \frac{U_0}{I_0}$ *(input impedance);* $X_1 = \frac{U_1}{I_1}$; $X_2 = \frac{U_2}{I_2}$; $X_3 = \frac{U_3}{I_3}$;...; $X_{N-1} = \frac{U_{N-1}}{I_{N-1}}$; $X_N = \frac{U_N}{I_N} = \infty$

6.3.2. *Network admittance*

6.3.2.1. *Expression of the admittance* $1/X_0(s)$ *versus the admittance* $1/X_1(s)$

Given Figure 6.10, the impedance $X_0(s)$ is written as:

$$X_0(s) = Z_1(s) + \frac{Z_1^{'}(s)X_1(s)}{Z_1^{'}(s) + X_1(s)}, \tag{6.211}$$

the corresponding admittance thus admitting the following expression:

$$\frac{1}{X_0(s)} = \frac{1}{Z_1(s) + \frac{Z_1^{'}(s)X_1(s)}{Z_1^{'}(s) + X_1(s)}}, \tag{6.212}$$

or also:

$$\frac{1}{X_0(s)}=\frac{1}{Z_1(s)+\dfrac{Z_1^{'}(s)}{1+Z_1^{'}(s)\dfrac{1}{X_1(s)}}}, \qquad [6.213]$$

or even:

$$\frac{1}{X_0(s)}=\frac{1}{Z_1(s)+\dfrac{Z_1^{'}(s)}{1+Z_1^{'}(s)\dfrac{1}{X_1(s)}}}. \qquad [6.214]$$

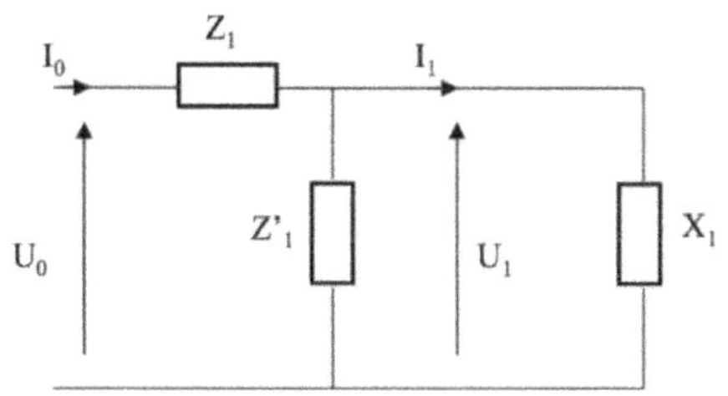

Figure 6.10. *Equivalent configuration*

6.3.2.2. *Expressions of the admittances* $1/X_1(s)$ *and* $1/X_2(s)$

The transposition of the results so obtained to the following two configurations (Figure 6.11) immediately leads to the corresponding admittances.

$$\frac{1}{X_1(s)}=\frac{1/Z_2(s)}{1+\dfrac{Z_2^{'}(s)/Z_2(s)}{1+Z_2^{'}(s)\dfrac{1}{X_2(s)}}} \qquad [6.215]$$

and

$$\frac{1}{X_2(s)}=\frac{1/Z_3(s)}{1+\dfrac{Z_3^{'}(s)/Z_3(s)}{1+Z_3^{'}(s)\dfrac{1}{X_3(s)}}}. \qquad [6.216]$$

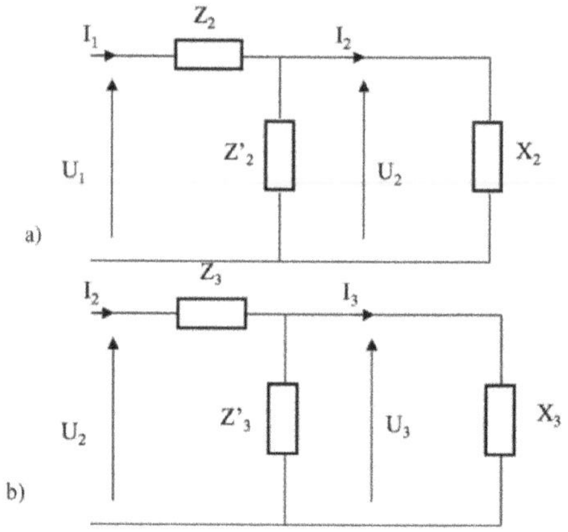

Figure 6.11. *a) First configuration and b) second configuration*

6.3.2.3. *Expression of the admittance* $1/X_0(s)$ *versus* $1/X_2(s)$ *then* $1/X_3(s)$*: extension to the rank* N

Putting the expression of $1/X_1(s)$ into equation [6.214] leads to the expression of $1/X_0(s)$ versus $1/X_2(s)$, namely:

$$\frac{1}{X_0(s)}=\frac{1/Z_1(s)}{1+\cfrac{Z_1'(s)/Z_1(s)}{1+\cfrac{Z_1'(s)/Z_2(s)}{1+\cfrac{Z_2'(s)/Z_2(s)}{1+Z_2'(s)\cfrac{1}{X_2(s)}}}}}. \qquad [6.217]$$

By now putting the expression of $1/X_2(s)$ into this equation, we get the expression of $1/X_0(s)$ versus $1/X_3(s)$, namely:

$$\frac{1}{X_0(s)}=\frac{1/Z_1(s)}{1+\cfrac{Z_1'(s)/Z_1(s)}{1+\cfrac{Z_1'(s)/Z_2(s)}{1+\cfrac{Z_2'(s)/Z_2(s)}{1+\cfrac{Z_2'(s)/Z_3(s)}{1+\cfrac{Z_3'(s)/Z_3(s)}{1+Z_3'(s)\cfrac{1}{X_3(s)}}}}}}}. \qquad [6.218]$$

The extension of the rank N is carried out:

– by expressing $1/X_0(s)$ versus $1/X_N(s)$;

– by putting $1/X_N(s)=0$ as $X_N(s)=\infty$;

– by introducing the rank N in the expression of $1/X_0(s)$ suggesting to denote this admittance by $Y_N(s)$ (admittance of the set of the N gamma cells), so namely:

$$Y_N(s)=\cfrac{1/Z_1(s)}{1+\cfrac{Z_1'(s)/Z_1(s)}{1+\cfrac{Z_1'(s)/Z_2(s)}{1+\cfrac{Z_2'(s)/Z_2(s)}{1+\cfrac{Z_2'(s)/Z_3(s)}{1+\ldots\cfrac{}{1+\cfrac{Z_{N-1}'(s)/Z_N(s)}{1+Z_N'(s)/Z_N(s)}}}}}}}. \qquad [6.219]$$

The form so obtained is nothing but the form of a *continued fraction* or a *continued fraction expansion.*

By particularizing $Z_i(s)$ and $Z_i'(s)$ in conformity with relations [6.208] and [6.209], $Y_N(s)$ becomes:

$$Y_N(s)=\cfrac{1/(R_1+L_1s)}{1+\cfrac{1/C_1s(R_1+L_1s)}{1+\cfrac{1/C_1s(R_2+L_2s)}{1+\cfrac{1/C_2s(R_2+L_2s)}{1+\cfrac{1/C_2s(R_3+L_3s)}{1+\ldots\cfrac{}{1+\cfrac{1/C_{N-1}s(R_N+L_Ns)}{1+1/C_Ns(R_N+L_Ns)}}}}}}}, \qquad [6.220]$$

or, given the recursive factors as defined by relation [6.210]:

$$Y_N(s)=\cfrac{1/(R_1+L_1s)}{1+\cfrac{1/C_1s(R_1+L_1s)}{1+\cfrac{1/C_1s(\alpha R_1+\lambda L_1s)}{1+\cfrac{1/\eta C_1s(\alpha R_1+\lambda L_1s)}{1+\cfrac{1/\eta C_1s(\alpha^2 R_1+\lambda^2 L_1s)}{1+\dots\cfrac{}{1+\cfrac{1/\eta^{N-2}C_1s(\alpha^{N-1}R_1+\lambda^{N-1}L_1s)}{1+1/\eta^{N-1}C_1s(\alpha^{N-1}R_1+\lambda^{N-1}L_1s)}}}}}}}. \quad [6.221]$$

For the frequencies to which the inductive effects can be neglected compared to the resistive (or dissipative) effects, namely, $Z_i(s)=R_i+L_is$ being written

$$Z_i(s)=\alpha^{i-1}\left[R_i+\left(\frac{\lambda}{\alpha}\right)^{i-1}L_is\right], \quad [6.222]$$

for the frequencies such that

$$\left(\frac{\lambda}{\alpha}\right)^{i-1}L_1|s|<<R_1 \quad \forall i\geq 1, \quad [6.223]$$

therefore verifying

$$\text{(for i=1) } L_1|s|<<R\text{, namely } |s|<<\frac{R}{L_1}, \quad [6.224]$$

$$\text{with (for i>1) } \frac{\lambda}{\alpha}\leq 1\text{, namely } \lambda\leq\alpha, \quad [6.225]$$

the admittance $Y_N(s)$ can be reduced to the following expression:

$$Y_N(s)=\cfrac{1/R_1}{1+\cfrac{1/R_1C_1s}{1+\cfrac{1/\alpha R_1C_1s}{1+\cfrac{1/\alpha\eta R_1C_1s}{1+\cfrac{1/\alpha^2\eta R_1C_1s}{1+\cfrac{1/\alpha^2\eta^2 R_1C_1s}{1+\dots\cfrac{}{1+\cfrac{1/\alpha^{N-1}\eta^{N-2}R_1C_1s}{1+1/\alpha^{N-1}\eta^{N-1}R_1C_1s}}}}}}}}. \quad [6.226]$$

By putting $W(s) = 1/R_1C_1\,s$, $Y_N(s)$ can be written as:

$$Y_N(s) = \frac{1/R_1}{1+g(W(s),\alpha,\eta)}, \qquad [6.227]$$

where $g(W(s),\alpha,\eta)$ is a new continued fraction such that:

$$g(W(s),\alpha,\eta) = \cfrac{W(s)}{1+\cfrac{W(s)/\alpha}{1+\cfrac{W(s)/\alpha\eta}{1+\cfrac{W(s)/\alpha^2\eta}{1+\cfrac{W(s)/\alpha^2\eta^2}{\ddots}}}}}\,. \qquad [6.228]$$

6.3.3. *Non-integer order of the admittance*

Using the Pringsheim notation, the continued fraction $g(W(s),\alpha,\eta)$ is written as:

$$\begin{aligned} g(W(s),\alpha,\eta) = {} & \frac{W(s)\,|}{|\;1} + \frac{W(s)/\alpha\,|}{|\;1} + \frac{W(s)/\alpha\eta\,|}{|\;1} \\ & + \frac{W(s)/\alpha^2\eta\,|}{|\;1} + \frac{W(s)/\alpha^2\eta^2\,|}{|\;1} + \cdots. \end{aligned} \qquad [6.229]$$

Through a triple variable change that consists of replacing $W(s)$ by $W(s)/\alpha$, α by η and η by α, $g(W(s),\alpha,\eta)$ becomes:

$$\begin{aligned} g(\frac{W(s)}{\alpha},\alpha,\eta) = {} & \frac{W(s)/\alpha\,|}{|\;1} + \frac{W(s)/\alpha\eta\,|}{|\;1} \\ & + \frac{W(s)/\alpha^2\eta\,|}{|\;1} + \frac{W(s)/\alpha^2\eta^2\,|}{|\;1} + \cdots. \end{aligned} \qquad [6.230]$$

The comparison of these two expansions shows that:

$$g(W(s),\alpha,\eta) = \frac{W(s)\,|}{|\;1} + g(\frac{W(s)}{\alpha},\alpha,\eta), \qquad [6.231]$$

namely:

$$g(W(s),\alpha,\eta) = \frac{W(s)}{1+g(\frac{W(s)}{\alpha},\alpha,\eta)}, \qquad [6.232]$$

from which we draw, in a first approximation, at the frequencies for which

$$\left| g(\frac{W(s)}{\alpha},\alpha,\eta) \right| >> 1: \qquad [6.233]$$

$$g(W(s),\alpha,\eta)\ g(\frac{W(s)}{\alpha},\eta,\alpha) = W(s). \qquad [6.234]$$

6.3.3.1. *Hypothesis of an approximated equation of the function* g

For this, we can then put, by assumption:

$$g(W(s),\alpha,\eta) \approx K(\alpha,\eta)W(s)^{n(\alpha,\eta)}, \qquad [6.235]$$

an expression that put into [6.234] leads to:

$$K(\alpha,\eta)W(s)^{n(\alpha,\eta)} K(\eta,\alpha)\frac{W(s)^{n(\eta,\alpha)}}{\alpha^{n(\eta,\alpha)}} = W(s), \qquad [6.236]$$

namely:

$$\frac{K(\alpha,\eta)\ K(\eta,\alpha)}{\alpha^{n(\eta,\alpha)}}\ W(s)^{n(\alpha,\eta)+n(\eta,\alpha)} = W(s), \qquad [6.237]$$

from which we draw through a term-to-term identification:

$$n(\alpha,\eta)+n(\eta,\alpha) = 1 \qquad [6.238]$$

and

$$K(\alpha,\eta)K(\eta,\alpha) = \alpha^{n(\eta,\alpha)}, \qquad [6.239]$$

the double variable change such that $\alpha \to \eta$ and $\eta \to \alpha$ (permutation of α and η) and which allows us to write

$$K(\eta,\alpha)\ K(\alpha,\eta) = \eta^{n(\alpha,\eta)}, \qquad [6.240]$$

showing in addition that:

$$\alpha^{n(\eta,\alpha)} = \eta^{n(\alpha,\eta)}. \tag{6.241}$$

By taking the decimal logarithm of equation [6.241], we get:

$$n(\eta,\alpha)\log\alpha = n(\alpha,\eta)\log\eta, \tag{6.242}$$

namely, given [6.238]:

$$(1-n(\alpha,\eta))\log\alpha = n(\alpha,\eta)\,\log\eta, \tag{6.243}$$

from which we draw:

$$n(\alpha,\eta) = \frac{\log\alpha}{\log\alpha + \log\eta}. \tag{6.244}$$

CONCLUSION.–

Finally, for the frequencies to which

$$\left|g(W(s),\alpha,\eta)\right| >> 1, \tag{6.245}$$

assembling relations [6.227] and [6.235] where $n(\alpha,\eta)$ is defined by [6.244] makes it possible to express the admittance $Y_N(s)$ as:

$$Y_N(s) = \frac{1/R_1}{K(\alpha,\eta)W^n(s)} = \frac{1/R_1}{K(\alpha,\eta)}(R_1C_1s)^n, \tag{6.246}$$

a form that symbolically reveals a differentiation to a non-integer order such that:

$$n = n(\alpha,\eta) = \frac{\log\alpha}{\log(\alpha\eta)}. \tag{6.247}$$

6.3.3.2. *Hypothesis of an exact equation of the function g*

After making the assumption of an approximated equation of the function g through relation [6.235], we now make the (stronger) assumption of an exact equation of the function g by putting:

$$g(W(s),\alpha,\eta) = h(W(s),\alpha,\eta)W(s)^{n(\alpha,\eta)}. \tag{6.248}$$

Putting this expression into relation [6.234] leads to:

$$h\left(W(s),\alpha,\eta\right)W(s)^{n(\alpha,\eta)}\,h\left(\frac{W(s)}{\alpha},\eta,\alpha\right)\frac{W(s)^{n(\eta,\alpha)}}{\alpha^{n(\eta,\alpha)}}=W(s), \qquad [6.249]$$

namely:

$$\frac{h\left(W(s),\alpha,\eta\right)h\left(W(s)/\alpha,\eta,\alpha\right)}{\alpha^{n(\eta,\alpha)}}W(s)^{n(\alpha,\eta)+n(\eta,\alpha)}=W(s), \qquad [6.250]$$

from which we draw through a term-to-term identification:

$$n(\alpha,\eta)+n(\eta,\alpha)=1 \qquad [6.251]$$

and

$$h\left(W(s),\alpha,\eta\right)h\left(\frac{W(s)}{\alpha},\eta,\alpha\right)=\alpha^{n(\eta,\alpha)}. \qquad [6.252]$$

The triple variable change such that $W(s)\to\eta W(s)$, $\alpha\to\eta$ and $\eta\to\alpha$ enables us to rewrite relation [6.252] according to:

$$h\left(\eta W(s),\eta,\alpha\right)h\left(W(s),\alpha,\eta\right)=\eta^{n(\alpha,\eta)}, \qquad [6.253]$$

an expression that, compared to [6.252], leads to:

$$h\left(W(s),\alpha,\eta\right)=\frac{\alpha^{n(\eta,\alpha)}}{h\left(\frac{W(s)}{\alpha},\eta,\alpha\right)}=\frac{\eta^{n(\alpha,\eta)}}{h\left(\eta W(s),\eta,\alpha\right)}. \qquad [6.254]$$

By choosing $n(\alpha,\eta)$ and $n(\eta,\alpha)$ such that

$$\alpha^{n(\eta,\alpha)}=\eta^{n(\alpha,\eta)}, \qquad [6.255]$$

namely

$$n(\eta,\alpha)\log\alpha=n(\alpha,\eta)\log\eta, \qquad [6.256]$$

or (from [6.251])

$$(1-n(\alpha,\eta))\log\alpha = n(\alpha,\eta)\log\eta\,, \qquad [6.257]$$

from which we draw

$$n(\alpha,\eta) = \frac{\log\alpha}{\log\alpha+\log\eta}\,, \qquad [6.258]$$

relation [6.254] shows that the function h verifies an equation of the form:

$$h\left(\frac{W(s)}{\alpha},\eta,\alpha\right) = h(\eta W(s),\eta,\alpha)\,, \qquad [6.259]$$

or, by replacing $W(s)$ by $\alpha W(s)$:

$$h(W(s),\eta,\alpha) = h(\alpha\eta W(s),\eta,\alpha)\,, \qquad [6.260]$$

or even, α becoming η and reciprocally:

$$h(W(s),\alpha,\eta) = h(\lambda W(s),\alpha,\eta) \text{ with } \lambda = \alpha\eta\,, \qquad [6.261]$$

which expresses the periodicity of the variations of $h(W(j\omega),\alpha,\eta)$ around the constant $K(\alpha,\eta)$, the period being given by

$$\log\lambda = \log(\alpha\eta) \qquad [6.262]$$

in a semi-logarithmic representation.

Stemmed from equation [6.259], this well-known periodicity of the gain and phase undulations validates this equation and so justifies (*a posteriori*) the choice of equality [6.255].

CONCLUSION.–

Finally, for the same condition as previously, namely (relation [6.245])

$$|g(W(s),\alpha,\eta)| >> 1\,, \qquad [6.263]$$

the admittance defined by relation [6.246] becomes, in the case of the new assumption on g (relation [6.248]):

$$Y_N(s) = \frac{1/R_1}{h(W(s),\alpha,\eta)W^n(s)} = \frac{1/R_1}{h(W(s),\alpha,\eta)}(R_1C_1s)^n, \qquad [6.264]$$

a form that reveals, up to the undulations of h around the constant $K(\alpha,\eta)$, a differentiation to the non-integer order

$$n = n(\alpha,\eta) = \frac{\log\alpha}{\log(\alpha\eta)}, \qquad [6.265]$$

a differentiation that so appears as the result of a frequency filtering that reduces $h(W(j\omega),\alpha,\eta)$ to $K(\alpha,\eta)$, the first assumption on g (relation [6.235]) integrating this filtering idea.

6.3.4. *Frequency band corresponding to the non-integer behavior*

The non-integer behavior occurs only in a frequency band covered by the spectrum of the transitional frequencies of the various cells, this spectrum being delimited by the extremal transitional frequencies $1/R_1C_1$ and $1/R_NC_N$.

Knowing that the frequency areas where the boundary effects occur cannot present a non-integer behavior, only the frequency band defined by

$$\frac{1}{R_NC_N} << |s| << \frac{1}{R_1C_1}, \qquad [6.266]$$

is likely to present a non-integer behavior.

In order to increase this frequency band, and even to extend it to the whole frequency domain, it suffices:

– to decrease the time constant R_1C_1, *the high transitional frequency* $1/R_1C_1$ *can even be infinite for* $R_1C_1 = 0$;

– to increase the time constant R_NC_N (through an increasing of the cell number N), *the low transitional frequency* $1/R_NC_N$ *can even be nil for* $R_NC_N = \infty$.

For the limit so defined,

$$|W(s)| = 1/R_1C_1|s| \quad [6.267]$$

then being infinite for any finite frequency ($|s| = \omega$ finite), the functions

$$g(W(s),\alpha,\eta) = h(W(s),\alpha,\eta)W(s)^{n(\alpha,\eta)} \quad [6.268]$$

and

$$g\left(\frac{W(s)}{\alpha},\eta,\alpha\right) = h\left(\frac{W(s)}{\alpha},\eta,\alpha\right)\left(\frac{W(s)}{\alpha}\right)^{n(\eta,\alpha)}, \quad [6.269]$$

whose moduli are written as

$$|g(W(s),\alpha,\eta)| = |h(W(s),\alpha,\eta)||W(s)|^{n(\alpha,\eta)} \quad [6.270]$$

and

$$\left|g\left(\frac{W(s)}{\alpha},\eta,\alpha\right)\right| = \left|h\left(\frac{W(s)}{\alpha},\eta,\alpha\right)\right|\left(\frac{|W(s)|}{\alpha}\right)^{n(\eta,\alpha)}, \quad [6.271]$$

show that, given

$$|h(W(s),\alpha,\eta)|, \left|h\left(\frac{W(s)}{\alpha},\eta,\alpha\right)\right|, n(\alpha,\eta) \; et \; n(\eta,\alpha) > 0: \quad [6.272]$$

$$|g(W(s),\alpha,\eta)| = \infty \quad [6.273]$$

and

$$\left|g\left(\frac{W(s)}{\alpha},\eta,\alpha\right)\right| = \infty, \quad [6.274]$$

relations which express that the study conditions [6.245], [6.263] and [6.233], namely

$$|g(W(s),\alpha,\eta)| >> 1 \quad [6.275]$$

and

$$\left| g\left(\frac{W(s)}{\alpha}, \eta, \alpha \right) \right| >> 1, \qquad [6.276]$$

are then well verified whatever the frequency.

6.4. Unified study of recursive parallel arrangements of RL, RC and RLC cells

This section presents some heuristic aspects through the proposition of pertinent assumptions that quickly lead to exact results with remarkable simplicity.

Notably, the non-integer order and the "frequency periodicity" of the admittance and also the non-integer power and the "time periodicity" of the impulse response inscribe their determination in these heuristic approaches.

6.4.1. *Recursive parallel arrangement of series RL cells*

6.4.1.1. *Definition*

An *indefinite recursive parallel arrangement of series RL cells* is a ladder network such that, for recursive factors α and η greater than 1:

– the cell of rank 0 is the *median (or central)* cell defined by a resistance R in series with an inductance L;

– the cell of rank 1 is defined by a resistance αR in series with an inductance L/η;

– the cell of rank –1 is defined by a resistance $\alpha^{-1}R$ in series with an inductance L/η^{-1};

– the cell of rank i is defined by a resistance $R_i = \alpha^i R$ in series with an inductance $L_i = L/\eta^i$, $i \in \mathbb{Z}$.

6.4.1.2. *Impulse response*

The network admittance is expressed by the indefinite sum:

$$Y(s) = \sum_{i \in \mathbb{Z}} Y_i(s) = \sum_{i=-\infty}^{+\infty} Y_i(s), \qquad [6.277]$$

where $Y_i(s)$ is the admittance of the cell $R_i L_i$, namely:

$$Y_i(s) = \frac{1}{R_i + L_i s} = \frac{\frac{1}{L_i}}{s + \frac{R_i}{L_i}}, \qquad [6.278]$$

or, knowing that $R_i = \alpha^i R$ and $L_i = L / \eta^i$ lead to

$$\frac{1}{L_i} = \frac{\eta^i}{L} \text{ and } \frac{R_i}{L_i} = \frac{\alpha^i R}{\frac{L}{\eta^i}} = (\alpha\eta)^i \frac{R}{L} = \lambda^i \omega_0, \qquad [6.279]$$

with $\lambda = \alpha\eta$ and $\omega_0 = \frac{R}{L}$, transitional frequency of the median cell:

$$Y_i(s) = \frac{\eta^i}{L} \frac{1}{s + \lambda^i \omega_0}. \qquad [6.280]$$

The original (or inverse transform) $y_i(t)$ of $Y_i(s)$, namely

$$y_i(t) = \mathcal{L}^{-1}\left[Y_i(s)\right] = \mathcal{L}^{-1}\left[\frac{\eta^i}{L} \frac{1}{s + \lambda^i \omega_0}\right], \qquad [6.281]$$

is written as:

$$y_i(t) = \frac{\eta^i}{L} e^{-\lambda^i \omega_0 t} u(t), \qquad [6.282]$$

or, for $t \geq 0$:

$$y_i(t) = \frac{\eta^i}{L} e^{-\lambda^i \omega_0 t}, \qquad [6.283]$$

the network impulse response (for $t \geq 0$) then being given by:

$$y(t) = \sum_{i=-\infty}^{+\infty} y_i(t) = \sum_{i=-\infty}^{+\infty} \frac{\eta^i}{L} e^{-\lambda^i \omega_0 t}. \qquad [6.284]$$

6.4.1.3. *Recursivity of relaxation modes*

For $t > 0$, the relaxation mode of rank $i+1$, $y_{i+1}(t)$, can be written as:

$$y_{i+1}(t) = \frac{\eta^{i+1}}{L} e^{-\lambda^{i+1}\omega_0 t} = \eta \frac{\eta^i}{L} e^{-\lambda^i \omega_0 (\lambda t)}, \qquad [6.285]$$

namely, from the expression of $y_i(t)$ established for $t \geq 0$:

$$y_{i+1}(t) = \eta y_i(\lambda t), \qquad [6.286]$$

where η is the *recursive factor along the amplitude* and λ is the *recursive factor along the time.*

6.4.1.4. *Operational approach of non-integer integration*

Given the rank i cell admittance $Y_i(s)$, the network admittance $Y(s)$ is given by:

$$Y(s) = \sum_{i=-\infty}^{+\infty} \frac{\eta^i}{L} \frac{1}{s + \lambda^i \omega_0}. \qquad [6.287]$$

By putting $i = j+1$, $Y(s)$ is rewritten in conformity with:

$$\begin{aligned} Y(s) &= \sum_{j=-\infty}^{+\infty} \frac{\eta^{j+1}}{L} \frac{1}{s + \lambda^{j+1}\omega_0} \\ &= \eta \sum_{j=-\infty}^{+\infty} \frac{\eta^j}{L} \frac{1}{s + \lambda^j \lambda \omega_0} \\ &= \frac{\eta}{\lambda} \sum_{j=-\infty}^{+\infty} \frac{\eta^j}{L} \frac{1}{\lambda^{-1} s + \lambda^j \omega_0}, \end{aligned} \qquad [6.288]$$

or, from [6.287]:

$$Y(s) = \alpha^{-1} Y(\lambda^{-1} s). \qquad [6.289]$$

n denoting a non-integer order and $X(s)$ an operational function taking into account the gain and phase undulations, let us put (by assumption):

$$Y(s) = s^n X(s), \qquad [6.290]$$

from which we draw, by putting [6.290] into [6.289], the following equation:

$$s^n X(s) = \alpha^{-1}\left(\lambda^{-1}s\right)^n X\left(\lambda^{-1}s\right), \tag{6.291}$$

therefore:

$$X(s) = \alpha^{-1}\lambda^{-n} X\left(\lambda^{-1}s\right). \tag{6.292}$$

By choosing $\alpha^{-1}\lambda^{-n} = 1$, that is to say

$$n = -\frac{\log\alpha}{\log\lambda} < 0, \tag{6.293}$$

which translates a *non-integer integration*, $X(s)$ verifies the equation:

$$X(s) = X(\lambda^{-1}s), \tag{6.294}$$

or, in frequency domain:

$$X(j\omega) = X(j\lambda^{-1}\omega), \tag{6.295}$$

which expresses *the periodicity of* $X(j\omega)$ *in a semi-logarithmic representation* (these results are commented upon in section 6.4.4.3).

6.4.1.5. *Time approach of non-integer integration*

Let us take again the network impulse response expression (for $t \geq 0$), namely

$$y(t) = \sum_{i=-\infty}^{+\infty} \frac{\eta^i}{L} e^{-\lambda^i \omega_0 t}, \tag{6.296}$$

then let us put $i = j+1$, $y_{(t)}$ then being rewritten:

$$\begin{aligned} y(t) &= \sum_{j=-\infty}^{+\infty} \frac{\eta^{j+1}}{L} e^{-\lambda^{j+1}\omega_0 t} \\ &= \eta \sum_{j=-\infty}^{+\infty} \frac{\eta^j}{L} e^{-\lambda^j \omega_0 (\lambda t)}, \end{aligned} \tag{6.297}$$

namely, from [6.296]:

$$y(t) = \eta y(\lambda t). \tag{6.298}$$

γ denoting a non-integer power and $x(t)$ a time function taking into account the gain and phase undulations, let us pose (by assumption):

$$y(t) = t^{\gamma} x(t), \quad [6.299]$$

from which we draw, by putting [6.299] into [6.298], the following equation:

$$t^{\gamma} x(t) = \eta (\lambda t)^{\gamma} x(\lambda t), \quad [6.300]$$

therefore:

$$x(t) = \eta \lambda^{\gamma} x(\lambda t). \quad [6.301]$$

By choosing $\eta \lambda^{\gamma} = 1$, that is to say

$$\gamma = -\frac{\log \eta}{\log \lambda} = -\frac{\log(\alpha\eta) - \log\alpha}{\log \lambda} = -n - 1, \quad [6.302]$$

$x(t)$ verifies the equation:

$$x(t) = x(\lambda t), \quad [6.303]$$

which expresses *the periodicity of* $x(t)$ *in a semi-logarithmic representation* (section 6.4.4.4).

6.4.2. *Recursive parallel arrangement of series RC cells*

6.4.2.1. *Definition*

An *indefinite recursive parallel arrangement of series RC cells* is a ladder network such that (for recursive factors α and η greater than 1):

– the cell of rank 0 is the *median (or central)* cell defined by a resistance R in series with a capacitance C;

– the cell of rank 1 is defined by a resistance R/α in series with a capacitance C/η;

– the cell of rank –1 is defined by a resistance R/α^{-1} in series with a capacitance C/η^{-1};

– the cell of rank i is defined by a resistance $R_i = R/\alpha^i$ in series with a capacitance $C_i = C/\eta^i$, $i \in \mathbb{Z}$.

6.4.2.2. *Impulse response*

The network admittance is expressed by the indefinite sum:

$$Y(s) = \sum_{i \in \mathbb{Z}} Y_i(s) = \sum_{i=-\infty}^{+\infty} Y_i(s), \qquad [6.304]$$

where $Y_i(s)$ is the admittance of the cell R_iC_i, namely:

$$Y_i(s) = \frac{1}{R_i + \dfrac{1}{C_i s}} = \frac{C_i s}{R_i C_i s + 1}, \qquad [6.305]$$

or:

$$Y_i(s) = \frac{\dfrac{C_i s}{R_i C_i}}{s + \dfrac{1}{R_i C_i}} = \frac{1}{R_i} \frac{s}{s + \dfrac{1}{R_i C_i}}, \qquad [6.306]$$

or even, knowing that $R_i = R/\alpha^i$ and $C_i = C/\eta^i$ lead to

$$\frac{1}{R_i} = \frac{\alpha^i}{R} \text{ and } \frac{1}{R_i C_i} = \frac{1}{\dfrac{R}{\alpha^i}\dfrac{C}{\eta^i}} = \frac{(\alpha\eta)^i}{RC} = \lambda^i \omega_0, \qquad [6.307]$$

with $\lambda = \alpha\eta$ and $\omega_0 = \dfrac{1}{RC}$, transitional frequency of the median cell:

$$Y_i(s) = \frac{\alpha^i}{R} \frac{s}{s + \lambda^i \omega_0}. \qquad [6.308]$$

The original $y_i(t)$ of $Y_i(s)$, namely

$$y_i(t) = \mathcal{L}^{-1}\left[\frac{\alpha^i}{R} \frac{s}{s + \lambda^i \omega_0}\right] = \frac{\alpha^i}{R} \mathcal{L}^{-1}\left[s \frac{1}{s + \lambda^i \omega_0}\right], \qquad [6.309]$$

is written as:

$$\begin{aligned} y_i(t) &= \frac{\alpha^i}{R}\frac{d}{dt}\mathcal{L}^{-1}\left[\frac{1}{s+\lambda^i\omega_0}\right] \\ &= \frac{\alpha^i}{R}\frac{d}{dt}\mathcal{L}^{-1}\left[e^{-\lambda^i\omega_0 t}u(t)\right] \\ &= \frac{\alpha^i}{R}\left[e^{-\lambda^i\omega_0 t}\delta(t)-\lambda^i\omega_0 e^{-\lambda^i\omega_0 t}u(t)\right] \\ &= \frac{\alpha^i}{R}\left[\delta(t)-\lambda^i\omega_0 e^{-\lambda^i\omega_0 t}u(t)\right], \end{aligned} \qquad [6.310]$$

namely, for $t \geq 0$:

$$y_i(t)=\frac{\alpha^i}{R}\left[\delta(t)-\lambda^i\omega_0 e^{-\lambda^i\omega_0 t}\right], \qquad [6.311]$$

the network impulse response (for $t \geq 0$) then being given by:

$$y(t)=\sum_{i=-\infty}^{+\infty}\frac{\alpha^i}{R}\left[\delta(t)-\lambda^i\omega_0 e^{-\lambda^i\omega_0 t}\right], \qquad [6.312]$$

or:

$$y(t)=\left(\frac{1}{R}\sum_{i=-\infty}^{+\infty}\alpha^i\right)\delta(t)-\frac{\omega_0}{R}\sum_{i=-\infty}^{+\infty}(\alpha\lambda)^i e^{-\lambda^i\omega_0 t}, \qquad [6.313]$$

or even, given that $\sum_{i=-\infty}^{+\infty}\alpha^i=\infty$:

$$y(t)=(\infty)\delta(t)-\frac{\omega_0}{R}\sum_{i=-\infty}^{+\infty}(\alpha\lambda)^i e^{-\lambda^i\omega_0 t}, \qquad [6.314]$$

a result that expresses that *the impulse response presents an indefinite number of Dirac impulsions at initial instant* $t=0$.

6.4.2.3. *Recursivity of relaxation modes*

For $t>0$, the relaxation mode of rank i, $y_i(t)$, is written as:

$$y_i(t)=-\frac{\omega_0}{R}(\alpha\lambda)^i e^{-\lambda^i\omega_0 t}, \qquad [6.315]$$

from which we deduce the relaxation mode of rank $i+1$:

$$\begin{aligned} y_{i+1}(t) &= -\frac{\omega_0}{R}(\alpha\lambda)^{i+1} e^{-\lambda^{i+1}\omega_0 t} \\ &= -(\alpha\lambda)\frac{\omega_0}{R}(\alpha\lambda)^{i} e^{-\lambda^{i}\omega_0(\lambda t)}, \end{aligned} \quad [6.316]$$

namely, from the expression of $y_i(\lambda t)$ established for $t>0$:

$$y_{i+1}(t) = (\alpha\lambda)\, y_i(\lambda t), \quad [6.317]$$

where $\alpha\lambda$ is the *recursive factor along the amplitude* and λ is the *recursive factor along the time.*

6.4.2.4. *Operational approach of non-integer differentiation*

Given the rank i cell admittance $Y_i(s)$, the network admittance $Y(s)$ is given by:

$$Y(s) = \sum_{i=-\infty}^{+\infty} \frac{\alpha^i}{R}\frac{s}{s+\lambda^i\omega_0}. \quad [6.318]$$

By putting $i = j+1$, $Y(s)$ is rewritten in conformity with:

$$\begin{aligned} Y(s) &= \sum_{j=-\infty}^{+\infty} \frac{\alpha^{j+1}}{R}\frac{s}{s+\lambda^{j+1}\omega_0} \\ &= \alpha\sum_{j=-\infty}^{+\infty} \frac{\alpha^{j}}{R}\frac{s}{s+\lambda^{j}\lambda\omega_0} \\ &= \alpha\sum_{j=-\infty}^{+\infty} \frac{\alpha^{j}}{R}\frac{\lambda^{-1}s}{\lambda^{-1}s+\lambda^{j}\omega_0}, \end{aligned} \quad [6.319]$$

or, from [6.318]:

$$Y(s) = \alpha Y(\lambda^{-1}s). \quad [6.320]$$

n denoting a non-integer order and $X(s)$ an operational function taking into account the gain and phase undulations, let us put (by assumption):

$$Y(s) = s^n X(s), \quad [6.321]$$

from which we draw, by putting [6.321] into [6.320], the following equation:

$$s^n X(s) = \alpha\left(\lambda^{-1}s\right)^n X(\lambda^{-1}s), \qquad [6.322]$$

therefore:

$$X(s) = \alpha\lambda^{-n} X(\lambda^{-1}s). \qquad [6.323]$$

By choosing $\alpha\lambda^{-n} = 1$, that is to say

$$n = \frac{\log\alpha}{\log\lambda} > 0, \qquad [6.324]$$

which translates a *non-integer differentiation,* $X(s)$ verifies the equation:

$$X(s) = X(\lambda^{-1}s), \qquad [6.325]$$

or, in frequency domain:

$$X(j\omega) = X(j\lambda^{-1}\omega), \qquad [6.326]$$

which expresses *the periodicity of* $X(j\omega)$ *in a semi-logarithmic representation* (these results are commented upon in section 6.4.4.3).

6.4.2.5. *Time approach of non-integer differentiation*

Let us again take the network impulse response expression (for $t \geq 0$), namely

$$y(t) = \sum_{i=-\infty}^{+\infty} \frac{\alpha^i}{R}\left[\delta(t) - \lambda^i\omega_0 e^{-\lambda^i\omega_0 t}\right], \qquad [6.327]$$

that we consider here for $t > 0$, namely

$$y(t) = -\frac{\omega_0}{R}\sum_{i=-\infty}^{+\infty} (\alpha\lambda)^i e^{-\lambda^i\omega_0 t}, \qquad [6.328]$$

then let us put $i = j+1$, $y(t)$ then being rewritten:

$$\begin{aligned} y(t) &= -\frac{\omega_0}{R}\sum_{j=-\infty}^{+\infty} (\alpha\lambda)^{j+1} e^{-\lambda^{j+1}\omega_0 t} \\ &= -(\alpha\lambda)\frac{\omega_0}{R}\sum_{j=-\infty}^{+\infty} (\alpha\lambda)^j e^{-\lambda^j\omega_0(\lambda t)}, \end{aligned} \qquad [6.329]$$

namely, from [6.328]:

$$y(t) = (\alpha\lambda)\, y(\lambda t). \qquad [6.330]$$

γ denoting a non-integer power and $x(t)$ a time function taking into account the gain and phase undulations, let us pose (by assumption):

$$y(t) = t^{\gamma} x(t), \qquad [6.331]$$

from which we draw, by putting [6.331] into [6.330], the following equation:

$$t^{\gamma} x(t) = (\alpha\lambda)(\lambda t)^{\gamma}\, x(\lambda t), \qquad [6.332]$$

therefore:

$$x(t) = (\alpha\lambda)\, \lambda^{\gamma} x(\lambda t). \qquad [6.333]$$

By choosing $(\alpha\lambda)\lambda^{\gamma} = 1$, that is to say

$$\gamma = -\frac{\log(\alpha\lambda)}{\log\lambda} = -\frac{\log\alpha + \log\lambda}{\log\lambda} = -n - 1, \qquad [6.334]$$

$x(t)$ verifies the equation:

$$x(t) = x(\lambda t), \qquad [6.335]$$

which expresses *the periodicity of* $x(t)$ *in a semi-logarithmic representation* (section 6.4.4.4).

6.4.3. *Recursive parallel arrangement of series RLC cells*

As an introduction, it is convenient to mention that such an arrangement is liable to represent the porous face model (Chapter 1 and Appendix 2) when *the mass of the motion water in each pore is taken into account at the energy level.*

Indeed, by taking into account the mass of the motion water in each of the porous face pores, the *energy balance* of a pore is such that:

– through the orifice which laminates water, the pore is the place where a *dissipative energy* is located;

– through the mass of the motion water, the pore is the place where a *kinetic energy* is located;

– through the cavity which traps air (compressible by the motion water), the pore is the place where an *elasticity potential energy* is located.

This leads us to state that a pore can be represented by a purely resistive dissipative element (*resistance*) in series with a purely inductive non-dissipative element (*inductance*) and a purely capacitive non-dissipative element (*capacitance*), actually a *series RLC cell.*

6.4.3.1. *Definition*

An *indefinite recursive parallel arrangement of series RLC cells* is a ladder network such that (for recursive factors α and η greater than 1):

– the cell of rank 0 is the *median (or central)* cell defined by a resistance R in series with an inductance L and a capacitance C;

– the cell of rank 1 is defined by a resistance $R(\eta / \alpha)$ in series with an inductance L / α^2 and a capacitance C / η^2 ;

– the cell of rank –1 is defined by a resistance $R(\eta / \alpha)^{-1}$ in series with an inductance $L / (\alpha^2)^{-1}$ and a capacitance $C / (\eta^2)^{-1}$;

– the cell of rank i is defined by a resistance $R_i = R(\eta / \alpha)^i$ in series with an inductance $L_i = L / (\alpha^2)^i$ and a capacitance $C_i = C / (\eta^2)^i$, $i \in \mathbb{Z}$.

6.4.3.2. *Impulse response*

The network admittance is expressed by the indefinite sum:

$$Y(s) = \sum_{i \in \mathbb{Z}} Y_i(s) = \sum_{i=-\infty}^{+\infty} Y_i(s), \qquad [6.336]$$

where $Y_i(s)$ is the admittance of the cell $R_i L_i C_i$, namely:

$$Y_i(s) = \frac{1}{R_i + L_i s + \dfrac{1}{C_i s}} = \frac{C_i s}{1 + R_i C_i s + L_i C_i s^2}, \qquad [6.337]$$

or, under a canonical form:

$$Y_i(s) = \frac{C_i s}{1 + 2\zeta_i \dfrac{s}{\omega_i} + \dfrac{s^2}{\omega_i^2}}, \qquad [6.338]$$

with

$$\omega_i = \frac{1}{\sqrt{L_i C_i}} \text{ and } \zeta_i = \frac{R_i}{2}\sqrt{\frac{C_i}{L_i}}, \qquad [6.339]$$

or even, knowing that $R_i = R(\eta/\alpha)^i$, $L_i = L/(\alpha^2)^i$ and $C_i = C/(\eta^2)^i$ lead to

$$\omega_i = \lambda^i \omega_0 \text{ and } \zeta_i = \zeta_0, \qquad [6.340]$$

with $\lambda = \alpha\eta$ and also $\omega_0 = \frac{1}{\sqrt{LC}}$ and $\zeta_0 = \frac{R}{2}\sqrt{\frac{C}{L}}$, transitional frequency and damping ratio of the median cell:

$$Y_i(s) = \frac{\frac{C}{(\eta^2)^i}s}{1 + 2\zeta_0\frac{s}{\lambda^i\omega_0} + \frac{s^2}{(\lambda^i\omega_0)^2}}. \qquad [6.341]$$

The original $y_i(t)$ of $Y_i(s)$, namely

$$\begin{aligned} y_i(t) &= \mathcal{L}^{-1}\left[\frac{C}{(\eta^2)^i}\frac{s}{1 + 2\zeta_0\frac{s}{\lambda^i\omega_0} + \frac{s^2}{(\lambda^i\omega_0)^2}}\right] \\ &= \frac{C}{(\eta^2)^i}\mathcal{L}^{-1}\left[s\frac{1}{1 + 2\zeta_0\frac{s}{\lambda^i\omega_0} + \frac{s^2}{(\lambda^i\omega_0)^2}}\right], \end{aligned} \qquad [6.342]$$

is written as:

$$y_i(t) = \frac{C}{(\eta^2)^i}\frac{d}{dt}\mathcal{L}^{-1}\left[\frac{1}{1 + 2\zeta_0\frac{s}{\lambda^i\omega_0} + \frac{s^2}{(\lambda^i\omega_0)^2}}\right], \qquad [6.343]$$

or, for the *critical damping* defined by $\zeta_0 = 1$:

$$\begin{aligned} y_i(t) &= \frac{C}{\left(\eta^2\right)^i}\frac{d}{dt}\left[\left(\lambda^i\omega_0\right)^2 te^{-\lambda^i\omega_0 t}u(t)\right] \\ &= \frac{C}{\left(\eta^2\right)^i}\left(\lambda^i\omega_0\right)^2\frac{d}{dt}\left[e^{-\lambda^i\omega_0 t}\left(tu(t)\right)\right] \\ &= C\left(\alpha^2\right)^i\omega_0^{\ 2}\left[-\lambda^i\omega_0 e^{-\lambda^i\omega_0 t}tu(t)+e^{-\lambda^i\omega_0 t}u(t)\right] \\ &= C\left(\alpha^2\right)^i\omega_0^{\ 2}e^{-\lambda^i\omega_0 t}\left[1-\lambda^i\omega_0 t\right]u(t), \end{aligned} \qquad [6.344]$$

or, for $t \geq 0$:

$$y_i(t) = C\omega_0^{\ 2}\left(\alpha^2\right)^i\left[1-\lambda^i\omega_0 t\right]e^{-\lambda^i\omega_0 t}, \qquad [6.345]$$

the network impulse response (for $t \geq 0$) then being given by:

$$y(t) = \sum_{i=-\infty}^{+\infty} C\omega_0^{\ 2}\left(\alpha^2\right)^i\left[1-\lambda^i\omega_0 t\right]e^{-\lambda^i\omega_0 t}. \qquad [6.346]$$

6.4.3.3. *Recursivity of relaxation modes*

For $t > 0$, the relaxation mode of rank $i+1$, $y_{i+1}(t)$, can be written as:

$$\begin{aligned} y_{i+1}(t) &= C\omega_0^{\ 2}\left(\alpha^2\right)^{i+1}\left[1-\lambda^{i+1}\omega_0 t\right]e^{-\lambda^{i+1}\omega_0 t} \\ &= \left(\alpha^2\right)C\omega_0^{\ 2}\left(\alpha^2\right)^i\left[1-\lambda^i\omega_0\left(\lambda t\right)\right]e^{-\lambda^i\omega_0(\lambda t)}, \end{aligned} \qquad [6.347]$$

namely, from the expression of $y_i(\lambda t)$ established for $t \geq 0$:

$$y_{i+1}(t) = \left(\alpha^2\right)y_i(\lambda t), \qquad [6.348]$$

where α^2 is the *recursive factor along the amplitude* and λ is the *recursive factor along the time*.

6.4.3.4. *Operational approach of non-integer differentiation or integration*

Given the rank i cell admittance $Y_i(s)$, the network admittance $Y(s)$ is given by:

$$Y(s)=\sum_{i=-\infty}^{+\infty}\frac{C}{\left(\eta^2\right)^i}\frac{s}{1+2\zeta_0\dfrac{s}{\lambda^i\omega_0}+\left(\dfrac{s}{\lambda^i\omega_0}\right)^2}. \qquad [6.349]$$

By putting $i=j+1$, $Y(s)$ is rewritten in conformity with:

$$\begin{aligned}Y(s) &= \sum_{j=-\infty}^{+\infty}\frac{C}{\left(\eta^2\right)^{j+1}}\frac{s}{1+2\zeta_0\dfrac{s}{\lambda^{j+1}\omega_0}+\left(\dfrac{s}{\lambda^{j+1}\omega_0}\right)^2}\\ &= \frac{1}{\eta^2}\sum_{j=-\infty}^{+\infty}\frac{C}{\left(\eta^2\right)^{j}}\frac{s}{1+2\zeta_0\dfrac{s}{\lambda^{j}\lambda\omega_0}+\left(\dfrac{s}{\lambda^{j}\lambda\omega_0}\right)^2}\\ &= \frac{\lambda}{\eta^2}\sum_{j=-\infty}^{+\infty}\frac{C}{\left(\eta^2\right)^{j}}\frac{\lambda^{-1}s}{1+2\zeta_0\dfrac{\lambda^{-1}s}{\lambda^{j}\omega_0}+\left(\dfrac{\lambda^{-1}s}{\lambda^{j}\omega_0}\right)^2},\end{aligned} \qquad [6.350]$$

or, from [6.349]:

$$Y(s)=\frac{\alpha}{\eta}Y\left(\lambda^{-1}s\right). \qquad [6.351]$$

n denoting a non-integer order and $X(s)$ an operational function taking into account the gain and phase undulations, let us put (by assumption):

$$Y(s)=s^nX(s), \qquad [6.352]$$

from which we draw, by putting [6.352] into [6.351], the following equation:

$$s^nX(s)=\frac{\alpha}{\eta}\left(\lambda^{-1}s\right)^nX\left(\lambda^{-1}s\right), \qquad [6.353]$$

therefore:

$$X(s)=\frac{\alpha}{\eta}\lambda^{-n}X\left(\lambda^{-1}s\right). \qquad [6.354]$$

We choose n such that $(\alpha/\eta)\lambda^{-n}=1$, namely

$$n=\frac{\log\alpha-\log\eta}{\log\lambda}, \qquad [6.355]$$

the non-integer order so determined translating *a non-integer differentiation for* $\alpha > \eta$ $(n > 0)$ *and a non-integer integration for* $\alpha < \eta$ $(n < 0)$.

$X(s)$ then verifies the equation:

$$X(s) = X\left(\lambda^{-1}s\right), \quad [6.356]$$

or, in the frequency domain:

$$X(j\omega) = X\left(j\lambda^{-1}\omega\right), \quad [6.357]$$

which expresses *the periodicity of* $X(j\omega)$ *in a semi-logarithmic representation.*

As in the case of RL or RC cells, the results so obtained as for the order n and the frequency response $X(j\omega)$ are the subject of commentaries in section 6.4.4.3.

6.4.3.5. *Time approach of non-integer differentiation or integration*

Let us again take the network impulse response expression (for $t \geq 0$), namely

$$y(t) = \sum_{i=-\infty}^{+\infty} C\omega_0^2\left(\alpha^2\right)^i\left[1-\lambda^i\omega_0 t\right]e^{-\lambda^i\omega_0 t}, \quad [6.358]$$

then let us put $i = j+1$, $y(t)$ then being rewritten:

$$\begin{aligned} y(t) &= \sum_{j=-\infty}^{+\infty} C\omega_0^2\left(\alpha^2\right)^{j+1}\left[1-\lambda^{j+1}\omega_0 t\right]e^{-\lambda^{j+1}\omega_0 t} \\ &= \left(\alpha^2\right)\sum_{j=-\infty}^{+\infty} C\omega_0^2\left(\alpha^2\right)^j\left[1-\lambda^j\omega_0\left(\lambda t\right)\right]e^{-\lambda^j\omega_0(\lambda t)}, \end{aligned} \quad [6.359]$$

namely, from [6.358]:

$$y(t) = \left(\alpha^2\right)y\left(\lambda t\right). \quad [6.360]$$

γ denoting a non-integer power and $x(t)$ a time function taking into account the gain and phase undulations, let us pose (by assumption):

$$y(t) = t^\gamma x(t), \quad [6.361]$$

from which we draw, by putting [6.361] into [6.360], the following equation:

$$t^\gamma x(t) = \left(\alpha^2\right)\left(\lambda t\right)^\gamma x\left(\lambda t\right), \quad [6.362]$$

therefore:

$$x(t) = \left(\alpha^2\right)\lambda^{\gamma} x\left(\lambda t\right). \qquad [6.363]$$

By choosing $\left(\alpha^2\right)\lambda^{\gamma} = 1$, that is to say

$$\begin{aligned} \gamma = -2\frac{\log\alpha}{\log\lambda} &= \frac{-\log\alpha - \log\alpha + \log\eta - \log\eta}{\log\lambda} \\ &= \frac{-\log\alpha + \log\eta - \log\left(\alpha\eta\right)}{\log\lambda} \\ &= \frac{-\left(\log\alpha - \log\eta\right) - \log\lambda}{\log\lambda} \\ &= -\frac{\log\alpha - \log\eta}{\log\lambda} - 1 = -n - 1, \end{aligned} \qquad [6.364]$$

$x(t)$ verifies the equation:

$$x(t) = x\left(\lambda t\right), \qquad [6.365]$$

which expresses, as in the case of RL or RC cells, *the periodicity of* $x(t)$ *in a semi-logarithmic representation* (section 6.4.4.4).

6.4.3.6. *Graphical verification of differentiation or integration non-integer order*

A calculation simplified by the consideration of recursive factors α and η large compared to unit, as detailed in [OUS 95, Chapter 5], enables us to plot the Bode asymptotic diagrams of the admittance $Y(j\omega)$ (Figure 6.12) for which:

$$\frac{\omega_i}{\omega_i^{'}} = \frac{\omega_{i+1}}{\omega_{i+1}^{'}} = \alpha \qquad [6.366]$$

and

$$\frac{\omega_{i+1}^{'}}{\omega_i} = \frac{\omega_{i+2}^{'}}{\omega_{i+1}} = \eta\,. \qquad [6.367]$$

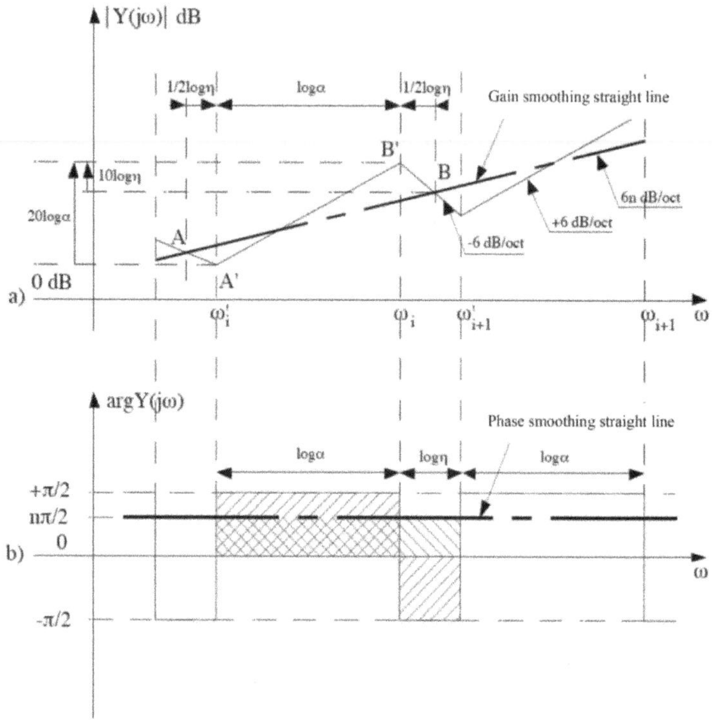

Figure 6.12. *Gain and phase asymptotic diagrams of the admittance Y(jω)*

As in the case of the recursive parallel arrangement of a large number of series RC cells (Chapter 2), the smoothing of these diagrams can be materialized by a *gain smoothing straight line* of slope $6ndB/oct$ and a *phase smoothing straight line* of ordinate $n\pi/2$, the value of n being determined in conformity with the following developments.

The slopes of segments AB and $A'B'$ (Figure 6.12(a)) are, respectively, given by the equations:

$$6n\text{dB/oct} = \frac{20\log\alpha - 20\log\eta}{\log\alpha + \log\eta} \qquad [6.368]$$

and

$$6\,\text{dB/oct} = \frac{20\log\alpha}{\log\alpha} = 20, \qquad [6.369]$$

from which we draw, by carrying out their ratio, the expression of the non-integer order n versus the recursive factors α and η:

$$n = \frac{\log\alpha - \log\eta}{\log\alpha + \log\eta} = \frac{\log\alpha - \log\eta}{\log\lambda}. \qquad [6.370]$$

The identity of the hatched surfaces (Figure 6.12(b)) is conveyed by the equation:

$$n\frac{\pi}{2}(\log\alpha + \log\eta) = \frac{\pi}{2}\log\alpha - \frac{\pi}{2}\log\eta, \qquad [6.371]$$

from which n is immediately deduced, namely:

$$n = \frac{\log\alpha - \log\eta}{\log\alpha + \log\eta} = \frac{\log\alpha - \log\eta}{\log\lambda}, \qquad [6.372]$$

an expression which is identical to the one determined from the gain.

6.4.4. *Commented synthesis of the main results*

6.4.4.1. *On the impulse response and its non-definition at* $t = 0$ *in non-integer differentiation*

For $t \geq 0$, the impulse response of an *indefinite recursive parallel arrangement* is given by:

$$y(t) = \sum_{i=-\infty}^{+\infty} \frac{\eta^i}{L} e^{-\lambda^i \omega_0 t}, \text{ with } \omega_0 = \frac{R}{L}, \qquad [6.373]$$

for series RL cells;

$$y(t) = \sum_{i=-\infty}^{+\infty} \frac{\alpha^i}{R}\left[\delta(t) - \lambda^i \omega_0 e^{-\lambda^i \omega_0 t}\right], \text{ with } \omega_0 = \frac{1}{RC}, \qquad [6.374]$$

for series RC cells;

$$y(t) = \sum_{i=-\infty}^{+\infty} C\omega_0 \left(\alpha^2\right)^i \left[1 - \lambda^i \omega_0 t\right] e^{-\lambda^i \omega_0 t}, \text{ with } \omega_0 = \frac{1}{\sqrt{LC}}, \qquad [6.375]$$

for series RLC cells.

Among all these responses, the impulse response of a series RC cell arrangement is distinguished by *an infinity of Dirac impulsions at initial instant* $t = 0$.

It is true that such an arrangement approximately achieves a non-integer differentiator, of reduced transmittance s^n with $n>0$, for which it is well known (notably by mathematicians) that *the impulse response is not defined at* $t=0$.

On the contrary, a series RLC cell arrangement that also approximates (for $\alpha>\eta$) a non-integer differentiator, of reduced transmittance s^n with $n>0$, presents *an impulse response without Dirac impulsion at* $t=0$, because of the inductances that are highly opposed to the current fast variations.

It then seems that for the same ideal transmittance s^n with $n>0$, the initial impulse behavior is different according to the (approximate) achievement of this transmittance, thus reinforcing (by physics) *the non-defined character at* $t=0$ *of the non-integer differentiator impulse response.*

6.4.4.2. *On the recursivity of relaxation modes*

For $t>0$, the relaxation modes of ranks i and $i+1$ of an indefinite recursive parallel arrangement verify a *recursivity relation* expressed by the general formula:

$$y_{i+1}(t)=py_i(qt), \qquad [6.376]$$

where p and q, respectively, denote:

– the *recursive (or homothety) factor along the amplitude*;

– the *recursive (or homothety) factor along the time*;

p and q being, according to the cell nature, particularized in conformity with:

$$p=\eta \text{ and } q=\lambda \qquad [6.377]$$

for series RL cells;

$$p=\alpha\lambda \text{ and } q=\lambda \qquad [6.378]$$

for series RC cells;

$$p=\alpha^2 \text{ and } q=\lambda \qquad [6.379]$$

for series RLC cells.

6.4.4.3. *On the operational approach of non-integer differentiation or integration*

The admittance of an indefinite recursive parallel arrangement is given by:

$$Y(s)=\sum_{i=-\infty}^{+\infty}\frac{\eta^i}{L}\frac{1}{s+\lambda^i\omega_0}\text{, with }\omega_0=\frac{R}{L}, \qquad [6.380]$$

for series RL cells;

$$Y(s)=\sum_{i=-\infty}^{+\infty}\frac{\alpha^i}{R}\frac{s}{s+\lambda^i\omega_0}\text{, with }\omega_0=\frac{1}{RC}, \qquad [6.381]$$

for series RC cells;

$$Y(s)=\sum_{i=-\infty}^{+\infty}\frac{C}{\left(\eta^2\right)^i}\frac{s}{1+2\zeta_0\dfrac{s}{\lambda^i\omega_0}+\left(\dfrac{s}{\lambda^i\omega_0}\right)^2}\text{, with }\omega_0=\frac{1}{\sqrt{LC}}, \qquad [6.382]$$

for series RLC cells.

The admittance $Y(s)$ verifies a general relation of the form:

$$Y(s)=\mu Y\left(\lambda^{-1}s\right), \qquad [6.383]$$

μ being particularized by the cell type in conformity with:

$\mu=\alpha^{-1}$ for series RL cells; [6.384]

$\mu=\alpha$ for series RC cells; [6.385]

$\mu=\alpha/\eta$ for series RLC cells. [6.386]

Attributing to the admittance $Y(s)$ a general expression

$$Y(s)=s^nX(s), \qquad [6.387]$$

and then by putting this expression into the general relation that $Y(s)$ verifies, namely

$$Y(s)=\mu Y\left(\lambda^{-1}s\right), \qquad [6.388]$$

we obtain the general relation that $X(s)$ verifies, namely

$$X(s)=\mu\lambda^{-n}X\left(\lambda^{-1}s\right), \qquad [6.389]$$

a relation that is reduced to

$$X(s) = X\left(\lambda^{-1}s\right), \qquad [6.390]$$

or

$$X(j\omega) = X\left(\lambda^{-1}j\omega\right), \qquad [6.391]$$

by choosing n such that

$$\mu\lambda^{-n} = 1, \qquad [6.392]$$

namely

$$n = \frac{\log \mu}{\log \lambda}, \qquad [6.393]$$

the non-integer order n being particularized by the cell type in conformity with:

$$n = -\frac{\log \alpha}{\log \lambda} \text{ for series RL cells } (\mu = \alpha^{-1}); \qquad [6.394]$$

$$n = \frac{\log \alpha}{\log \lambda} \text{ for series RC cells } (\mu = \alpha); \qquad [6.395]$$

$$n = \frac{\log \alpha - \log \eta}{\log \lambda} \text{ for series RLC cells } (\mu = \alpha / \eta). \qquad [6.396]$$

For series RL cells, *the order* n *is negative*, which expresses that *the arrangement of such cells approximates a non-integer integrator (of order* $-n$ *).*

For series RC cells, *the order* n *is positive*, which expresses that *the arrangement of such cells approximates a non-integer differentiator (of order* n *).*

For series RLC cells, *the order* n *is positive or negative* whether α is greater or less than η (or whether α / η is greater or less than 1), thus expressing that *the arrangement of such* cells *approximates both a non-integer differentiator (of order* n*) and a non-integer integrator (of order* $-n$*).*

The general relation that governs the frequency function $X(j\omega)$ taking into account the gain and phase undulations (around the smoothing straight lines), namely

$$X(j\omega) = X\left(\lambda^{-1} j\omega\right), \qquad [6.397]$$

expresses the periodicity of $X(j\omega)$ in a semi-logarithmic representation where the frequencies are carried on a logarithmic scale (scale in $\log\omega$), the period then being $\log\lambda = \log(\alpha\eta)$.

Furthermore, an annex calculation founded on a Fourier series decomposition shows that $X(j\omega)$ tends toward a constant K when the period $\log\lambda$ tends toward zero, λ then tending toward unit.

Thus, in a structuration of $X(j\omega)$ in conformity with

$$X(j\omega) = KZ(j\omega), \qquad [6.398]$$

where $Z(j\omega)$ tends toward 1 when λ tends toward 1, $Z(j\omega)$ can be interpreted as a *(multiplicative) corrective term* that affects the ideal frequency response, $K(j\omega)^n$, which is no other than the frequency reponse arising from the smoothing as defined in Chapter 1.

REMARK.–

In the synthesis of non-integer differentiation such as led in the CRONE approach, the well-known periodicity of the gain and phase undulations around the smoothing straight lines, can here be used (as a piece of knowledge) to determine the general expression of n without having to choose the condition $\mu\lambda^{-n} = 1$. Indeed, the periodicity of $X(j\omega)$ that is expressed by $X(j\omega) = X\left(\lambda^{-1} j\omega\right)$ straightaway dictates this condition, in which case the resulting non-integer order n is no longer consecutive to a choice, but to a true proof.

6.4.4.4. *On the time approach of non-integer differentiation or integration*

For $t \geq 0$ (or strictly $t > 0$ for series RC cells), the impulse response of an indefinite recursive parallel arrangement fulfills a general relation of the form:

$$y(t) = \mu' y(\lambda t), \qquad [6.399]$$

μ' being particularized by the cell type in conformity with:

$$\mu' = \eta \text{ for series RL cells;} \quad [6.400]$$

$$\mu' = \alpha\eta \text{ for series RC cells;} \quad [6.401]$$

$$\mu' = \alpha^2 \text{ for series RLC cells.} \quad [6.402]$$

Attributing to the impulse response $y(t)$ a general expression of the form

$$y(t) = t^{\gamma} x(t), \quad [6.403]$$

then putting this expression into the general relation that $y(t)$ verifies, namely

$$y(t) = \mu' y(\lambda t), \quad [6.404]$$

one obtains the general relation that $x(t)$ verifies, namely

$$x(t) = \mu' \lambda^{\gamma} x(\lambda t), \quad [6.405]$$

a relation that is reduced to

$$x(t) = x(\lambda t), \quad [6.406]$$

by choosing γ such that

$$\mu' \lambda^{\gamma} = 1, \quad [6.407]$$

namely

$$\gamma = -\frac{\log \mu'}{\log \lambda}, \quad [6.408]$$

the non-integer power γ being particularized by the cell type in conformity with:

$$\gamma = -\frac{\log \eta}{\log \lambda} = -n - 1 \text{ for series RL cells } (\mu' = \eta); \quad [6.409]$$

$$\gamma = -\frac{\log(\alpha\lambda)}{\log \lambda} = -n - 1 \text{ for series RC cells } (\mu' = \alpha\lambda); \quad [6.410]$$

$$\gamma = -\frac{\log(\alpha^2)}{\log \lambda} = -n-1 \text{ for series RLC cells } (\mu' = \alpha^2). \qquad [6.411]$$

For $t > 0$, the impulse response approximately varies in t^{-n-1} for the three studied arrangements, of ideal transmittance Ks^n, with n positive or negative

– according to the (series RC or RL) cell type

– or according to the ratio α / η (greater than or less than one) for series RLC cells.

The general relation that governs the time function $x(t)$ taking into account the gain and phase undulations (around the smoothing straight lines), namely

$$x(t) = x(\lambda t), \qquad [6.412]$$

expresses the periodicity of $x(t)$ in a semi-logarithmic representation where the instants are carried on a logarithmic scale (scale in $\log t$), the period then being $\log \lambda = \log(\alpha\eta)$.

Furthermore, an annex calculation shows that $x(t)$ tends toward a constant k when the period $\log \lambda$ tends toward zero, λ then tending toward unit.

Thus, in a structuration of $x(t)$ in conformity with

$$x(t) = kz(t), \qquad [6.413]$$

where $z(t)$ tends toward 1 when λ tends toward 1, $z(t)$ can be interpreted as a *(multiplicative) corrective term* that affects the ideal impulse response, kt^γ, which is no other than the impulse response arising from the smoothing as defined in Chapter 1.

Finally, by exploiting the *time periodicity* discussed in this section as the *frequency periodicity* discussed in the previous section:

– we could finish this section with a remark analogous to that of the previous section;

– except that the time periodicity that affects the impulse response is less known, even if it is predictable as a consequence of the frequency periodicity that affects the admittance through which the non-integer differentiation synthesis is carried out.

6.5. A common presentation of results turning on eight RC and RL cell recursive arrangements

6.5.1. *Systemic recursivity*

The (indefinite or not) recursive parallel arrangement of series RC cells, which is widely studied in this book, constitutes the recursive system example *par excellence* liable to illustrate and materialize the notion of *systemic recursivity*. But in conformity with what follows, there exist of course other recursive system examples.

By still considering (among the passive components) resistances and capacitors, the entirety of the possible elementary combinations of a resistance and a capacitance determines four different RC cells (Figure 6.13): the series RC cell, the parallel RC cell, the gamma RC cell and the gamma CR cell.

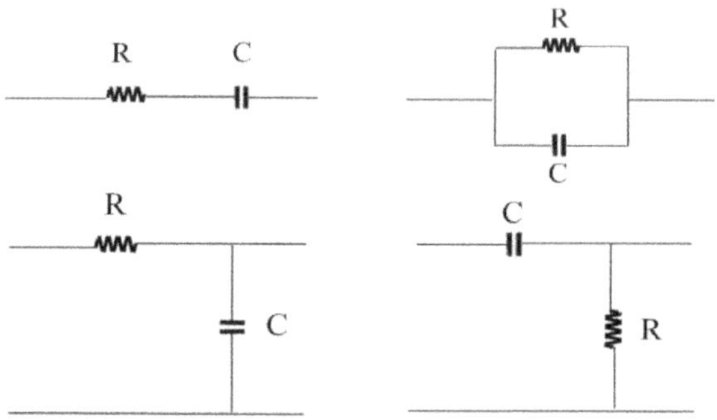

Figure 6.13. *Four possible combinations of a resistance and a capacitance*

Each type of RC cell dictating the nature of the arrangement that suits, the four cells so defined induce four appropriate arrangements (Figure 6.14):

– a recursive parallel arrangement of series RC cells;

– a recursive series arrangement of parallel RC cells;

– a recursive cascade arrangement of gamma RC cells;

– a recursive cascade arrangement of gamma CR cells.

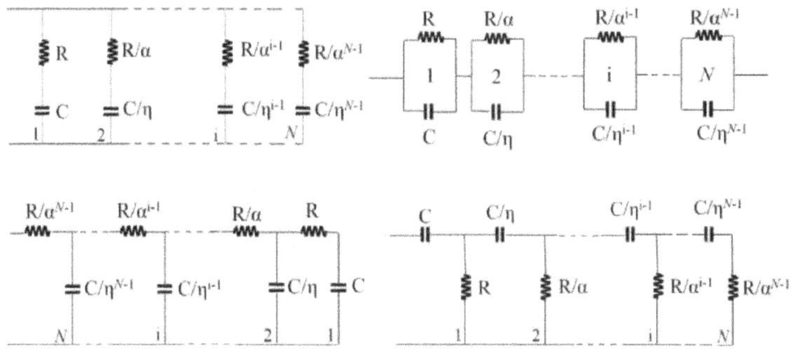

Figure 6.14. *Four possible recursive arrangements of RC cells*

By now considering coils instead of capacitors, the four RL cells that result from the possible elementary combinations of a resistance and an inductance lead to four appropriate arrangements (Figure 6.15):

– a recursive parallel arrangement of series RL cells;

– a recursive series arrangement of parallel RL cells;

– a recursive cascade arrangement of gamma RL cells;

– a recursive cascade arrangement of gamma LR cells.

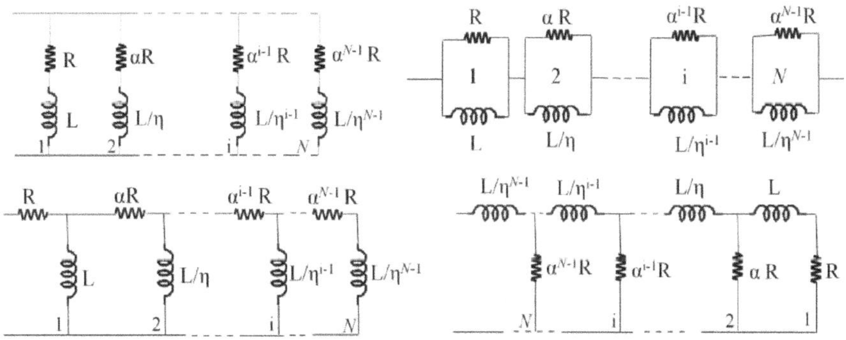

Figure 6.15. *Four possible recursive arrangements of RL cells*

6.5.2. *Frequency recursivity*

The aim of this section is to specify the distribution of the transitional frequencies ω_i' and ω_i, respectively, corresponding to the zeros z_i and to the poles

p_i of the admittance or impedance of the recursive arrangements as defined in the previous section.

At medium frequencies, which here are defined as frequencies that are:

– high compared to the transitional frequency of the cell of rank 1, $1/RC$ or R/L;

– and low compared to the transitional frequency of the cell of rank N, $(\alpha\eta)^{N-1} RC$ or $(\alpha\eta)^{N-1} R/L$;

the transitional frequencies of the admittance $Y(\mathrm{j}\omega)$ of the RC cell arrangements and of the impedance $Z(\mathrm{j}\omega)$ of the RL cell arrangements are *recursively* distributed according to the ratios:

$$\frac{\omega'_{i+1}}{\omega'_i} = \frac{\omega_{i+1}}{\omega_i} = \alpha\eta\ , \ \frac{\omega_i}{\omega'_i} = \alpha \text{ and } \frac{\omega'_{i+1}}{\omega_i} = \eta\ , \qquad [6.414]$$

which express the notion of *frequency recursivity*, the *recursive factors* α and η being defined by:

$$\frac{R_i}{R_{i+1}} = \alpha > 1 \text{ and } \frac{C_i}{C_{i+1}} = \eta \qquad [6.415]$$

for the RC cell arrangements, and

$$\frac{R_{i+1}}{R_i} = \alpha > 1 \text{ and } \frac{L_i}{L_{i+1}} = \eta > 1 \qquad [6.416]$$

for the RL cell arrangements [OUS 81c].

This frequency distribution, which is common to the eight recursive arrangements, clearly expresses the same behavior (or a *common behavior*) of these arrangements, on the condition of considering the admittance $Y(\mathrm{j}\omega)$ for the RC cell arrangements and the impedance $Z(\mathrm{j}\omega)$ for the RL cell arrangements [OUS 83].

The gain and phase smoothing straight lines that smooth the Bode asymptotic diagrams resulting from the frequency distribution so specified determine a differentiation non-integer order

$$n = \frac{\log(\alpha)}{\log(\alpha\eta)} . \qquad [6.417]$$

This differentiation order for the admittance $Y(\mathrm{j}\omega)$ of the RC cell arrangements, which is also a differentiation order for the impedance $Z(\mathrm{j}\omega)=1/Y(\mathrm{j}\omega)$ of the RL cell arrangements, is actually *an integration order for the admittance* $Y(\mathrm{j}\omega)$ *of the RL cell arrangements*. It is true that the RC cell arrangements approximate an *order n non-integer differentiator*, whereas the RL cell arrangements approximate an *order n non-integer integrator*; this assertion has, of course, to be inverted in impedance terms.

6.5.3. *Same recursivity on the components*

From the section title, it is expressed that our interest turns again to systemic recursivity.

When the distributions of resistances and capacitances or inductances present the same recursivity, two cases should be distinguished:

1) the case when the recursive factors are identical and superior to unit, namely

$$\alpha=\eta>1; \qquad [6.418]$$

2) the case when the recursive factors are identical and equal to unit, namely

$$\alpha=\eta=1, \qquad [6.419]$$

the components then being the same.

6.5.3.1. *Recursive factors identical and superior to unit*

When $\alpha=\eta>1$, the eight RC or RL cell recursive arrangements are still characterized by the differentiation non-integer order

$$n=\frac{\log(\alpha)}{\log(\alpha\eta)}, \qquad [6.420]$$

which is then reduced to

$$n=\frac{\log\alpha}{\log(\alpha^2)}=\frac{1}{2}, \qquad [6.421]$$

thus expressing a *differentiation half-integer order* for all of the eight arrangements (by still taking $Y(\mathrm{j}\omega)$ for the RC cell arrangements and $Z(\mathrm{j}\omega)$ for the RL cell arrangements).

6.5.3.2. *Recursive factors identical and equal to 1: same components*

When $\alpha = \eta = 1$, the resistances and capacitances or inductances then being the same:

– the two series RC or RL cell parallel arrangements come down to *setting identical series cells in parallel*, the structure of the admittance or impedance of these arrangements then being reduced to that of a cell (*therefore to the structure of a first order*);

– the two parallel RC or RL cell series arrangements come down to *setting identical parallel cells in series*, the structure of the admittance or impedance of these arrangements then being reduced to that of a cell (*therefore to the structure of a first order*).

For these four arrangements, the differentiation non-integer order, $n = \log\alpha / \log(\alpha\eta)$, therefore has no more meaning.

On the contrary, for the four gamma RC or RL cell cascade arrangements, the structure of the admittance or impedance is not reduced to that of a cell (even if all the cells are identical):

– the differentiation non-integer order, $n = \log\alpha / \log(\alpha\eta)$, indeed remains valid;

– even if its particularization in conformity with $\alpha = 1$ and $\eta = 1$ leads to an indetermination such as

$$n = \frac{\log 1}{\log 1} = \frac{0}{0} \qquad [6.422]$$

– this indetermination being immediately removable by making $\alpha = \eta$, namely

$$n = \frac{\log \alpha}{\log(\alpha^2)}, \qquad [6.423]$$

then by tending α toward 1, namely

$$n = \lim_{\alpha \to 1} \frac{\log(\alpha)}{\log(\alpha^2)} = \lim_{\alpha \to 1} \frac{\log(\alpha)}{2\log(\alpha)} = \frac{1}{2}, \qquad [6.424]$$

thus expressing a *differentiation half-integer order* (by still bringing the same care to the choice of $Y(\mathrm{j}\omega)$ or $Z(\mathrm{j}\omega)$).

This last result can be verified by a (more physical) approach from a cascade arrangement example, notably the one of an *identical gamma RdxCdx cell cascade arrangement* (Figure 6.16) which results from the cutting of a semi-infinite thermal rod into infinitesimal slices of thickness dx (where *R* and *C* represent the rod thermal resistance and capacitance per unit length).

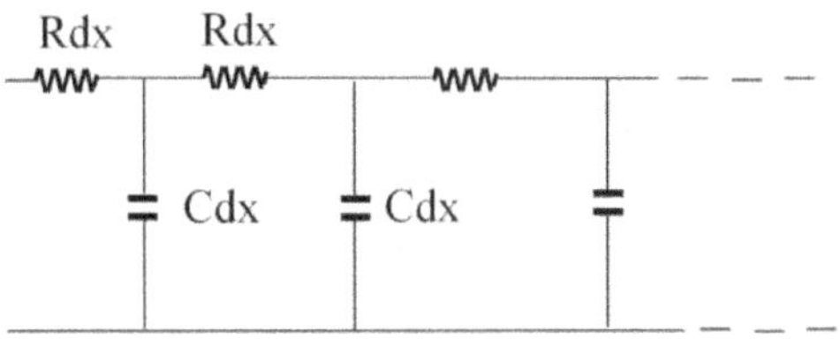

Figure 6.16. *Thermal representation of a semi-infinite rod cut into infinitesimal slices*

The approach consists of writing that the rod (thermal) impedance is the same up to a cell that vanishes when dx tends toward zero, namely:

$$Z(\mathrm{j}\omega) = R\,\mathrm{dx} + \frac{\dfrac{1}{\mathrm{j}C\mathrm{dx}\omega} Z(\mathrm{j}\omega)}{\dfrac{1}{\mathrm{j}C\mathrm{dx}\omega} + Z(\mathrm{j}\omega)}, \qquad [6.425]$$

or:

$$Z(\mathrm{j}\omega) = R\,\mathrm{dx} + \frac{Z(\mathrm{j}\omega)}{1 + \mathrm{j}Z(\mathrm{j}\omega)C\mathrm{dx}\omega}, \qquad [6.426]$$

or even:

$$Z(\mathrm{j}\omega)\left[1 - \frac{1}{1 + \mathrm{j}Z(\mathrm{j}\omega)C\mathrm{dx}\omega}\right] = R\,\mathrm{dx}, \qquad [6.427]$$

or else:

$$Z(\mathrm{j}\omega)\frac{\mathrm{j}Z(\mathrm{j}\omega)\mathrm{dx}\omega}{1 + \mathrm{j}Z(\mathrm{j}\omega)C\mathrm{dx}\omega} = R\,\mathrm{dx}, \qquad [6.428]$$

from which we draw, for dx tending toward zero $(Z(\mathrm{j}\omega)C\mathrm{dx}\omega << 1)$, the following:

$$Z(\mathrm{j}\omega) = \left(\frac{R}{\mathrm{j}C\omega}\right)^{\frac{1}{2}}, \qquad [6.429]$$

an impedance that determines an admittance of the form:

$$Y(\mathrm{j}\omega)=\left(\frac{C}{R}\mathrm{j}\omega\right)^{\frac{1}{2}}. \text{ QED} \qquad [6.430]$$

6.6. On unit gain frequency in non-integer differentiation or integration

For λ very close to 1, let us again take the expressions of the admittance and the impulse response of an indefinite recursive parallel arrangement of series RC cells, namely (relations [6.76] and [6.101]):

$$Y(s)=\frac{C(RC)^{n-1}}{\lambda-1}\frac{\pi}{\sin(\pi n)}s^n \qquad [6.431]$$

and

$$y(t)=\frac{C(RC)^{n-1}}{\lambda-1}\frac{\pi}{\sin(\pi n)}\frac{t^{-n-1}}{\Gamma(-n)},\ t>0, \qquad [6.432]$$

with $\lambda=\alpha\eta>1$ and $n=\dfrac{\log\alpha}{\log\lambda}$, $n\in\,]0,1[$.

For medium frequencies and times, $Y(s)$ and $y(t)$, respectively, tend toward infinity when λ tends toward 1. This is due to the impedance $Z(s)=1/Y(s)$ which tends toward 0 when λ tends toward 1, knowing that the increase of the parallel cell number when λ tends toward 1 is not accompanied by an increase of the impedance of each of the cells.

To answer this phenomenon, it is convenient to again take the impedance $Z_i(s)$ of the cell of rank i, namely

$$R_i+\frac{1}{C_i s}=\frac{R}{\alpha^i}+\frac{\eta^i}{Cs}, \qquad [6.433]$$

and then replace it with the new impedance (which tends toward infinity when λ tends toward 1):

$$\frac{1}{\lambda-1}\left(\frac{R}{\alpha^i}+\frac{\eta^i}{Cs}\right)=\frac{1}{\alpha^i}\frac{R}{\lambda-1}+\frac{\eta^i}{(\lambda-1)Cs}, \qquad [6.434]$$

which amounts to replacing R by $R/(\lambda-1)$ and C by $(\lambda-1)C$, $Y(s)$ and $y(t)$ then becoming:

$$Y(s)=C\left(RC\right)^{n-1}\frac{\pi}{\sin\left(\pi n\right)}s^{n} \quad [6.435]$$

and

$$y(t)=C\left(RC\right)^{n-1}\frac{\pi}{\sin\left(\pi n\right)}\frac{t^{-n-1}}{\Gamma\left(-n\right)}, \quad [6.436]$$

expressions that show well the disappearance of the term $\lambda-1$ from the denominator.

The admittance $Y(s)$ can be rewritten as:

$$Y(s)=\left(\left(C\left(RC\right)^{n-1}\frac{\pi}{\sin\left(\pi n\right)}\right)^{\frac{1}{n}}s\right)^{n}, \quad [6.437]$$

namely:

$$Y(s)=\left(\frac{s}{\omega_u}\right)^{n}, \quad [6.438]$$

by putting

$$\omega_u=\left(\frac{1}{C\left(RC\right)^{n-1}}\frac{\sin\left(\pi n\right)}{\pi}\right)^{\frac{1}{n}}, \quad [6.439]$$

as *unit gain frequency* (or *transition frequency*) of the non-integer order n differentiator that the studied arrangement achieves.

Similarly, for an indefinite recursive parallel arrangement of series RL cells, it is convenient to replace the impedance $Z_i(s)$ of the cell of rank i, namely

$$R_i+L_i s=\alpha^{i}R+\frac{L}{\eta^{i}}s, \quad [6.440]$$

by the new impedance (which tends toward infinity when λ tends toward 1):

$$\frac{1}{\lambda-1}\left(\alpha^i R+\frac{L}{\eta^i}s\right)=\alpha^i\frac{R}{\lambda-1}+\frac{1}{\eta^i}\frac{L}{\lambda-1}s, \qquad [6.441]$$

which amounts to replacing R by $R/(\lambda-1)$ and L by $L/(\lambda-1)$. The expressions of $Y(s)$ and $y(t)$ for R and L, namely

$$Y(s)=\frac{(L/R)^n}{R(\lambda-1)}\frac{\pi}{\sin((1+n)\pi)}s^n,\ n=-\frac{\log\alpha}{\log\lambda}\in\left]-1,0\right[, \qquad [6.442]$$

and

$$y(t)=\frac{(L/R)^n}{R(\lambda-1)}\frac{\pi}{\sin((1+n)\pi)}\frac{t^{-n-1}}{\Gamma(-n)},\ t\geq 0, \qquad [6.443]$$

indeed become for $R/(\lambda-1)$ and $L/(\lambda-1)$:

$$Y(s)=\frac{(L/R)^n}{R}\frac{\pi}{\sin((1+n)\pi)}s^n \qquad [6.444]$$

and

$$y(t)=\frac{(L/R)^n}{R}\frac{\pi}{\sin((1+n)\pi)}\frac{t^{-n-1}}{\Gamma(-n)}, \qquad [6.445]$$

the term $\lambda-1$ having disappeared from the denominator.

The admittance $Y(s)$ can be rewritten as:

$$Y(s)=\left(\left(\frac{(L/R)^n}{R}\frac{\pi}{\sin((1+n)\pi)}\right)^{\frac{1}{n}}s\right)^n, \qquad [6.447]$$

namely:

$$Y(s)=\left(\frac{s}{\omega_u}\right)^n=\left(\frac{\omega_u}{s}\right)^{-n}, \qquad [6.448]$$

by putting

$$\omega_u = \left(\frac{R}{(L/R)^n} \frac{\sin((1+n)\pi)}{\pi} \right)^{\frac{1}{n}}, \qquad [6.449]$$

as *unit gain frequency* (or *transition frequency*) of the non-integer order $-n$ integrator that the studied arrangement achieves.

6.7. On stored energy in non-integer differentiation or integration

The arrangements being seen under the admittance angle, we now know that if λ tends toward 1, an indefinite recursive arrangement of RC cells truly achieves a non-integer differentiator, as an indefinite recursive arrangement of RL cells truly achieves a non-integer integrator.

So, in terms of energy stored by capacitors or coils, we will successively study an indefinite recursive parallel arrangement of series RC cells for non-integer differentiation and an indefinite recursive series arrangement of parallel RL cells for non-integer integration. This choice is not mere coincidence. It in fact results from the *duality* of energetic approaches that these arrangements ensure.

6.7.1. *Non-integer differentiator*

The energy stored by all the capacitances of the indefinite recursive parallel arrangement of series RC cells, submitted to a direct (or constant) voltage U, is expressed by an indefinite sum of the form:

$$W = \frac{1}{2}(\lambda - 1) C U^2 \sum_{i=-\infty}^{+\infty} \frac{1}{\eta^i}, \qquad [6.450]$$

an expression that can be rewritten as:

$$W = \lim_{N \to \infty} W_N, \qquad [6.451]$$

with

$$W_N = \frac{1}{2}(\lambda - 1) C U^2 \sum_{i=-N}^{N} \frac{1}{\eta^i}. \qquad [6.452]$$

The sum from $i=-N$ to $i=N$ can be decomposed according to two sums:

$$\begin{aligned}\sum_{i=-N}^{N}\frac{1}{\eta^i} &= \sum_{i=0}^{N}\frac{1}{\eta^i}+\sum_{i=-N}^{-1}\frac{1}{\eta^i}\\ &= \sum_{i=0}^{N}\left(\frac{1}{\eta}\right)^i+\sum_{i=-N}^{-1}\eta^{-i}\\ &= \sum_{i=0}^{N}\left(\frac{1}{\eta}\right)^i+\sum_{i=1}^{N}\eta^{i}\\ &= \sum_{i=0}^{N}\left(\frac{1}{\eta}\right)^i+\eta\sum_{i=0}^{N-1}\eta^{i},\end{aligned} \quad [6.453]$$

or, given the general formula

$$\sum_{i=0}^{m}q^i=\frac{1-q^{m+1}}{1-q}: \quad [6.454]$$

$$\begin{aligned}\sum_{i=-N}^{N}\frac{1}{\eta^i} &= \frac{1-\dfrac{1}{\eta^{N+1}}}{1-\dfrac{1}{\eta}}+\eta\frac{1-\eta^N}{1-\eta}\\ &= \frac{\eta-\dfrac{1}{\eta^N}}{\eta-1}+\frac{\eta^{N+1}-\eta}{\eta-1}\\ &= \frac{\eta^{N+1}-\dfrac{1}{\eta^N}}{\eta-1}.\end{aligned} \quad [6.455]$$

W_N then becomes:

$$W_N=\frac{1}{2}CU^2\frac{\lambda-1}{\eta-1}\left(\eta^{N+1}-\frac{1}{\eta^N}\right), \quad [6.456]$$

or, knowing that

$$n=\frac{\log\alpha}{\log\lambda} \text{ leads to } \alpha=\lambda^n \quad [6.457]$$

and that

$$\eta = \frac{\lambda}{\alpha} = \frac{\lambda}{\lambda^n} = \lambda^{1-n}: \qquad [6.458]$$

$$W_N = \frac{1}{2} C U^2 \frac{\lambda - 1}{\lambda^{1-n} - 1} \left(\eta^{N+1} - \frac{1}{\eta^N} \right). \qquad [6.459]$$

Concerning the ratio

$$\frac{\lambda - 1}{\lambda^{1-n} - 1}, \qquad [6.460]$$

the study of its limit when λ tends toward 1 leads to an indetermination of the form $0/0$ that can be removed by computing the limit when $\lambda \to 1$ of the ratio of the numerator and denominator derivatives with respect to λ, namely

$$\lim_{\lambda \to 1} \frac{\lambda - 1}{\lambda^{1-n} - 1} = \lim_{\lambda \to 1} \frac{1}{(1-n)\lambda} = \frac{1}{1-n}. \qquad [6.461]$$

W_N thus becomes:

$$W_N = \frac{1}{2} C U^2 \frac{1}{1-n} \left(\eta^{N+1} - \frac{1}{\eta^N} \right). \qquad [6.462]$$

Finally, as for the energy W which is given by the limit of W_N when $N \to \infty$, the limits of η^{N+1} and $1/\eta^N$, namely

$$\lim_{N \to \infty} \eta^{N+1} = \infty \text{ and } \lim_{N \to \infty} \frac{1}{\eta^N} = 0, \qquad [6.463]$$

enable us to write:

$$W = \lim_{N \to \infty} W_N = \infty, \qquad [6.464]$$

a result that expresses an *infinite energy* stored by the whole of the capacitances that actually constitutes an *infinite capacity*.

But this capacitance is not that of a genuine capacitor. It is indeed indissociable from the arrangement resistances that condition the energy storage and release phases. The stored energy release can be studied during the arrangement *relaxation* that $U=0$ defines, which amounts to shunting the arrangement with a short-circuit. The short-circuit current is then the sum of the discharge currents of each of the capacitances. Each discharge current is easily computable since, for each RC cell, the resistance is both in series and in parallel with the capacitance (through a same current and a same voltage). The energy released by a capacitance is then dissipated by the corresponding resistance. The relaxation thus puts at stake infinitely fast discharges due to infinitely small time constants and infinitely slow discharges due to infinitely high time constants that are besides at the origin of the *long memory* (in this case *infinite*) which is well known on the subject.

6.7.2. *Non-integer integrator*

An *indefinite recursive series arrangement of parallel RL cells* is such that (for recursive factors α and η greater than 1):

– the cell of rank 0 is the *median* (or *central*) *cell* defined by a resistance R in parallel with an inductance L;

– the cell of rank 1 is defined by a resistance αR in parallel with an inductance L/η;

– the cell of rank –1 is defined by a resistance $\alpha^{-1}R$ in parallel with an inductance L/η^{-1};

– the cell of rank i is defined by a resistance $R_i=\alpha^i R$ in parallel with an inductance $L_i=L/\eta^i$, $i\in\mathbb{Z}$.

Not yet established, the impedance $Z(s)$ of such an arrangement can be very simply determined from the admittance $Y(s)$ of the indefinite recursive parallel arrangement of series RC cells. Indeed, the impedance $Z_i(s)$ of the rank i parallel RL cell of the first arrangement, namely

$$Z_i(s)=\frac{R_iL_is}{R_i+L_is}=\frac{\alpha^i Ls}{\dfrac{L}{R}s+\lambda^i}, \qquad [6.465]$$

presents the same structure as the admittance $Y_i(s)$ of the rank i series RC cell of the second arrangement, namely (relation [6.2])

$$Y_i(s)=\frac{\alpha^i Cs}{RCs+\lambda^i}. \qquad [6.466]$$

The comparison of the parameters of $Z_i(s)$ and $Y_i(s)$ shows that, by replacing C by L and RC by L/R in the admittance expression $Y(s)$ given by [6.76], we can directly obtain the impedance expression $Z(s)$ of the first arrangement, namely:

$$Z(s) = \frac{L\left(L/R\right)^{n-1}}{\lambda - 1} \frac{\pi}{\sin(\pi n)} s^n \text{ with } n = \frac{\log \alpha}{\log \lambda} \in \left]0,1\right[. \quad [6.467]$$

In order to get rid of the term $\lambda - 1$ from the denominator of $Z(s)$, it is convenient to express the admittance $Y_i(s)$ of the rank i parallel RL cell, namely

$$\frac{1}{R_i} + \frac{1}{L_i s} = \frac{1}{\alpha^i R} + \frac{\eta^i}{Ls}, \quad [6.468]$$

and then replace it with the new admittance (which tends toward infinity when λ tends toward 1):

$$\frac{1}{\lambda - 1}\left(\frac{1}{\alpha^i R} + \frac{\eta^i}{Ls}\right) = \frac{1}{\alpha^i (\lambda - 1) R} + \frac{\eta^i}{(\lambda - 1) Ls}, \quad [6.469]$$

which amounts to replacing R by $(\lambda - 1)R$ and L by $(\lambda - 1)L$, $Z(s)$ then becoming:

$$Z(s) = L\left(L/R\right)^{n-1} \frac{\pi}{\sin(\pi n)} s^n . \quad [6.470]$$

The energy stored by all the inductances of the indefinite recursive series arrangement of parallel RL cells, submitted to a direct (or constant) current I, is expressed by an indefinite sum of the form:

$$W = \frac{1}{2}(\lambda - 1) L I^2 \sum_{i=-\infty}^{+\infty} \frac{1}{\eta^i}, \quad [6.471]$$

an expression that can be rewritten as:

$$W = \lim_{N \to \infty} W_N , \quad [6.472]$$

with:

$$W_N = \frac{1}{2}(\lambda - 1) L I^2 \sum_{i=-N}^{N} \frac{1}{\eta^i}, \qquad [6.473]$$

or, given the calculation developed in section 6.7.1 and in which it suffices to replace CU^2 by LI^2:

$$W_N = \frac{1}{2} L I^2 \frac{1}{1-n}\left(\eta^{N+1} - \frac{1}{\eta^N}\right), \qquad [6.474]$$

from which we draw:

$$W = \lim_{N\to\infty} W_N = \infty, \qquad [6.475]$$

a result that expresses an *infinite energy* stored by the whole of the inductances that actually constitutes an *infinite inductance*.

But this inductance is not that of a genuine coil, being indeed indissociable from the arrangement resistances that condition the energy storage and release phases. The stored energy release can be studied during the arrangement *relaxation* that $I = 0$ defines, which amounts to opening the circuit. The voltage between the arrangement terminals is then the sum of the discharge voltages of each of the inductances. Each discharge voltage is easily computable since, for each RL cell, the resistance is both in series and in parallel with the inductance (through a same current and a same voltage). The energy released by an inductance is then dissipated by the corresponding resistance. The relaxation thus puts at stake infinitely fast discharges due to infinitely small time constants and infinitely slow discharges due to infinitely high time constants, the latter being the cause of the *infinite memory* phenomenon.

6.8. Bibliography

[ION 09] IONESCU C., OUSTALOUP A., LEVRON F., *et al.*, "A model of the lungs based on fractal geometrical and structural properties", *15th IFAC Symposium on System Identification (SYSID '09)*, Saint Malo, 2009.

[ION 09] IONESCU C., SEGERS P., DE KEYSER R., "Mechanical properties of the respiratory system derived from morphologic insight", *IEEE Transactions on Biomedical Engineering*, vol. 56, no. 4, pp. 949–959, 2009.

[ION 10] IONESCU C., DEROM E., DE KEYSER R., "Assessment of respiratory mechanical properties with constant-phase models in healthy and COPD lungs", *Computer Methods and Programs in Biomedicine*, vol. 97, no. 1, pp. 78–85, 2010.

[ION 11] IONESCU C., DE KEYSER R., SABATIER J., *et al.*, "Low frequency constant-phase behavior in the respiratory impedance", *Biomedical Signal Processing and Control*, vol. 6, no. 2, pp. 197–208, April 2011.

[ION 13] IONESCU C., *The Human Respiratory System: An Analysis of the Interplay between Anatomy, Structure, Breathing and Fractal Dynamics*, Series in BioEngineering, Springer, 2013.

[OUS 81a] OUSTALOUP A., "Fractional order sinusoidal oscillators: optimization and their use in highly linear FM modulation", *IEEE Transactions on Circuits and Systems*, vol. 28, no. 10, pp. 1007–1009, 1981.

[OUS 81b] OUSTALOUP A., "Linear feedback control systems of fractional order between 1 and 2", *IEEE International Symposium on Circuits and Systems*, Chicago, IL, 27–29 April 1981.

[OUS 81c] OUSTALOUP A., Systèmes asservis linéaires d'ordre fractionnaire, PhD Thesis, University of Bordeaux 1, 1981.

[OUS 82] OUSTALOUP A., PISTRE J.D., MORA A., "Fractional order systems with localized and distributed parameters", *IASTED Conference*, Daros, Switzerland, 2–5 March 1982.

[OUS 83] OUSTALOUP A., *Systèmes asservis linéaires d'ordre fractionnaire*, Masson, Paris, 1983.

[OUS 91] OUSTALOUP A., *La commande CRONE*, Hermès, Paris, 1991.

[OUS 95] OUSTALOUP A., *La dérivation non entière: théorie, synthèse et applications*, Hermès, Paris, 1995.

[SAB 98] SABATIER J., La dérivation non entière en modélisation des systèmes à paramètres distribués récursifs et en commande robuste des procédés non stationnaires, PhD Thesis, University of Bordeaux 1, 1998.

Appendix 1

Damping of a Usual Automotive Suspension

A1.1. The usual automotive suspension

Let a suspension be of type mass-spring-dashpot configured, through its variables, according to two operating states: the *free state* (Figure A1.1(a)) and the *forced state* (Figure A1.1(b), where M represents the bodywork mass supported per wheel, k denotes the spring stiffness and C denotes the viscous friction coefficient of the dashpot.

In the first configuration, through the wheel (on the ground) without vertical motion, the suspension is a *free system* with one variable, i.e. the output variable $y(t)$ that represents the vertical displacement of the bodywork.

In the second configuration, through the wheel in vertical motion, the suspension is a *forced (or controlled) system* with two variables: the input variable $x(t)$ that represents the wheel vertical displacement and the output variable $y(t)$ that represents the bodywork vertical displacement (in response to the one of the wheel).

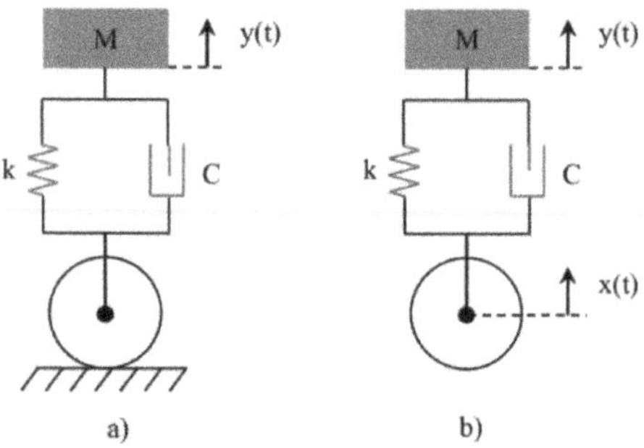

Figure A1.1. *Vehicle quarter usual suspension configured a) as a free system (wheel on the ground) and b) as a forced system (wheel submitted to a vertical motion)*

A1.2. Damping that the damping ratio measures

A1.2.1. *From the suspension as a free system*

Intrinsic to the system and therefore independent of the solicitation, the damping is determined from the free state defined by $x(t)=0$. In this case, the suspension second configuration (Figure A1.1(b)) comes down to the first configuration (Figure A1.1(a)), only the output variable $y(t)$ then being effective.

Thus, applying Newton's second law to the motion mass M enables us to formulate the differential equation:

$$M\frac{d^2y(t)}{dt^2} = -\,ky(t) - C\frac{dy(t)}{dt}\,, \qquad \text{[A1.1]}$$

namely:

$$M\frac{d^2y(t)}{dt^2} + C\frac{dy(t)}{dt} + ky(t) = 0\,, \qquad \text{[A1.2]}$$

a homogenous differential equation that admits for *characteristic equation*:

$$Ms^2 + Cs + k = 0\,, \qquad \text{[A1.3]}$$

or:

$$1 + \frac{C}{k}s + \frac{M}{k}s^2 = 0\,, \qquad \text{[A1.4]}$$

or even, under a canonical form:

$$1+2\zeta\frac{s}{\omega_n}+\frac{s^2}{\omega_n^{\ 2}}=0\,, \tag{A1.5}$$

by putting:

$$2\frac{\zeta}{\omega_n}=\frac{C}{k} \text{ and } \frac{1}{\omega_n^{\ 2}}=\frac{M}{k}, \tag{A1.6}$$

a system with two equations from which we successively draw:

– the *natural frequency without damping*, namely

$$\omega_n=\sqrt{\frac{k}{M}} \tag{A1.7}$$

– and the *damping ratio* (or *reduced damping coefficient*), namely

$$\zeta=\frac{\omega_n}{2}\frac{C}{k}=\frac{1}{2}\sqrt{\frac{k}{M}}\frac{C}{k}=\frac{1}{2}\frac{C}{\sqrt{k}}\frac{1}{\sqrt{M}}, \tag{A1.8}$$

or else

$$\zeta=\frac{K}{\sqrt{M}} \text{ with } K=\frac{C}{2\sqrt{k}}. \tag{A1.9}$$

Furthermore, as for the *natural frequency* (or *pseudo-frequency*), namely

$$\omega_p=\omega_n\sqrt{1-\zeta^2}\,, \tag{A1.10}$$

such that

$$\left[\omega_p\right]_{\zeta=0}=\omega_n \text{ (\textit{natural frequency without damping})}, \tag{A1.11}$$

it is particularized in conformity with:

$$\omega_p=\sqrt{\frac{k}{M}\left(1-\frac{K^2}{M}\right)}. \tag{A1.12}$$

A1.2.2. *From the suspension as a forced system*

From Newton's second law, Figure A1.1(b) enables us to write the differential equation:

$$M\frac{d^2y(t)}{dt^2}=-k\left[y(t)-x(t)\right]-C\left(\frac{d}{dt}\right)\left[y(t)-x(t)\right], \qquad \text{[A1.13]}$$

namely:

$$M\frac{d^2y(t)}{dt^2}+C\frac{dy(t)}{dt}+ky(t)=C\frac{dx(t)}{dt}+kx(t), \qquad \text{[A1.14]}$$

from which we draw the *transfer function* (or *transmittance*):

$$H(s)=\frac{Y(s)}{X(s)}=\frac{k+Cs}{k+Cs+Ms^2}, \qquad \text{[A1.15]}$$

where $Y(s)$ and $X(s)$, respectively, denote the Laplace transforms of $y(t)$ and $x(t)$.

In contrast to the previous case of the free system, the case of the forced system is distinguished by the existence of a differential equation with a second member and, consequently, of a transfer function (in this case $H(s)$ defined by [A1.15]).

Given that the forced system damping is the same as the one of the free system (since intrinsic of the system), it is determined, from the transfer function:

– by equaling the denominator to zero (or *characteristic polynomial*);

– the characteristic polynomial equaled to zero being no other than the *characteristic equation*, namely in this case:

$$Ms^2+Cs+k=0, \qquad \text{[A1.16]}$$

an equation which is well in conformity with [A1.3] and which leads then to identical expressions for ω_n, ζ and ω_p (relations [A1.7], [A1.9] and [A1.12]).

A1.3. On the importance of the transmittance denominator

The transmittance $H(s)$ can be rewritten as:

$$H(s) = \frac{1+(C/k)s}{1+\frac{C}{k}s+\frac{M}{k}s^2}, \quad [A1.17]$$

namely, under the canonical form:

$$H(s) = \frac{1+(C/k)s}{1+2\zeta\frac{s}{\omega_n}+\frac{s^2}{\omega_n^2}}, \quad [A1.18]$$

by putting:

$$2\frac{\zeta}{\omega_n} = \frac{C}{k} \text{ and } \frac{1}{\omega_n^2} = \frac{M}{k}, \quad [A1.19]$$

from which we draw:

$$\omega_n = \sqrt{\frac{k}{M}} \text{ and } \zeta = \frac{K}{\sqrt{M}} \text{ with } K = \frac{C}{2\sqrt{k}}. \quad [A1.20]$$

By rewriting canonical form [A1.18] as

$$H(s) = \omega_n^2 \frac{1+(C/k)s}{s^2+2\zeta\omega_n s+\omega_n^2}, \quad [A1.21]$$

$H(s)$ admits the new canonical form:

$$H(s) = \omega_n^2 \frac{1+(C/k)s}{(s-a)^2+b^2}, \quad [A1.22]$$

by putting:

$$-2a = 2\zeta\omega_n \text{ and } a^2+b^2 = \omega_n^2, \quad [A1.23]$$

from which is drawn:

$$a = -\zeta\omega_n \text{ and } b = \omega_n\sqrt{1-\zeta^2}\,, \qquad [A1.24]$$

or even, knowing that $\zeta\omega_n$ is the *damping coefficient* denoted by α, namely $\zeta\omega_n = \alpha$, and that $\omega_n\sqrt{1-\zeta^2}$ is the *natural frequency* denoted by ω_p, namely $\omega_n\sqrt{1-\zeta^2} = \omega_p$ (see [A1.10]):

$$a = -\alpha \text{ and } b = \omega_p. \qquad [A1.25]$$

From the Laplace transform tables that show that

$$\frac{1}{(s-a)^2+b^2} = L\left[\frac{e^{at}}{b}\sin(bt)\right] \qquad [A1.26]$$

and

$$\frac{s}{(s-a)^2+b^2} = L\left[e^{at}\left(\cos(bt)+\frac{a}{b}\sin(bt)\right)\right], \qquad [A1.27]$$

the transmittance $H(s)$ admits the original function (which is the suspension impulse response):

$$h(t) = \omega_n^{\,2}\frac{e^{at}}{b}\sin(bt) + \omega_n^{\,2}(C/k)e^{at}\left(\cos(bt)+\frac{a}{b}\sin(bt)\right), \qquad [A1.28]$$

namely, by replacing a and b with their respective expressions:

$$h(t) = \frac{\omega_n^{\,2}}{\omega_p}e^{-\alpha t}\sin(\omega_p t) + \omega_n^{\,2}(C/k)e^{-\alpha t}\left(\cos(\omega_p t) - \frac{\alpha}{\omega_p}\sin(\omega_p t)\right). \qquad [A1.29]$$

A1.3.1. *Conclusion*

The form of this impulse response expresses a remarkable result that shows that each term of $h(t)$ is characterized by the same damping ratio (or reduced damping coefficient) $\zeta = \alpha/\omega_n$, the same natural frequency ω_p and the same natural frequency without damping ω_n :

– each of these (dynamic) characteristics being imposed by the denominator of the transmittance $H(s)$;

– and this, independently of the numerator (which remains true for a numerator having any number of terms).

Nevertheless, note that the numerator has an influence on the impulse or step response, notably through the *reduced first overshoot* which is another dynamic characteristic and which indeed depends on the numerator.

A1.4. From denominator zeros to dynamics damping and rapidity

A1.4.1. *On dynamics and its performances*

The dynamics of a linear system is a conceptual translation of the step response transient. It can be characterized by:

– the overshoot of the established state that is measured by the *reduced first overshoot*, D_1', a time dynamic performance defined as the ratio between the first overshoot D_1 and the step response final value $y(\infty)$;

– damping that is measured by the *damping ratio*, ζ, a time dynamic performance defined by the reduced damping coefficient α / ω_n of the step response oscillatory mode (the dominant one if there are several modes);

– rapidity that is measured by the *natural frequency*, ω_p, a time dynamic performance defined by the pseudo-frequency of this mode.

A1.4.2. *Denominator zeros or characteristic equation roots*

The *zeros* of the denominator are the *roots* of the characteristic equation

$$1+2\zeta\frac{s}{\omega_n}+\frac{s^2}{\omega_n^2}=0, \qquad \text{[A1.30]}$$

namely, for $0<\zeta<1$ (underdamped state):

$$s_{1,2}=-\zeta\omega_n \pm j\omega_n\sqrt{1-\zeta^2}, \qquad \text{[A1.31]}$$

conjugate complex roots that form a center angle 2θ (Figure A1.2).

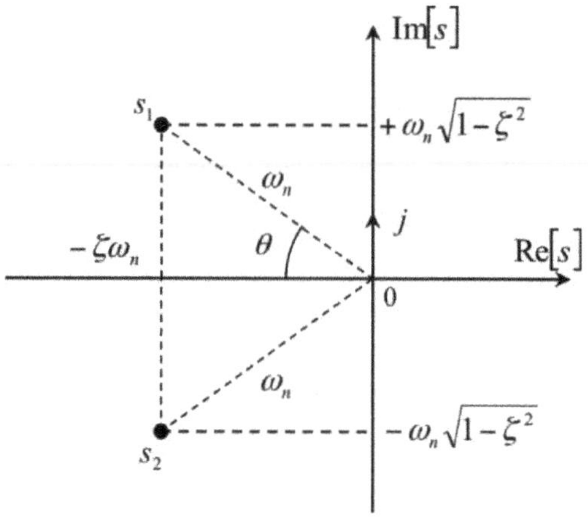

Figure A1.2. *θ is the half-center angle formed by the conjugate complex root pair*

A1.4.3. *Damping and rapidity performances of dynamics*

The aim of this section is to precisely answer the question:

– "What about the damping and rapidity performances of the oscillatory modes resulting from the *same pair* of conjugate complex roots?"

knowing that the modes, which are at stake, correspond to the two modes defined by [A1.26] and [A1.27].

The calculation of the cosine of the half-center angle θ (Figure A1.2) makes it possible to write:

$$\cos\theta = \frac{\zeta\omega_n}{\sqrt{\zeta^2\omega_n^2 + \omega_n^2(1-\zeta^2)}} = \frac{\zeta\omega_n}{\omega_n}, \qquad [A1.32]$$

therefore:

$$\cos\theta = \zeta, \qquad [A1.33]$$

a result which expresses that *the damping ratio ζ is given by the cosine of the half-center angle θ which is formed by the conjugate complex root pair*, namely:

$$\zeta = \cos\theta. \qquad [A1.34]$$

Furthermore, knowing that

$$\omega_n \sqrt{1-\zeta^2} = \omega_p \,, \qquad \text{[A1.35]}$$

the observation of Figure A1.2 shows that *the natural frequency* ω_p *is given by the projection on the imaginary axis of the positive imaginary part root*.

In other words (given the expression of this projection), *the natural frequency* ω_p *is given by the modulus* ω_n *of the conjugate complex root pair, multiplied by the sine of the half-center angle* θ it forms, namely:

$$\omega_p = \omega_n \sin\theta \,. \qquad \text{[A1.36]}$$

Appendix 2

Relaxation of Water on a Porous Dyke

A2.1. Context and study assumptions

Dyke builders have long been aware of the damping properties of very irregular dykes, in particular those with cavities or depressions imprisoning pockets of air which can be compressed by the oncoming water.

Besides, as everyone can see on the ebb and flow consecutive to the damping of water on fluvial or coastal dykes (where the motion water masses are different):

– the natural frequency of the water relaxation seems to be different whether the dyke is fluvial or coastal;

– the damping of the water relaxation seems to be independent of the dyke whether it is fluvial or coastal.

Given that nature is an inexhaustible source of solutions, we have studied the attenuation of the motion of water on this type of dyke.

To tackle such a study which concerns the physics of the complex media, we aimed to formulate the differential equation that governs a water mass M, of speed

$V(t)$, in horizontal[1] relaxation on a dyke, said to be porous, exerting a reaction force $F(t)$ (Figure A2.1)

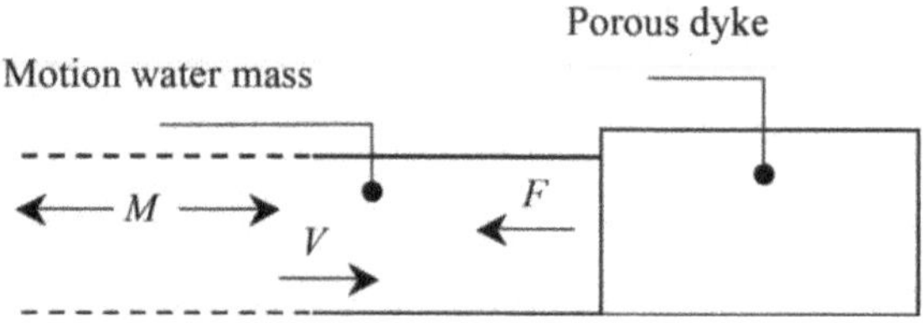

Figure A2.1. *Study system*

The dyke is then interpreted as a porous medium liable to conceive a model replying to a good compromise between reality and simplicity. The porosity assumption is, indeed, motivated by a set of properties that favors this compromise [OUS 99, OUS 05].

The indimensional and unordered nature of porosity implies an indefinite and indeterminate distribution of different size pores. This distribution justifies the uniformity of the dynamic pressure and the flow density on the *cross-section of flow*, which defines the water-dyke interface (just upstream of the dyke). Such a study configuration permits us to characterize the interface by a hydraulic admittance.

Porous rock is not necessarily permeable, which suggests considering pores not linked between them which gives them an alveolar character (dead-end).

1 A propagating wave carries, horizontally, a mass of water; the greater the wave height and length, the greater the value of the mass. The comparative observation of coastal, harbor or river waves is sufficiently explicit in this respect, and the dyke builders integrate this phenomenon into the dimensioning of their achievements.

In the literature, the motion water mass is interpreted as a "flow of mass" depending on the density of water as well as the speed and average height of the wave. This motion water mass is more perceptible as the wave is propagated by a surge, a phenomenon that results from the conversion of kinetic energy into potential energy when the depth decreases (which is the case near the coasts). In the absence of surge, the flow of mass is explained partly by the horizontal speed in the direction of the propagation of the water particles at the top of the wave.

In 1890, the hydraulicians already evaluated the pressure of the waves on vertical walls by the ratio of the "momentum", $HLC/2g$, over the duration of the movement, $L/2C$, where H is the height of the wave, L is the length of the wave, C is the celerity of the wave and g denotes the gravitational acceleration. Defined as such, this concept of momentum justifies well considering a horizontal motion water mass.

Such a study consideration allows us to describe a pore by its constituent elements, which cause two distinct physical phenomena:

– an orifice or canal that laminates water, which can be characterized by a *hydraulic resistance*, and where a dissipative energy is located;

– a cavity or alveolus that imprisons air compressible by the motion water, which can be characterized by a *pneumatic capacitance*, and where an elasticity potential energy is located; the corresponding elasticity reduces the dynamic pressure peaks at the water-dyke interface.

Due to the additivity of the flows in the pores (Figure A2.2), namely $Q = \sum_i Q_i$, the admittance of the water-dyke interface is that of a parallel arrangement of resistance–capacitance hydropneumatic cells. Each cell illustrates both, for each pore, the energy dissipation through viscosity (and turbulence) and the elasticity potential energy through air compression (in the case of the non-saturated porous media).

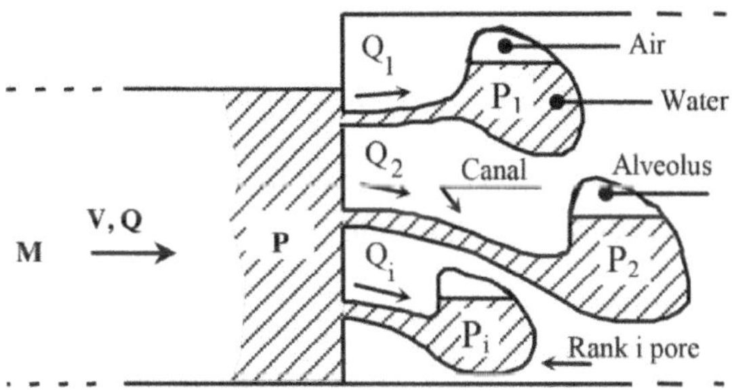

Figure A2.2. *Dyke interpretation (as an indeterminate distribution of an indefinite number of pores of different sizes not linked with each other)*

The fractal character attributed by Benoît Mandelbrot to porosity because of its property of self-similarity [MAN 75, MAN 82] suggests the (simplifying) assumption of recursivity on the distribution of resistances and capacitances, thus leading ***to** a recursive parallel arrangement of series RC cells*, with constant ratios α and η between the resistances and capacitances of two consecutive cells, α and η being greater than 1 and called *recursive factors* (Figure A2.3). It should be noted that:

– the fractal character of porosity as seen by Mandelbrot is that of the random fractals, as the different size pores are randomly distributed;

– the recursive parallel arrangement of series RC cells obtained apparently results from a determinist fractal approach because of the ordered decreasing, versus the rank, both of the resistances and capacitances.

In fact, the impermeability assumption of the porous medium (unlinked pores) allows us, given the additivity of the flows, an ordered reorganization of the pores and, therefore, of the RC cells according to the recursive parallel arrangement of series RC cells. The corresponding deterministic fractal approach thus fits with the random fractal approach to porosity as proposed by Mandelbrot.

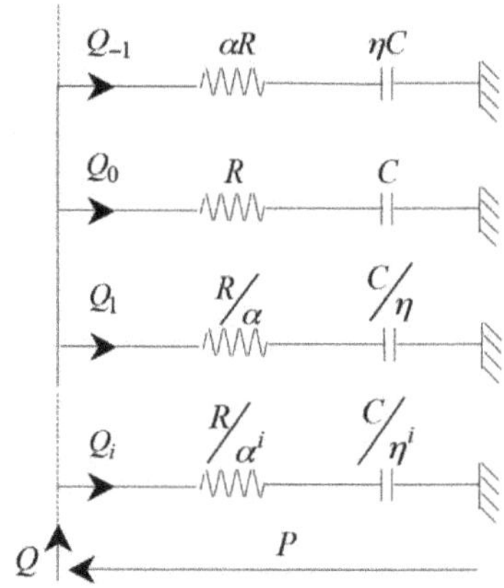

Figure A2.3. *Recursive parallel arrangement of series RC cells*

A2.2. Flow-pressure differential equation

The application of the fundamental law of dynamics to the motion water mass (which is here assumed to be undeformable) leads to the differential equation (Figure A2.1):

$$M \frac{dV(t)}{dt} + F(t) = 0 . \qquad \text{[A2.1]}$$

If S represents the flow cross-section of water, the speed V(t) is expressed as a function of the flow Q(t) according to

$$V(t) = \frac{Q(t)}{S} . \qquad \text{[A2.2]}$$

Moreover, the force $F(t)$ is expressed as a function of the dynamic pressure *P(t)* at the water-dyke interface, namely

$$F(t) = P(t)S . \qquad \text{[A2.3]}$$

Putting expressions [A2.2] and [A2.3] into relation [A2.1], we obtain a new form of the differential equation [A2.1]:

$$\frac{M}{S^2}\frac{dQ(t)}{dt}+P(t)=0\,. \qquad \text{[A2.4]}$$

A2.3. Model of the water-dyke interface

A2.3.1. *Admittance continuous form: introduction of a hypergeometric function*

The purpose of this section is to determine a non-integer order of differentiation from the introduction of a hypergeometric function in the continuous form of the admittance of the recursive parallel arrangement of series RC cells.

The series RC cells recursive parallel arrangement (Figure A2.3) admits an admittance of the form:

$$Y(s)=s\sum_i \frac{C_i}{1+\frac{C_i}{G_i}s}, \qquad \text{[A2.5]}$$

where:

$$G_i=\frac{\alpha^i}{R} \text{ and } C_i=\frac{C}{\eta^i}, \qquad \text{[A2.6]}$$

G_i and C_i denoting the conductance and the capacitance of the rank i cell, respectively.

Given the undimensional character of the arrangement:

– the discrete sum extended to infinity of terms

$$\frac{C_i}{1+\frac{C_i}{G_i}s} \qquad \text{[A2.7]}$$

– can be replaced by the continuous sum of the infinitesimal terms

$$\frac{C(z)\,dz}{1+\frac{C(z)}{G(z)}s}, \qquad \text{[A2.8]}$$

where $G(z)$ and $C(z)$ represent the lineic conductance and capacitance of the equivalent continuous medium, respectively.

Because of the recursivity of the parameters, the ratios between the conductances and capacitances of two consecutive infinitesimal slices are independent of the slice position and are thus expressed by:

$$\frac{G(z+dz)dz}{G(z)dz}=1+\frac{G'(z)}{G(z)}dz=\alpha(dz)=1+Bdz \qquad \text{[A2.9]}$$

and

$$\frac{C(z+dz)dz}{C(z)dz}=1+\frac{C'(z)}{C(z)}dz=\frac{1}{\eta(dz)}=1-Adz\,, \qquad \text{[A2.10]}$$

with:

$$B=\frac{G'(z)}{G(z)} \text{ and } A=-\frac{C'(z)}{C(z)}, \qquad \text{[A2.11]}$$

from which the following equations can be derived (through an integration):

$$G(z)=G(0)e^{Bz} \text{ and } C(z)=C(0)e^{-Az}. \qquad \text{[A2.12]}$$

Carried out as such, this approach makes it possible to establish an admittance continuous form, which is adapted for the introduction of a *hypergeometric function* as defined in [ERD 55, OUS 95].

Having done all the calculations, we thus obtained a canonical form of the admittance which is the transmittance of a *non-integer differentiator*, namely:

$$Y(s)=\left(\frac{s}{\omega_0}\right)^m, \qquad \text{[A2.13]}$$

where the non-integer order m is expressed by:

$$m=\frac{B}{A+B}. \qquad \text{[A2.14]}$$

From relations [A2.9] and [A2.10], it is possible to write:

$$\operatorname{Log}\alpha(dz) = \operatorname{Log}(1 + Bdz) = Bdz \qquad \text{[A2.15]}$$

and

$$-\operatorname{Log}\eta(dz) = \operatorname{Log}(1 - Adz) = -Adz\,, \qquad \text{[A2.16]}$$

from which can be drawn:

$$B = \frac{\operatorname{Log}\alpha(dz)}{dz} \text{ and } A = \frac{\operatorname{Log}\eta(dz)}{dz}. \qquad \text{[A2.17]}$$

Putting these parameters into [A2.14] leads to:

$$m = \frac{\operatorname{Log}\alpha(dz)}{\operatorname{Log}\alpha(dz) + \operatorname{Log}\eta(dz)} = \frac{\operatorname{Log}\alpha(dz)}{\operatorname{Log}(\alpha(dz)\eta(dz))}, \qquad \text{[A2.18]}$$

namely, in decimal logarithm:

$$m = \frac{\log\alpha(dz)}{\log(\alpha(dz)\eta(dz))}. \qquad \text{[A2.19]}$$

A2.3.2. *Admittance discrete form: smoothing of the Bode asymptotic diagrams*

The aim of this section is to determine the differentiation non-integer order from the Bode asymptotic diagram smoothing of the discrete form of the admittance of the recursive parallel arrangement of series RC cells.

The study of the admittance [A2.5] close to the transitional frequencies ω_i and ω_{i+1} of the cells of ranks i and $i+1$ [OUS 83] introduces new transitional frequencies ω'_i and ω'_{i+1} such that:

$$\frac{\omega_{i+1}}{\omega_i} = \frac{\omega'_{i+1}}{\omega'_i} = \alpha\eta\,, \qquad \text{[A2.20]}$$

$$\frac{\omega_i}{\omega' i} = \alpha \text{ and } \frac{\omega'_{i+1}}{\omega_i} = \eta\,. \qquad \text{[A2.21]}$$

These constant ratios express a recursive distribution of the transitional frequencies ω_i and $\omega_i^{'}$, and condition the form of the Bode asymptotic diagrams of $Y(j\omega)$ (Figure A2.4):

– the gain asymptotic diagram results from a regular sequence of steps (Figure A2.4(a));

– the phase asymptotic diagram results from a regular sequence of crenels (Figure A2.4(b)).

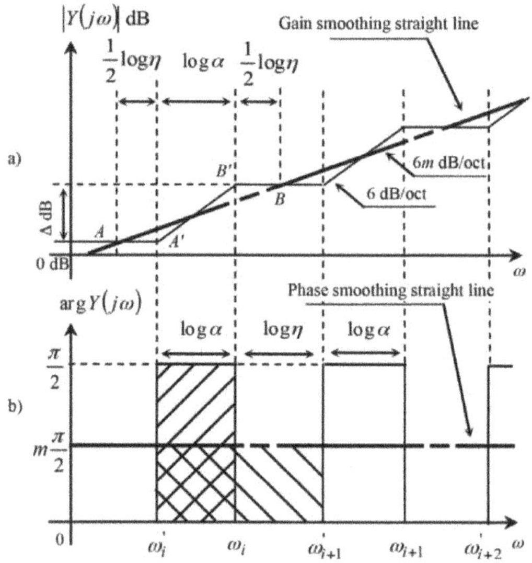

Figure A2.4. *Admittance asymptotic diagrams*

The smoothing of the steps of the gain asymptotic diagram can be illustrated by a straight line, called *gain smoothing straight line*, the slope of which is less than $6\ \text{dB/oct}$, namely 6m dB/oct (m being between 0 and 1). The slopes of segments AB and A'B' (Figure A2.4. (a)) are, respectively, given by the equations:

$$6m\ \text{dB/oct} = \frac{\Delta\, dB}{\log\alpha + \log\eta} \qquad \text{[A2.22]}$$

and

$$6\,\text{dB/oct} = \frac{\Delta\, dB}{\log\alpha}, \qquad \text{[A2.23]}$$

from which we draw, carrying out their ratio, the expression of the non-integer order m versus the recursive factors α and η:

$$m = \frac{\log \alpha}{\log(\alpha\eta)}. \qquad \text{[A2.24]}$$

The smoothing of the crenels of the phase asymptotic diagram can be illustrated by a straight line, called a *phase smoothing straight line*, the ordinate of which is less than $\pi/2$, namely $m\pi/2$ (m being between 0 and 1), and it is equal to the average of the phase asymptotic variation. The identity of the shaded surfaces (Figure A2.4. (b)) is translated by the equation:

$$m\frac{\pi}{2}(\log\alpha + \log\eta) = \frac{\pi}{2}\log\alpha, \qquad \text{[A2.25]}$$

from which m is directly deduced:

$$m = \frac{\log \alpha}{\log(\alpha\eta)}. \qquad \text{[A2.26]}$$

This expression is identical to relation [A2.24] obtained from the gain and is in conformity with relation [A2.19] obtained from the hypergeometric function.

The gain and phase smoothing straight lines so defined are, in fact, the Bode diagrams of a *non-integer differentiator* of frequency response

$$Y(j\omega) = \left(j\frac{\omega}{\omega_0}\right)^m, \qquad \text{[A2.27]}$$

where ω_0 denotes the frequency for which the gain smoothing straight line intersects the axis 0 dB, called *unit gain frequency* or *transition frequency*.

A2.3.3. *Non-integer differentiation as a model of the interface*

Finally, relation [A2.13] or [A2.27] determines the hydraulic admittance of the water-dyke interface:

$$Y(s) = \frac{Q(s)}{P(s)} = \left(\frac{s}{\omega_0}\right)^m, \qquad \text{[A2.28]}$$

from which the symbolic equation is deduced:

$$Q(s) = \left(\frac{s}{\omega_0}\right)^m P(s), \qquad \text{[A2.29]}$$

which, in the time domain, admits the concrete equation:

$$Q(t) = \frac{1}{\omega_0^m}\left(\frac{d}{dt}\right)^m P(t), \qquad \text{[A2.30]}$$

This equation represents the model of the interface. It expresses that the water flow Q(t) is proportional to the non-integer derivative of the dynamic pressure P(t) at the interface (at least in a medium frequency range, given that the frequency interval which regroups the whole of the ω_i and ω_i' (Figure A2.4) does not cover all the frequency domains).

The non-integer approach thus permits us to replace:

– a model of infinite integer order;

– by a model of finite non-integer order.

A2.3.4. *A remark on pore dispersion*

For recursive factors α and η sufficiently close to the unit such as

$$\alpha = 1 + \varepsilon_\alpha \text{ and } \eta = 1 + \varepsilon_\eta, \qquad \text{[A2.31]}$$

with ε_α and $\varepsilon_\eta << 1$, the non-integer order m, namely

$$m = \frac{\log \alpha}{\log \alpha + \log \eta}, \qquad \text{[A2.32]}$$

is particularized in conformity with:

$$m = \frac{\log(1+\varepsilon_\alpha)}{\log(1+\varepsilon_\alpha) + \log(1+\varepsilon_\eta)} = \frac{\varepsilon_\alpha}{\varepsilon_\alpha + \varepsilon_\eta}, \qquad \text{[A2.33]}$$

from which we draw:

$$\frac{\varepsilon_\alpha}{\varepsilon_\eta} = \frac{m}{1-m}, \qquad [A2.34]$$

or else, for $m<<1$, which is a sufficient condition to ensure a good damping ratio of the relaxation since $\zeta = -\cos\left(\pi/(1+m)\right)$ (Chapters 1 and 2):

$$\frac{\varepsilon_\alpha}{\varepsilon_\eta} \approx m << 1. \qquad [A2.35]$$

This ratio shows that practically identical pores from one rank to the other (ε_α and $\varepsilon_\eta << 1$), with a greater dispersion on the alveoli than on the orifices ($\varepsilon_\alpha / \varepsilon_\eta << 1$), make it possible to obtain a non-integer differentiation order in conformity with a good damping ($\zeta \approx 1$), a natural dispersion of the pores as little as possible responding to this condition.

A2.4. Non-integer order differential equation as a model governing the water relaxation

A2.4.1. *Pressure as a variable of the differential equation*

Putting the expression of $Q(t)$ given by [A2.30] into relation [A2.4] determines a differential equation of non-integer order $n=1+m$ between 1 and 2 (since m is between 0 and 1), namely:

$$\frac{M}{S^2}\frac{1}{\omega_0^m}\left(\frac{d}{dt}\right)^n P(t) + P(t) = 0, \qquad [A2.36]$$

or, under a canonical form:

$$\tau^n \left(\frac{d}{dt}\right)^n P(t) + P(t) = 0, \qquad [A2.37]$$

putting

$$\tau = \left(\frac{M}{S^2}\frac{1}{\omega_0^m}\right)^{\frac{1}{n}}, \qquad [A2.38]$$

which is called *transitional time constant.*

A2.4.2. *Flow as a variable of the differential equation*

The order $-m$ derivative of equation [A2.30] is written as:

$$\left(\frac{d}{dt}\right)^{-m} Q(t) = \frac{1}{\omega_0{}^m} P(t), \qquad \text{[A2.39]}$$

from which the following equation is drawn:

$$P(t) = \omega_0{}^m \left(\frac{d}{dt}\right)^{-m} Q(t). \qquad \text{[A2.40]}$$

This expression when put into [A2.37] leads to:

$$\tau^n \left(\frac{d}{dt}\right)^{1} Q(t) + \left(\frac{d}{dt}\right)^{-m} Q(t) = 0, \qquad \text{[A2.41]}$$

namely, through an order m differentiation:

$$\tau^n \left(\frac{d}{dt}\right)^{n} Q(t) + Q(t) = 0. \qquad \text{[A2.42]}$$

This differential equation that governs the flow has indeed the same form as the one governing the pressure (relation [A2.37]).

A2.5. Bibliography

[ERD 55] ERDELYI A., *Higher transcendental functions*, California Institute of Technology, MacGraw-Hill, vol. 1, 1955.

[MAN 75] MANDELBROT B., *Les Fractals*, Flammarion, 1975.

[MAN 82] MANDELBROT B., *The fractal geometry of nature*, Freeman, San Francisco (USA), 1982.

[OUS 95] OUSTALOUP A., *La dérivation non entière: théorie, synthèse et applications*, Hermès, Paris, 1995.

[OUS 99] OUSTALOUP A., SABATIER J., LANUSSE P., "From fractal robustness to the CRONE control", *Fractional Calculus and Applied Analysis (FCAA): An International Journal for Theory and Applications*, vol. 2, no. 1, pp. 1–30, January 1999.

[OUS 05] OUSTALOUP A., COIS O., LANUSSE P., *et al.*, "A survey on the CRONE approach", Plenary lecture, *IEEE International Conference on Systems, Signals, Devices SSD*, Sousse, Tunisia, 21–24 March 2005.

Appendix 3

Systems with Explicit and Implicit Generalized Derivative

A3.1. Explicit and implicit generalized differentiation

A3.1.1. *Explicit generalized derivative*

The generalized derivative of a function $f(t)$ is said to be *explicit* when it directly turns on $f(t)$, namely:

$$\left(\frac{d}{dt}\right)_{\exp l}^{n} f(t) = \left(\frac{d}{dt}\right)^{n} f(t), \qquad [A3.1]$$

a relation that defines what we call the *order n explicit derivative of* $f(t)$.

A3.1.2. *Implicit generalized derivative*

The generalized derivative of a function $f(t)$ is said to be *implicit* when it does not directly turn on $f(t)$ but on the product of $f(t)$ with an increasing exponential of time constant τ, $\exp(t/\tau)$, namely:

$$\left(\frac{d}{dt}\right)_{impl}^{n} f(t) = \left(\frac{d}{dt}\right)^{n} [f(t)\exp(t/\tau)], \qquad \text{[A3.2]}$$

a relation defining what we call the *order n implicit derivative of* $f(t)$.

A3.2. System with explicit generalized derivative

A3.2.1. *Symbolic characterization*

In the operational domain, a fundamental system is said to be *with explicit generalized derivative* when it is described by a generalized transmittance of the form:

$$F(s) = \frac{1}{1+(\tau s)^{n}} \text{ with } \tau \in \mathbb{R}^{+} \text{ and } n \in \mathbb{C}, \qquad \text{[A3.3]}$$

an expression whose form well evokes the explicit presence of an order n differentiation.

A3.2.2. *Time characterization*

The introduction of the Laplace transforms of the input and output, namely $E(s)$ and $S(s)$, enables us to write equation [A3.3] as:

$$\tau^{n} s^{n} S(s) + S(s) = E(s), \qquad \text{[A3.4]}$$

a symbolic equation that, in the time domain, admits for original:

$$\tau^{n} \left(\frac{d}{dt}\right)^{n} s(t) + s(t) = e(t), \qquad \text{[A3.5]}$$

a relation that defines *a differential equation with explicit generalized derivative.*

A3.3. System with implicit generalized derivative

A3.3.1. *Symbolic characterization*

In the operational domain, a fundamental system is said to be *with implicit generalized derivative* when it is described by a generalized transmittance of the form:

$$F(s)=\frac{1}{(1+\tau s)^n} \text{ with } \tau \in \mathbb{R}^+ \text{ and } n \in \mathbb{C}, \quad [A3.6]$$

an expression whose form well evokes the implicit presence of an order n differentiation.

A3.3.2. *Time characterization*

Considering the Laplace transforms $E(s)$ and $S(s)$ enables us to write equation [A3.6] as:

$$(1+\tau s)^n S(s)=E(s), \quad [A3.7]$$

namely

$$\tau^n (s+1/\tau)^n S(s)=E(s), \quad [A3.8]$$

or, by making the variable change defined by

$$s'=s+1/\tau \text{ (namely } s=s'-1/\tau\text{):} \quad [A3.9]$$

$$\tau^n s'^n S(s'-1/\tau)=E(s'-1/\tau). \quad [A3.10]$$

Given the following transformations, namely

$$E(s')=\mathcal{L}[e(t)] \quad [A3.11]$$

$$E(s'-1/\tau)=\mathcal{L}[e(t)\exp(t/\tau)] \quad [A3.12]$$

$$S(s')=\mathcal{L}[s(t)] \quad [A3.13]$$

$$S(s'-1/\tau)=\mathcal{L}[s(t)\exp(t/\tau)] \quad [A3.14]$$

and

$$s'^n S(s'-1/\tau)=\mathcal{L}\left[\left(\frac{d}{dt}\right)^n [s(t)\exp(t/\tau)]\right], \quad [A3.15]$$

it is easy to deduce the original of symbolic equation [A3.10]:

$$\tau^n \left(\frac{d}{dt}\right)^n \left[s(t)\exp(t/\tau)\right] = e(t)\exp(t/\tau)\,, \qquad [A3.16]$$

namely, from [A3.2]:

$$\tau^n \left(\frac{d}{dt}\right)^n_{impl} s(t) = e(t)\exp(t/\tau)\,, \qquad [A3.17]$$

a relation that defines a *differential equation with implicit generalized derivative.*

Differentiating equation [A3.16] at the order $-n$, namely

$$\tau^n s(t)\exp(t/\tau) = \left(\frac{d}{dt}\right)^{-n} \left[e(t)\exp(t/\tau)\right], \qquad [A3.18]$$

and then dividing by $\tau^n \exp(t/\tau)$, we obtain the solution of differential equation [A3.17]:

$$s(t) = \tau^{-n} \exp(-t/\tau)\left(\frac{d}{dt}\right)^{-n} \left[e(t)\exp(t/\tau)\right], \qquad [A3.19]$$

or, from [A3.2]:

$$s(t) = \tau^{-n} \exp(-t/\tau)\left(\frac{d}{dt}\right)^{-n}_{impl} e(t)\,, \qquad [A3.20]$$

or even, under an integral form, given the generalized derivative definition:

$$s(t) = \tau^{-n} \int_0^t \frac{\theta^{n-1}}{\Gamma(n)} e(t-\theta)\exp\left(-\frac{\theta}{\tau}\right) d\theta \qquad [A3.21]$$

for $\mathrm{Re}[-n] \in \mathbb{R}^{-*}$, and

$$s(t) = \tau^{-n} \exp(-t/\tau) FP \int_0^t \frac{\theta^{n-1}}{\Gamma(n)} e(t-\theta)\exp\left(\frac{t-\theta}{\tau}\right) d\theta \qquad [A3.22]$$

for $\mathrm{Re}[-n] \in \mathbb{R}^+ - \mathbb{N}^*$.

A3.4. Change between systems with implicit and explicit generalized derivative

The change from a system with implicit generalized derivative to a system with explicit generalized derivative is based on the *high gain control principle,* which, in this case, consists of:

– introducing a *high gain,* K ($K >> 1$), in cascade with a system with implicit generalized derivative, namely

$$F(s) = \frac{1}{(1+\tau s)^n} \qquad \text{[A3.23]}$$

– then looping the action chain so obtained by a *unit feedback* that reduces the open-loop transmittance to the one of the action chain, namely

$$\beta(s) = KF(s) = \frac{K}{(1+\tau s)^n}. \qquad \text{[A3.24]}$$

The corresponding closed-loop transmittance is then written as:

$$F'(s) = \frac{\beta(s)}{1+\beta(s)} = \frac{K}{K+(1+\tau s)^n}, \qquad \text{[A3.25]}$$

or, given that K is sufficiently high to allow the following simplification:

$$F'(s) = \frac{K}{K+(\tau s)^n}, \qquad \text{[A3.26]}$$

or even:

$$F'(s) = \frac{1}{1+\left(\frac{\tau s}{K^{1/n}}\right)^n}, \qquad \text{[A3.27]}$$

or else:

$$F'(s) = \frac{1}{1+(\tau' s)^n}, \qquad \text{[A3.28]}$$

by putting

$$\tau' = \tau / K^{1/n} . \qquad [A3.29]$$

The comparison of relations [A3.23] and [A3.28] reveals a *remarkable result*, which expresses that *if the differentiation implicit character is linked to the open-loop concept, its explicit character is linked to the closed-loop concept.*

A3.5. Impulse response of a system with explicit generalized derivative

Given by the original of transmittance [A3.3], the unit impulse response is expressed by the general relation:

$$s(t) = \mathcal{L}^{-1}\left[\frac{1}{1+(\tau s)^n}\right] \text{ with } \tau \in \mathbb{R}^+ \text{ and } n \in \mathbb{R}^+ . \qquad [A3.30]$$

The calculation of such an inverse transform calls on the integration methods of multiform functions by residues. In order to simplify the mathematical developments, the non-integer order is considered real knowing that the results induced by this study consideration are directly generalizable to a complex non-integer order.

A3.5.1. *Search for singular points*

s^n with n non-integer has a meaning only if

$$s \in \mathbb{C} - \mathbb{R}^- , \qquad [A3.31]$$

namely:

$$s \notin \left]-\infty, 0\right[, \qquad [A3.32]$$

or even:

$$-\pi < \arg s < +\pi , \qquad [A3.33]$$

which expresses that the restriction of the argument of the operational variable s is nothing but

$$\left]-\pi, +\pi\right[. \qquad [A3.34]$$

The singular points satisfy the equation:

$$1+(\tau s)^n = 0, \qquad [A3.35]$$

namely:

$$(\tau s)^n = -1 = e^{j(\pi + 2k\pi)}, \qquad [A3.36]$$

from which we draw:

$$s_k = \tau^{-1} e^{j\theta_k} \text{ with } \theta_k = \frac{1+2k}{n}\pi . \qquad [A3.37]$$

The restriction $]-\pi,+\pi[$ on the argument of s allows us to write:

$$-\pi < \frac{1+2k}{n}\pi < +\pi , \qquad [A3.38]$$

namely:

$$-1 < \frac{1+2k}{n} < +1, \qquad [A3.39]$$

translating that the problem comes down to search for the values of k that satisfy this double inequality.

Considering $n > 0$ allows us to multiply this double inequation by n without changing the corresponding inequality direction:

$$-n < 1+2k < n, \qquad [A3.40]$$

or, by subtracting 1, and then dividing by 2:

$$-\frac{n+1}{2} < k < \frac{n-1}{2}, \qquad [A3.41]$$

a condition on k that leads to the following results.

There is no pole for $0 < n < 1$.

There are two poles for $1 < n < 3$, namely:

$$p_{-1} = \tau^{-1} e^{-j\pi/n} \text{ and } p_0 = \tau^{-1} e^{j\pi/n} . \qquad [A3.42]$$

There are four poles for $3 < n < 5$, namely:

$$\begin{aligned} p_{-2} &= \tau^{-1} e^{-j3\pi/n}; \quad p_{-1} = \tau^{-1} e^{-j\pi/n}; \\ p_0 &= \tau^{-1} e^{j\pi/n}; \quad p_1 = \tau^{-1} e^{j3\pi/n}. \end{aligned} \quad [A3.43]$$

The generalization is obvious:

6 poles for $5 < n < 7$,

8 poles for $7 < n < 9$, ...

A3.5.2. *Mellin–Fourier transform*

Given the Bromwich–Wagner contour, the original function $s(t)$ of $S(s)$ is expressed by the integral:

$$s(t) = \frac{1}{2\pi j} \int_{c-j\infty}^{c+j\infty} e^{ts} S(s) ds, \quad [A3.44]$$

where c is greater than the abscissas of the singular points of $S(s)$. As in our case the poles have a modulus equal to τ^{-1}, it is sufficient to choose an *integration contour* γ that respects $c > \tau^{-1}$ and makes $S(s)$ uniform by respecting the cutoff of the complex plane along $\mathbb{R}^-$ [OUS 95].

The integration along γ and the sum of the residues relative to the poles inside γ determine the general expression of the unit impulse response, namely:

$$s(t) = \left\{ \frac{\tau^n \sin n\pi}{\pi} \int_0^\infty \frac{x^n e^{-xt} dx}{1 + 2\tau^n x^n \cos n\pi + \tau^{2n} x^{2n}} - \frac{1}{n} \sum_k s_k e^{ts_k} \right\} u(t), \quad [A3.45]$$

the unit step function $u(t)$ being introduced in order to cancel $s(t)$ for $t < 0$.

Case 1: $0<n<1$

The pole absence enables us to reduce the general expression of the impulse response to:

$$s(t)=\frac{\tau^n \sin n\pi}{\pi}u(t)\int_0^\infty \frac{x^n e^{-xt}dx}{1+2(\tau x)^n \cos n\pi+(\tau x)^{2n}}; \qquad [A3.46]$$

as the integral that figures in this relation is a monotonous decreasing function, the response transient is aperiodic.

Case 2: $1<n<3$

The presence of two poles inside γ (corresponding to $k=-1$ and $k=0$) particularizes the general expression of the impulse response:

$$\begin{aligned} s(t) &= \left\{\frac{\tau^n \sin n\pi}{\pi}\int_0^\infty \frac{x^n e^{-xt}dx}{1+2(\tau x)^n \cos n\pi+(\tau x)^{2n}}\right\}u(t) \\ &-\left\{\frac{2}{n}\tau^{-1}e^{t\tau^{-1}\cos\frac{\pi}{n}}\cos\left(t\tau^{-1}\sin\frac{\pi}{n}+\frac{\pi}{n}\right)\right\}u(t). \end{aligned} \qquad [A3.47]$$

Three well distinct cases must be considered successively.

1) $1<n<2$

For $1<n<2$, namely $\frac{\pi}{2}<\frac{\pi}{n}<\pi$, $\cos\frac{\pi}{n}<0$, which defines a *positive damping state* (or *damped state*); the system is stable.

2) $n=2$

For $n=2$, namely $\frac{\pi}{n}=\frac{\pi}{2}$, $\cos\frac{\pi}{n}=0$, which corresponds to a *nil damping state* (or *undamped state*); the system is just oscillating.

3) $2<n<3$

For $2<n<3$, namely $\frac{\pi}{3}<\frac{\pi}{n}<\frac{\pi}{2}$, $\cos\frac{\pi}{n}>0$, which defines a *negative damping state*; the system is instable.

Case 3: $3 < n < 5$

The presence of four poles inside γ (corresponding to $k = -2, -1, 0$ and 1) particularizes the general expression of the impulse response:

$$s(t) = \left\{ \frac{\tau^n \sin n\pi}{\pi} \int_0^\infty \frac{x^n e^{-xt}\, dx}{1 + 2(\tau x)^n \cos n\pi + (\tau x)^{2n}} \right\} u(t)$$

$$- \left\{ \frac{2}{n} \tau^{-1} e^{t\tau^{-1} \cos\frac{\pi}{n}} \cos\left(t\tau^{-1} \sin\frac{\pi}{n} + \frac{\pi}{n} \right) \right\} u(t) \qquad \text{[A3.48]}$$

$$- \left\{ \frac{2}{n} \tau^{-1} e^{t\tau^{-1} \cos\frac{3\pi}{n}} \cos\left(t\tau^{-1} \sin\frac{3\pi}{n} + \frac{3\pi}{n} \right) \right\} u(t).$$

The interval of n corresponding to this response, namely $3 < n < 5$, is characterized by:

$$\frac{\pi}{5} < \frac{\pi}{n} < \frac{\pi}{3} \Rightarrow \cos\frac{\pi}{n} > 0 \text{ (negative damping)} \qquad \text{[A3.49]}$$

and

$$\frac{3\pi}{5} < \frac{3\pi}{n} < \pi \Rightarrow \cos\frac{3\pi}{n} < 0 \text{ (positive damping).} \qquad \text{[A3.50]}$$

Therefore, there is an *instable oscillatory mode* that corresponds to a negative damping, namely

$$-\frac{2}{n} \tau^{-1} e^{t\tau^{-1} \cos\frac{\pi}{n}} \cos\left(t\tau^{-1} \sin\frac{\pi}{n} + \frac{\pi}{n} \right) u(t), \qquad \text{[A3.51]}$$

and a *stable oscillatory mode* whose damping is positive, namely

$$-\frac{2}{n} \tau^{-1} e^{t\tau^{-1} \cos\frac{3\pi}{n}} \cos\left(t\tau^{-1} \sin\frac{3\pi}{n} + \frac{3\pi}{n} \right) u(t); \qquad \text{[A3.52]}$$

hence, the system is instable.

A3.6. Impulse response of a system with implicit generalized derivative

The Laplace transform of the unit impulse response is given by the transmittance [A3.6], namely:

$$S(s) = \mathcal{L}\left[s(t)\right] = \frac{1}{\left(1+\tau s\right)^n} \text{ with } \tau \in \mathbb{R}^+ \text{ and } n \in \mathbb{C} - \mathbb{Z}^- . \qquad \text{[A3.53]}$$

The determination of $s(t)$ is the subject of three calculation methods presented in the following three sections; each of which aims to familiarize the readers with its principle.

A3.6.1. *First method: implicit differentiation of the Dirac impulse*

From the response to any input of a system with implicit generalized derivative (relation [A3.20]), the impulse response is expressed as a function of the order $-n$ implicit derivative of $\delta(t)$, namely:

$$s(t) = \tau^{-n} \exp(\ t/\tau)\left(\frac{d}{dt}\right)_{impl}^{-n} \delta(t), \qquad \text{[A3.54]}$$

or, given definition [A3.2]:

$$s(t) = \tau^{-n} \exp\left(-\frac{t}{\tau}\right)\left(\frac{d}{dt}\right)^{-n}\left[\delta(t)\exp\left(\frac{t}{\tau}\right)\right]. \qquad \text{[A3.55]}$$

Now, for $\mathrm{Re}[n] > 0$, the order $-n$ explicit derivative of the product $\delta(t)\exp(t/\tau)$ is written as:

$$\left(\frac{d}{dt}\right)^{-n}\left[\delta(t)\exp\left(\frac{t}{\tau}\right)\right] = \frac{t^{n-1}}{\Gamma(n)} u(t) * \delta(t)\exp\left(\frac{t}{\tau}\right), \qquad \text{[A3.56]}$$

namely:

$$\left(\frac{d}{dt}\right)^{-n}\left[\delta(t)\exp\left(\frac{t}{\tau}\right)\right] = \int_{-\infty}^{+\infty} \frac{(t-\theta)^{n-1}}{\Gamma(n)} u(t-\theta)\delta(\theta)\exp\left(\frac{\theta}{\tau}\right) d\theta , \qquad \text{[A3.57]}$$

or, knowing that $\delta(t)$ exclusively exists at instant $\theta = 0$:

$$\left(\frac{d}{dt}\right)^{-n}\left[\delta(t)\exp\left(\frac{t}{\tau}\right)\right] = \int_{0^-}^{0^+}\frac{(t-\theta)^{n-1}}{\Gamma(n)}u(t-\theta)\delta(\theta)\exp\left(\frac{\theta}{\tau}\right)d\theta$$

$$= \int_{0^-}^{0^+}\frac{t^{n-1}}{\Gamma(n)}u(t)\delta(\theta)d\theta \qquad [A3.58]$$

$$= \frac{t^{n-1}}{\Gamma(n)}u(t)\int_{0^-}^{0^+}\delta(\theta)d\theta = \frac{t^{n-1}}{\Gamma(n)}u(t).$$

Finally, by putting this last result into equation [A3.55], we obtain the expression of $s(t)$, namely:

$$s(t) = \frac{1}{\tau}\frac{1}{\Gamma(n)}\left(\frac{t}{\tau}\right)^{n-1}\exp\left(-\frac{t}{\tau}\right)u(t) . \qquad [A3.59]$$

A3.6.2. *Second method: translation of the operational function* $1/s^n$

The expression of $S(s)$ that relation [A3.53] defines can be rewritten as:

$$S(s) = \frac{1}{\tau^n\left(s+1/\tau\right)^n} , \qquad [A3.60]$$

namely, by putting $H(s) = 1/s^n$:

$$S(s) = \frac{1}{\tau^n}H\left(s+1/\tau\right) , \qquad [A3.61]$$

or, given the transformation

$$H\left(s+1/\tau\right) = \mathcal{L}\left[\exp\left(-t/\tau\right)\mathcal{L}^{1}\left[H(s)\right]\right] : \qquad [A3.62]$$

$$S(s) = \frac{1}{\tau^n}\mathcal{L}\left[\exp\left(-t/\tau\right)\mathcal{L}^{1}\left[H(s)\right]\right] , \qquad [A3.63]$$

or even, for $\operatorname{Re}[n]>0$, then knowing that

$$\mathcal{L}^{-1}[H(s)]=\mathcal{L}^{-1}\left[1/s^{n}\right]=\frac{t^{n-1}}{\Gamma(n)}u(t): \qquad [A3.64]$$

$$S(s)=\frac{1}{\tau^{n}}\mathcal{L}\left[\exp\left(-\frac{t}{\tau}\right)\frac{t^{n-1}}{\Gamma(n)}u(t)\right], \qquad [A3.65]$$

from which we can directly draw the original:

$$s(t)=\frac{1}{\tau}\frac{1}{\Gamma(n)}\left(\frac{t}{\tau}\right)^{n-1}\exp\left(-\frac{t}{\tau}\right)u(t). \qquad [A3.66]$$

A3.6.3. *Third method: series expansion of* $(1+1/\tau s)^{-n}$

Expression [A3.53] can be rewritten as:

$$S(s)=\frac{1}{(\tau s)^{n}}\left(1+\frac{1}{\tau s}\right)^{-n}, \qquad [A3.67]$$

namely, by applying the Newton's binomial formula:

$$S(s)=\frac{1}{(\tau s)^{n}}\left[1-\frac{n}{\tau s}+\frac{n(n+1)}{2!}\frac{1}{(\tau s)^{2}}-\frac{n(n+1)(n+2)}{3!}\frac{1}{(\tau s)^{3}}+\cdots\right], \qquad [A3.68]$$

or:

$$S(s)=\frac{1}{(\tau s)^{n}}-\frac{n}{(\tau s)^{n+1}}+\frac{n(n+1)}{2!}\frac{1}{(\tau s)^{n+2}}-\frac{n(n+1)(n+2)}{3!}\frac{1}{(\tau s)^{n+3}}+\cdots, \qquad [A3.69]$$

from which we can draw, for $\operatorname{Re}[n]>0$, given the transformation

$$\frac{1}{s^{n}}=\mathcal{L}\left[\frac{t^{n-1}}{\Gamma(n)}u(t)\right]: \qquad [A3.70]$$

$$s(t) = u(t)\left[\frac{t^{n-1}}{\tau^n \Gamma(n)} - \frac{nt^n}{\tau^{n+1}\Gamma(n+1)} + \frac{n(n+1)}{2!}\frac{t^{n+1}}{\tau^{n+2}\Gamma(n+2)} - \frac{n(n+1)(n+2)}{3!}\frac{t^{n+2}}{\tau^{n+3}\Gamma(n+3)} + \cdots\right], \qquad [A3.71]$$

or else, knowing that $n\Gamma(n) = \Gamma(n+1)$:

$$s(t) = u(t)\frac{t^{n-1}}{\tau^n \Gamma(n)}\left[1 - \frac{1}{1!}\frac{t}{\tau} + \frac{1}{2!}\left(\frac{t}{\tau}\right)^2 - \frac{1}{3!}\left(\frac{t}{\tau}\right)^3 + \cdots\right], \qquad [A3.72]$$

or even:

$$s(t) = u(t)\frac{1}{\tau\,\Gamma(n)}\left(\frac{t}{\tau}\right)^{n-1} \sum_{k\geq 0} \frac{(-1)^k}{k!}\left(\frac{t}{\tau}\right)^k, \qquad [A3.73]$$

or finally, by remarking that the series sum is nothing but the exponential function $\exp(-t/\tau)$:

$$s(t) = \frac{1}{\tau\,\Gamma(n)}\left(\frac{t}{\tau}\right)^{n-1} \exp\left(-\frac{t}{\tau}\right) u(t). \qquad [A3.74]$$

A3.7. Impulse response of a frequency bounded generalized differentiator

The Laplace transform of the unit impulse response of a frequency bounded generalized differentiator is given by a transmittance of the form:

$$S(s) = \mathcal{L}\left[s(t)\right] = \left(\frac{1+\tau_1 s}{1+\tau_2 s}\right)^{\mathrm{n}} \text{ with } \tau_1,\ \tau_2 \in \mathbb{R}^+ \text{ and } n \in \mathbb{C} - \mathbb{Z}^-. \qquad [A3.75]$$

The determination of $s(t)$ is the subject of two calculation methods presented in the next two sections. A third method is presented after (section A3.7.3), which aims to familiarize the readers with its principle.

A3.7.1. *First method: change of variable*

The expression of $S(s)$ that relation [A3.75] defines can be reformulated by making the change of variable defined by

$$s'=1+\tau_1 s \text{ (namely } s=(s'-1)/\tau_1\text{):} \quad \text{[A3.76]}$$

$$S(s')=K\left(\frac{s'}{1+\tau s'}\right)^n, \quad \text{[A3.77]}$$

by putting:

$$K=\left(\frac{\tau_1}{\tau_1-\tau_2}\right)^n \quad \text{[A3.78]}$$

and

$$\tau=\left(\frac{\tau_2}{\tau_1-\tau_2}\right). \quad \text{[A3.79]}$$

Under this form, relation [A3.77] involves the transmittance of an implicit generalized derivative system (relation [A3.6]):

$$F(s')=\frac{1}{(1+\tau s')^n}. \quad \text{[A3.80]}$$

The inverse transform of $F(s')$, namely (relation [A3.1])

$$\mathcal{L}^1\left[F(s')\right]=\frac{1}{\tau\Gamma(n)}\left(\frac{t'}{\tau}\right)^{n-1}\exp\left(-\frac{t'}{\tau}\right)u(t'), \quad \text{[A3.81]}$$

can be rewritten by replacing $\exp(-t'/\tau)$ by its series expansion:

$$\mathcal{L}^1\left[F(s')\right]=\frac{1}{\tau\Gamma(n)}\sum_{k=0}^{\infty}\frac{(-1)^k}{k!}\left(\frac{t'}{\tau}\right)^{n+k-1}u(t'), \quad \text{[A3.82]}$$

or, by taking the Laplace transform of each of the members:

$$F(s')=\frac{1}{\tau\Gamma(n)}\sum_{k=0}^{\infty}\frac{(-1)^k}{k!\tau^{n+k-1}}\Gamma(n+k)(s')^{-n-k}\,. \quad [A3.83]$$

The expression of $F(s')$ so obtained enables us to express $S(s')$ as:

$$S(s')=\frac{K}{\tau\Gamma(n)}\sum_{k=0}^{\infty}\frac{(-1)^k}{k!\tau^{n+k-1}}\Gamma(n+k)(s')^{-k}, \quad [A3.84]$$

or, through the change of variable defined by

$$s'=1+\tau_1 s \text{ (namely } s=(s'-1)/\tau_1\text{):} \quad [A3.85]$$

$$S(s)=\frac{K}{\tau\Gamma(n)}\sum_{k=0}^{\infty}\frac{(-1)^k}{k!\tau^{n+k-1}}\Gamma(n+k)\frac{1}{\left(1+\tau_1 s\right)^k}, \quad [A3.86]$$

from which we can directly draw the original:

$$s(t)=\left(\frac{\tau_1}{\tau_2}\right)^n\left[\delta(t)+\sum_{k=1}^{\infty}\frac{(-1)^k}{k!(k-1)!}\frac{\Gamma(n+k)}{\Gamma(n)}\left(\frac{1}{\tau_2}-\frac{1}{\tau_1}\right)^k t^{k-1}e^{-\frac{t}{\tau_1}}u(t)\right]. \quad [A3.87]$$

A3.7.2. *Second method: series expansion*

The expression of $S(s)$ that relation [A3.75] defines can be rewritten by using the Euclidian division, namely:

$$S(s)=\left[\frac{\tau_1}{\tau_2}\left(1+\frac{\frac{\tau_2}{\tau_1}-1}{1+\tau_2 s}\right)\right]^n, \quad [A3.88]$$

or, by applying the Newton's binomial formula:

$$S(s)=\left(\frac{\tau_1}{\tau_2}\right)^n\sum_{k=0}^{\infty}\binom{n}{k}\left(\frac{\tau_2}{\tau_1}-1\right)^k\frac{1}{\left(1+\tau_2 s\right)^k}, \quad [A3.89]$$

from which we can draw, for $\mathrm{Re}[n] > 0$, given the transformation

$$\mathcal{L}^{1}\left[\frac{1}{(1+\tau_2 s)^k}\right] = \frac{t^{k-1}}{\tau_2{}^k (k-1)!} e^{-\frac{t}{\tau_2}}: \quad [A3.90]$$

$$s(t) = \left(\frac{\tau_1}{\tau_2}\right)^n \left[\delta(t) + \sum_{k=1}^{\infty} \frac{\binom{n}{k}}{(k-1)!} \left(\frac{1}{\tau_1} - \frac{1}{\tau_2}\right)^k t^{k-1} e^{-\frac{t}{\tau_2}} u(t)\right]. \quad [A3.91]$$

This expression is, indeed, identical to relation [A3.87] determined by the method of change of variable.

A3.7.3. *Third method: convolution product*

The expression of $S(s)$ defined by relation [A3.75] can be restructured according to the following equation:

$$S(s) = (1+\tau_1 s)^n (1+\tau_2 s)^{-n}. \quad [A3.92]$$

Under this form, $S(s)$ can be interpreted as resulting from the product of the transmittances of two implicit generalized derivative systems. Thus, in the time domain, the original $s(t)$ of $S(s)$ is nothing but the convolution product of the impulse responses of these systems, namely:

$$s(t) = s_1(t) * s_2(t), \quad [A3.93]$$

with, given [A3.59]:

$$s_1(t) = \frac{1}{\tau_1} \frac{1}{\Gamma(-n)} \left(\frac{t}{\tau_1}\right)^{-n-1} \exp\left(-\frac{t}{\tau_1}\right) u(t) \quad [A3.94]$$

and

$$s_2(t) = \frac{1}{\tau_2} \frac{1}{\Gamma(n)} \left(\frac{t}{\tau_2}\right)^{n-1} \exp\left(-\frac{t}{\tau_2}\right) u(t). \quad [A3.95]$$

The expression of $s(t)$ so formulated implies incompatible existence conditions of the convolution integral ($\mathrm{Re}[n]<0$ for $s_1(t)$ and $\mathrm{Re}[n]>0$ for $s_2(t)$). To avoid this problem, it is necessary first to calculate the step response in the following way.

Through the splitting of s^{-1}, it is possible to express $S(s)$ as:

$$S(s)=s^{-m}\left(1+\tau_1 s\right)^n s^{m-1}\left(1+\tau_2 s\right)^{-n}, \qquad \text{[A3.96]}$$

with $m\in \mathbb{Z}$ and $n\in \mathbb{C}-\mathbb{Z}$, $\mathrm{Re}[-m+n]<0$ and $\mathrm{Re}[m-1-n]<0$,

namely

$$\mathrm{Re}[n]<m<\mathrm{Re}[n]+1, \qquad \text{[A3.97]}$$

or even:

$$S(s)=\left(\frac{\tau_1}{\tau_2}\right)^n s^{-m+n}\left(1+\frac{1}{\tau_1 s}\right)^n s^{m-n-1}\left(1+\frac{1}{\tau_2 s}\right)^{-n}, \qquad \text{[A3.98]}$$

or, by applying Newton's binomial formula:

$$S(s)=\left(\frac{\tau_1}{\tau_2}\right)^n \sum_{k=0}^{\infty}\binom{n}{k}\frac{s^{n-m-k}}{\tau_1^k}\sum_{k=0}^{\infty}\binom{-n}{k}\frac{s^{m-n-k-1}}{\tau_2^k}. \qquad \text{[A3.99]}$$

The incompatibility linked to the symmetry $(n,-n)$ of the powers in t in the time domain (relations [A3.94] and [A3.95]) no longer appears here. Indeed, the new integral of the product $s_1(\theta)s_2(\theta-t)$, where θ is the mute variable, is now computed from the following two expressions:

$$s_1(t)=\sum_{k=0}^{\infty}\binom{n}{k}\frac{t^{m-n+k-1}}{\tau_1^{k-n}\Gamma(m-n+k)} \qquad \text{[A3.100]}$$

and

$$s_2(t)=\sum_{k=0}^{\infty}\binom{-n}{k}\frac{t^{n-m+k}}{\tau_1^{k+n}\Gamma(n+1-m+k)}. \qquad \text{[A3.101]}$$

This reformulation of $s_1(t)$ and $s_2(t)$ that introduces a dissymmetry of the powers ($m-n+k-1$, $n-m+k$) ensures the integral convergence. So obtained, the step response leads, through a simple differentiation, to the impulse response.

A3.8. On the singularities of a system with explicit generalized derivative

A3.8.1. *An infinity of hidden poles*

Whatever the positive real order n, the impulse response of a system with explicit generalized derivative admits a general expression of the form (relation [A3.45]):

$$s(t)=v(t)+w(t), \qquad \text{[A3.102]}$$

an expression that reveals two well distinct functions:

– a function $v(t)$, structurally dependent on n, which is the classical part of $s(t)$ and results from the poles obtained following the cutoff of the complex plane along $\mathbb{R}^-$;

– a function $w(t)$, structurally independent of n, which is the non-classical part of $s(t)$ and is an *integral function* of the form

$$w(t)=\frac{\tau^n \sin n\pi}{\pi}u(t)\int_0^\infty \frac{x^n e^{-xt}dx}{1+2(\tau x)^n \cos n\pi+(\tau x)^{2n}}. \qquad \text{[A3.103]}$$

But what about the origin of this integral function *in terms of poles*? In other words, does it result from "*hidden poles*" whose presence merits to be clarified?

This question finds its answer in the result of the following calculation led from the expression of $w(t)$ given by [A3.103].

The function $w(t)$ can be rewritten as:

$$w(t)=Au(t)\int_0^\infty e^{-xt}y(x)dx, \qquad \text{[A3.104]}$$

by putting:

$$A = \frac{\tau^n \sin n\pi}{\pi} \text{ and } y(x) = \frac{x^n}{1 + 2(\tau x)^n \cos n\pi + (\tau x)^{2n}}. \qquad [A3.105]$$

Its Laplace transform, namely

$$W(s) = \mathcal{L}[w(t)] = \int_0^\infty e^{-st} w(t)dt, \qquad [A3.106]$$

then admits an expression of the form:

$$W(s) = \int_0^\infty e^{-st} \left(A \int_0^\infty e^{-xt} y(x)dx \right) dt, \qquad [A3.107]$$

or:

$$W(s) = A \int_0^\infty e^{-st} \left(\int_0^\infty e^{-xt} y(x)dx \right) dt, \qquad [A3.108]$$

or even:

$$W(s) = A \int_0^\infty \int_0^\infty e^{-(s+x)t} y(x)dxdt, \qquad [A3.109]$$

or else, by inverting the order of the integrations from the Fubini theorem:

$$W(s) = A \int_0^\infty y(x) \left(\int_0^\infty e^{-(s+x)t} dt \right) dx, \qquad [A3.110]$$

namely:

$$W(s) = A \int_0^\infty y(x) \left[-\frac{e^{-(s+x)t}}{s+x} \right]_{t=0}^{t=\infty} dx, \qquad [A3.111]$$

or, given that $\lim_{t\to\infty} e^{-(s+x)t} = 0$ for $\text{Re}(s) > 0$:

$$W(s) = A\int_0^{\infty} \frac{y(x)}{s+x}\,dx, \qquad \text{[A3.112]}$$

a continuous sum that admits the discrete form:

$$W(s) = A\sum_{i=-\infty}^{+\infty} \frac{y(x_i)}{s+x_i}\Delta x_i \text{ with } \Delta x_i \text{ tending toward zero.} \qquad \text{[A3.113]}$$

The discrete sum defined as such reveals an infinity of *first-order stable aperiodic modes* of the form

$$A\frac{y(x_i)\Delta x_i}{s+x_i}, \qquad \text{[A3.114]}$$

each of them having a *negative real pole*, $s_i = -x_i$ (x_i being positive).

All the modes have the same sign, positive or negative according to the value of n that determines the sign of A and therefore the sign of the modes, knowing that $y(x_i)$ is positive in conformity with the following proof led on $y(x)$ (relation [A3.105]). As x is positive, the numerator x^n of $y(x)$ is positive. As for the denominator, it suffices to rewrite it by replacing the unit term by $\sin^2 n\pi + \cos^2 n\pi$, namely:

$$\sin^2 n\pi + \cos^2 n\pi + 2(\tau x)^n \cos n\pi + (\tau x)^{2n}, \qquad \text{[A3.115]}$$

or:

$$\sin^2 n\pi + \left(\cos n\pi + (\tau x)^n\right)^2, \qquad \text{[A3.116]}$$

a sum of two positive terms that well expresses the positivity of the denominator of $y(x)$.

Returning to the continuous form of $W(s)$ (relation [A3.112]) expresses that any point of the half-straight line $\mathbb{R}^-$ represents a pole of $W(s)$ and therefore of $F(s)$, $s = -x$ (x being positive), thus conveying that there exist:

– for $W(s)$, *an indefinite distribution of infinitely close poles on* $\mathbb{R}^-$;

– for $F(s)$, *the same indefinite distribution of poles* and also *the (discrete and in finite number) poles obtained following the cutoff of the complex plane along* $\mathbb{R}^-$.

An important remark: it is convenient to specify that this indefinite distribution of infinitely close poles on $\mathbb{R}^-$ (whose presence is often said implicit) also appears when, in $F(s)$, s^n is decomposed according to an *indefinite recursive distribution of infinitely close zeros and poles on* $\mathbb{R}^-$ [OUS 90, OUS 91].

A remark on the zeros: concerning the zeros of $F(s)$, there is no zero as the form of the expression of $F(s)$ shows (relation [A3.3]).

A3.8.2. *A stable aperiodic multimode*

The (implicit) presence of an infinity of infinitely close poles on $\mathbb{R}^-$, at the origin of an infinity of stable aperiodic modes of the first order, thus determines a *stable aperiodic multimode* [OUS 91].

In the case of the second generation CRONE control ($1 < n < 2$), this multimode, $w(t)$ or $W(s)$, is added to the *(stable) oscillatory mode*, $v(t)$ or $V(s)$, due to the explicit presence of the conjugate complex pole pair obtained following the cutoff of the complex plane along $\mathbb{R}^-$. This oscillatory mode is said to be *dominant* knowing that it practically determines the impulse response of the control. It is true that the multimode does not practically affect the general form of the impulse response [OUS 91].

A3.8.3. *On the singularity of* s^n *for* $s \in \mathbb{R}^-$

In the decomposition of the non-integer differentiation symbolic operator, s^n, according to an indefinite recursive distribution of infinitely close zeros and poles on $\mathbb{R}^-$, when a zero and a consecutive pole come infinitely closer, the limit is the fusion of the zero and the pole, namely $z_i = p_i$.

This leads us to state that in terms of limit, any point of the half-straight line $\mathbb{R}^-$ represents at the same time a zero and a pole, that is to say a zero-pole couple where the zero z_i and the pole p_i merge, the transmittance element corresponding to this couple $\{z_i, p_i\}$ being of the form:

$$\frac{1-\dfrac{s}{z_i}}{1-\dfrac{s}{p_i}} \text{ with } \lim|z_i - p_i| = 0 . \qquad [A3.117]$$

So, any s belonging to $\mathbb{R}^-$ such as $s = z_i = p_i$ leads, for the transmittance element so defined and therefore for the global transmittance s^n, to an indetermination of the form $0/0$, thus expressing the *indeterminate character* of s^n for $s \in \mathbb{R}^-$.

This paradoxical (or singular) character of s^n for $s \in \mathbb{R}^-$ does not contradict, or even reinforce, the property according to which *the operator* s^n *is not defined for* $s \in \mathbb{R}^-$, a property that, besides, is no stranger to the choice of a cutoff of the complex plane along $\mathbb{R}^-$.

To prove that s^n (with n non-integer) is not defined for $s \in \mathbb{R}^-$, it is convenient to express s^n under the form

$$s^n = e^{\ln(s^n)}, \qquad [A3.118]$$

namely:

$$s^n = e^{n\ln(s)}. \qquad [A3.119]$$

If $s \in \mathbb{R}^-$, then $s = -x$ with $x > 0$; therefore

$$s^n = e^{n\ln(-x)}. \qquad [A3.120]$$

Now, $\ln(-x)$ has no meaning, neither does s^n.

A3.9. Bibliography

[OUS 83] OUSTALOUP A., *Systèmes asservis linéaires d'ordre fractionnaire*, Masson, Paris, 1983.

[OUS 90] OUSTALOUP A., "A mathematical approach of the non integer derivation in control", *Colloque International sur l'Analyse des Systèmes Dynamiques Contrôlés*, CNRS, Lyon, France, 3–6 July 1990.

[OUS 91] OUSTALOUP A., *La commande CRONE*, Hermès, Paris, 1991.

[OUS 95] OUSTALOUP A., *La dérivation non entière: théorie, synthèse et applications*, Hermès, Paris, 1995.

Appendix 4

Generalized Differential Equation and Generalized Characteristic Equation

A4.1. Generalized differential equation time solving

A *generalized differential equation* (GDE) is defined by a general expression of the form:

$$\sum_{l=1}^{L} a_l \left(\frac{d}{dt}\right)^{n_l} s(t) = \sum_{q=1}^{Q} b_q \left(\frac{d}{dt}\right)^{m_q} e(t)\,, \qquad [A4.1]$$

where the coefficients a_l and b_q and also the differentiation orders n_l and m_q are integers or non-integers, real or complex.

Given that the developments in Chapter 3 enable us to determine the generalized derivative of a function, the right-hand member of equation [A4.1], namely

$$\sum_{q=1}^{Q} b_q \left(\frac{d}{dt}\right)^{m_q} e(t)\,, \qquad [A4.2]$$

is easily computable since the function $e(t)$ is given.

Solving equation [A4.1] then comes down to solving the new equation:

$$\sum_{l=1}^{L} a_l \left(\frac{d}{dt}\right)^{n_l} s(t) = f(t)\,, \qquad [A4.3]$$

where $f(t)$ is a complex function of the real variable t, determined according to relation [A4.2].

Through a sampling step h discretization, [A4.3] is rewritten as:

$$\sum_{l=1}^{L} a_l D^{n_l} s(Kh) = f(Kh). \quad \text{[A4.4]}$$

Given relation [3.11], it is possible to write:

$$D^{n_l} s(Kh) = \sum_{k=0}^{K} C(k) s\left(\left(K-k\right)h\right), \quad \text{[A4.5]}$$

with

$$C(k) = C_r(k) + C_i(k), \quad \text{[A4.6]}$$

$C_r(k)$ and $C_i(k)$ being given by:

$$C_r(k) = \frac{\left(k - \mathrm{Re}\left[n_l\right] - 1\right) C_r(k-1)}{k} + \frac{\mathrm{Im}\left[n_l\right] C_i(k-1)}{k} \quad \text{[A4.7]}$$

and

$$C_i(k) = \frac{\left(k - \mathrm{Re}\left[n_l\right] - 1\right) C_i(k-1)}{k} - \frac{\mathrm{Im}\left[n_l\right] C_r(k-1)}{k}. \quad \text{[A4.8]}$$

The values of $C_r(k)$ and $C_i(k)$ at rank 0 are defined by:

$$C_r(0) = \frac{H_r}{H_r^2 + H_i^2} \quad \text{[A4.9]}$$

and

$$C_i(0) = \frac{-H_i}{H_r^2 + H_i^2}. \quad \text{[A4.10]}$$

By putting expression [A4.5] into equality [A4.4], we get:

$$\sum_{l=1}^{L} a_l \left(\sum_{k=0}^{K} C(k) s\left(\left(K-k \right) h \right) \right) = f(Kh) . \qquad [A4.11]$$

By separating the real and imaginary parts, the equality so obtained is decomposed into two equations:

$$\begin{aligned} &\sum_{l=1}^{L} \operatorname{Re}\left[a_l\right]\left(C_{r_l}(0) s_r(Kh) - C_{i_l}(0) s_i(Kh) + S_{r_l}\right) \\ &-\sum_{l=1}^{L} \operatorname{Im}\left[a_l\right]\left(C_{i_l}(0) s_r(Kh) + C_{r_l}(0) s_i(Kh) + S_{i_l}\right) = f_r(Kh) \end{aligned} \qquad [A4.12]$$

and

$$\begin{aligned} &\sum_{l=1}^{L} \operatorname{Im}\left[a_l\right]\left(C_{r_l}(0) s_r(Kh) - C_{i_l}(0) s_i(Kh) + S_{r_l}\right) \\ &+\sum_{l=1}^{L} \operatorname{Re}\left[a_l\right]\left(C_{i_l}(0) s_r(Kh) + C_{r_l}(0) s_i(Kh) + S_{i_l}\right) = f_i(Kh), \end{aligned} \qquad [A4.13]$$

where:

$$S_{r_l} = \sum_{k-1}^{K} \left(C_{r_l}(k) s_r\left(\left(K-k \right) h \right) - C_{i_l}(k) s_i\left(\left(K-k \right) h \right) \right) \qquad [A4.14]$$

and

$$S_{i_l} = \sum_{k=1}^{K} \left(C_{i_l}(k) s_r\left(\left(K-k \right) h \right) - C_{r_l}(k) s_i\left(\left(K-k \right) h \right) \right) . \qquad [A4.15]$$

Reordering the terms of equations [A4.12] and [A4.13] modifies these equations in conformity with:

$$\begin{aligned} &\sum_{l=1}^{L} \left(\operatorname{Re}\left[a_l\right] C_{r_l}(0) - \operatorname{Im}\left[a_l\right] C_{i_l}(0) \right) s_r(Kh) \\ &-\sum_{l=1}^{L} \left(\operatorname{Re}\left[a_l\right] C_{i_l}(0) + \operatorname{Im}\left[a_l\right] C_{r_l}(0) \right) s_i(Kh) \\ &+\sum_{l=1}^{L} \left(\operatorname{Re}\left[a_l\right] S_{r_l} - \operatorname{Im}\left[a_l\right] S_{i_l} \right) = f_r(Kh) \end{aligned} \qquad [A4.16]$$

and

$$\sum_{l=1}^{L}\left(\mathrm{Im}[a_l]C_{r_l}(0)-\mathrm{Re}[a_l]C_{i_l}(0)\right)s_r(Kh)$$
$$+\sum_{l=1}^{L}\left(\mathrm{Re}[a_l]C_{r_l}(0)-\mathrm{Im}[a_l]C_{i_l}(0)\right)s_i(Kh) \qquad [A4.17]$$
$$+\sum_{l=1}^{L}\left(\mathrm{Im}[a_l]S_{r_l}+\mathrm{Re}[a_l]S_{i_l}\right)=f_i(Kh).$$

Let us put:

$$A=\sum_{l=1}^{L}\left(\mathrm{Re}[a_l]C_{r_l}(0)-\mathrm{Im}[a_l]C_{i_l}(0)\right) \qquad [A4.18]$$

and

$$B=-\sum_{l=1}^{L}\left(\mathrm{Re}[a_l]C_{i_l}(0)+\mathrm{Im}[a_l]C_{r_l}(0)\right), \qquad [A4.19]$$

then:

$$S_r=\sum_{l=1}^{L}\left(\mathrm{Re}[a_l]S_{r_l}-\mathrm{Im}[a_l]S_{i_l}\right) \qquad [A4.20]$$

and

$$S_i=\sum_{l=1}^{L}\left(\mathrm{Im}[a_l]S_{r_l}+\mathrm{Re}[a_l]S_{i_l}\right). \qquad [A4.21]$$

The problem then comes down to solving a system of two equations with two unknowns:

$$As_r(Kh)+Bs_i(Kh)+S_r=f_r(Kh) \qquad [A4.22]$$

and

$$-Bs_r(Kh)+As_i(Kh)+S_i=f_i(Kh). \qquad [A4.23]$$

The solution is almost immediate, namely:

$$s_r(Kh) = \frac{A\left(f_r(Kh) - S_r\right) - B\left(f_i(Kh) - S_i\right)}{A^2 + B^2} \quad [A4.24]$$

and

$$s_i(Kh) = \frac{B\left(f_r(Kh) - S_r\right) + A\left(f_i(Kh) - S_i\right)}{A^2 + B^2}. \quad [A4.25]$$

Thus we obtain, for each value of K, the values that numerically define the real and imaginary parts of the differential equation solution at the instant $t = Kh$, namely

$$s(t) = s_r(t) + \mathrm{i}\, s_i(t). \quad [A4.26]$$

A4.2. Generalized characteristic equation solving

A4.2.1. *Generalized characteristic equation*

The characteristic polynomial of GDE [A4.1] is defined by the denominator of the generalized transmittance deduced from the symbolic representation of GDE [A4.1], namely:

$$\sum_{l=1}^{L} a_l s^{n_l}. \quad [A4.27]$$

Characterized by real or complex non-integer powers, the characteristic polynomial so defined is said to be generalized. It is the same for the characteristic equation, defined as the characteristic polynomial equal to zero.

A4.2.2. *Search for zeros of a polynomial with real non-integer powers*

The principle of the method developed in this section is based on the approximation of the polynomial real non-integer powers with rational powers.

The real non-integer power of rank l being written in the form:

$$n_l = \frac{r}{m} + e_l, \quad [A4.28]$$

where r/m and e_l represent the rational power and the rationalization error of the same rank, respectively, the first step of the procedure is to determine the integer m that minimizes the rationalization error sum. The search for m is carried out by iteratively calculating the sum of e_l for m increasing, until the sum becomes the smallest possible in conformity with the admissible values of the parameter r.

Let us take a polynomial with real non-integer powers of the form:

$$P\left(s^{1/m}\right) = a_n s^{n/m} + a_{n-1} s^{(n-1)/m} + \cdots + a_r s^{r/m} + \cdots + a_1 s^{1/m} + a_0 , \qquad \text{[A4.29]}$$

where:

$s \in \mathbb{C}$; $a_r \in \mathbb{R}$; $m \geq 2$ and $n \geq 1$ (m and n being integers).

The problem to be solved consists of determining the zeros of such a polynomial. To that effect, let us consider the integer degree polynomial $P(s)$, namely:

$$P(s) = \alpha_n s^n + \alpha_{n-1} s^{n-1} + \cdots + \alpha_r s^r + \cdots + \alpha_1 s + \alpha_0 , \qquad \text{[A4.30]}$$

which can be rewritten under the factorized form as:

$$P(s) = \alpha_n (s - s_1)(s - s_2) \cdots (s - s_r) \cdots (s - s_n) , \qquad \text{[A4.31]}$$

where the s_r are the n real or complex zeros of $P(s)$ such that $s_r = R_r e^{j\theta_r}$, where $-\pi < \theta_r \leq +\pi$.

By replacing s by $s^{1/m}$ in equation [A4.31], we obtain a new form of polynomial [A4.29]:

$$P\left(s^{1/m}\right) = \alpha_n (s^{1/m} - s_1)(s^{1/m} - s_2) \cdots (s^{1/m} - s_r) \cdots (s^{1/m} - s_n) . \qquad \text{[A4.32]}$$

Let $s = \rho e^{j\theta}$ be a zero of $P\left(s^{1/m}\right)$, where $-\pi < \theta < +\pi$ knowing that $s^{1/m}$ has a meaning only if $s \in \mathbb{C} - \mathbb{R}$. Such a zero satisfies the equation:

$$s^{1/m} - s_r = 0 , \qquad \text{[A4.33]}$$

or:

$$\rho^{1/m} e^{j\theta/m} = R_r e^{j\theta_r} , \qquad \text{[A4.34]}$$

from where we draw, by identifying the modulus and arguments up to $2k\pi$:

$$\rho^{1/m} = R_r \qquad [A4.35]$$

and

$$\frac{\theta}{m} = \theta_r + 2k\pi\ , \qquad [A4.36]$$

namely:

$$\rho = R_r^m \qquad [A4.37]$$

and

$$\theta = m\theta_r + 2mk\pi\ . \qquad [A4.38]$$

The restriction $]-\pi,+\pi[$ on the argument of s makes it possible to formulate the double inequation:

$$-\pi < m\theta_r + 2mk\pi < +\pi\ , \qquad [A4.39]$$

namely:

$$-\pi - m\theta_r < 2mk\pi < \pi - m\theta_r\ , \qquad [A4.40]$$

or even:

$$-\frac{1}{2m} - \frac{\theta_r}{2\pi} < k < \frac{1}{2m} - \frac{\theta_r}{2\pi}\ . \qquad [A4.41]$$

It is then essential to determine the values of k that verify this double inequality in conformity with the condition on θ_r, namely $-\pi < \theta_r \le +\pi$, or in this case:

$$-\frac{1}{2} < \frac{\theta_r}{2\pi} \le \frac{1}{2}\ . \qquad [A4.42]$$

The lower bound of k (relation [A4.41]), namely

$$-\frac{1}{2m}-\frac{\theta_r}{2\pi}, \qquad [A4.43]$$

is minimal for $\theta_r/2\pi = 1/2$; therefore:

$$-\frac{1}{2m}-\frac{\theta_r}{2\pi} \geq -\frac{1}{2m}-\frac{1}{2}, \qquad [A4.44]$$

or, given that $m \geq 2$:

$$-\frac{1}{2m}-\frac{\theta_r}{2\pi} \geq -\frac{3}{4}. \qquad [A4.45]$$

The upper bound of k (relation [A4.41]), namely

$$\frac{1}{2m}-\frac{\theta_r}{2\pi}, \qquad [A4.46]$$

is maximal for $\theta_r/2\pi = (-1/2)^+$; therefore:

$$\frac{1}{2m}-\frac{\theta_r}{2\pi} < \frac{1}{2m}+\frac{1}{2}, \qquad [A4.47]$$

or, knowing that $m \geq 2$:

$$\frac{1}{2m}-\frac{\theta_r}{2\pi} < \frac{3}{4}. \qquad [A4.48]$$

Finally, by minimizing and maximizing the lower and upper bounds of k (relations [A4.45] and [A4.48]), double inequality [A4.41] is reduced to:

$$-\frac{3}{4} < k < \frac{3}{4}. \qquad [A4.49]$$

This result conveys that $k = 0$ is the only possible solution.

Equations [A4.37] and [A4.38] then become:

$$\rho = R_r^m \text{ and } \theta = m\theta_r . \qquad [A4.50]$$

The zero of a factor of the type $s^{1/m} - s_r$, namely $s = \rho e^{j\theta}$, can be written as:

$$s = R_r^m e^{jm\theta_r} = \left(R_r e^{j\theta_r}\right)^m = s_r^m . \qquad [A4.51]$$

Moreover, as double inequation [A4.39] is reduced to:

$$-\pi < m\theta_r < +\pi , \qquad [A4.52]$$

namely:

$$-\frac{\pi}{m} < \theta_r < +\frac{\pi}{m} , \qquad [A4.53]$$

the argument of s_r satisfies the condition:

$$|\theta_r| < \frac{\pi}{m} . \qquad [A4.54]$$

CONCLUSION.– From relations [A4.51] and [A4.54], the zeros of the polynomial $P\left(s^{1/m}\right)$ are finally:

$$\left\{ s_r^{\ m} \textit{ where } s_r = R_r e^{j\theta_r} \textit{ with } |\theta_r| < \frac{\pi}{m} \right\} . \qquad [A4.55]$$

This leads us to say that *the zeros of* $P\left(s^{1/m}\right)$ *are nothing but some zeros of* $P(s)$ *raised to the power* m*, the retained zeros being the ones whose argument is less than* π / m *in absolute value.*

At an algorithmic level, knowing that the zeros of $P(s)$ are those of an integer degree polynomial, their seeking can be carried out by traditional methods, such as those of Bairstow or Newton. So obtained, it just remains to select them through their argument and to raise them to the integer power m. The entirety of these operations defines the algorithmic procedure that the CRONE software uses for solving a non-integer polynomial equation.

A4.3. Bibliography

[LAN 11] LANUSSE P., MALTI R., MELCHIOR P., "CRONE control-system design toolbox for the control engineering community", *International Design Engineering Technical Conferences and Computers and Information in Engineering Conference (ASME '11)*, Washington, pp. 129–136, 2011.

[MAL 12] MALTI R., MELCHIOR P., LANUSSE P., *et al.*, "Object-oriented CRONE toolbox for fractional differential signal processing", *Signal, Image and Video Processing,* Springer, vol. 6, no. 3, pp. 393–400, 2012.

[OUS 95] OUSTALOUP A., *La dérivation non entière: théorie, synthèse et applications*, Hermès, Paris, 1995.

[OUS 00] OUSTALOUP A., MELCHIOR P., LANUSSE P., *et al.*, "The CRONE toolbox for Matlab", *IEEE International Symposium on Computer-Aided Control System Design (CACSD'00)*, pp. 190–195, 2000.

Appendix 5

CRONE Control Response with Initial Conditions

A5.1. Laplace transform of the generalized derivative of a time function in the presence of initial conditions

The application of the first definition of the generalized derivative to a function $f(t)$, which is continuous and infinitely differentiable on $\mathbb{R}$, is expressed by the equation:

$$D^n f(t) = \frac{1}{h^n} \sum_{k=0}^{\infty} (-1)^k \binom{n}{k} f(t-kh). \qquad [A5.1]$$

Its Laplace transform, namely:

$$\mathcal{L}\left[\mathrm{D}^n f(t)\right] = \int_0^{\infty} \left[\mathrm{D}^n f(t)\right] e^{-st}\, dt, \qquad [A5.2]$$

is written as (given [A5.1]):

$$\mathcal{L}\left[\mathrm{D}^n f(t)\right] = \int_0^{\infty} \frac{1}{h^n} \sum_{k=0}^{\infty} (-1)^k \binom{n}{k} f(t-kh) e^{-st}\, dt, \qquad [A5.3]$$

or:

$$\mathcal{L}\left[\mathrm{D}^n f(t)\right]=\frac{1}{h^n}\int_0^{\infty}\sum_{k=0}^{\infty}(-1)^k\binom{n}{k}f(t-kh)e^{-st}\,dt\,, \qquad \text{[A5.4]}$$

or even, by permuting the integral and sum signs:

$$\mathcal{L}\left[\mathrm{D}^n f(t)\right]=\frac{1}{h^n}\sum_{k=0}^{\infty}(-1)^k\binom{n}{k}\int_0^{\infty}f(t-kh)e^{-st}\,dt\,. \qquad \text{[A5.5]}$$

The change of variable defined by:

$$\theta=t-kh\,, \qquad \text{[A5.6]}$$

allows us to write:

$$\mathcal{L}\left[\mathrm{D}^n f(t)\right]=\frac{1}{h^n}\sum_{k=0}^{\infty}(-1)^k\binom{n}{k}e^{-skh}\int_{-kh}^{\infty}f(\theta)e^{-s\theta}\,d\theta\,, \qquad \text{[A5.7]}$$

namely:

$$\mathcal{L}\left[\mathrm{D}^n f(t)\right]=\frac{1}{h^n}\sum_{k=0}^{\infty}(-1)^k\binom{n}{k}e^{-skh}\left[\int_{-kh}^{0}f(\theta)e^{-s\theta}\,d\theta+\int_{0}^{\infty}f(\theta)e^{-s\theta}\,d\theta\right], \qquad \text{[A5.8]}$$

or else, given that the second integral is the Laplace transform $F(s)$ of $f(t)$:

$$\mathcal{L}\left[\mathrm{D}^n f(t)\right]=\frac{1}{h^n}\sum_{k=0}^{\infty}(-1)^k\binom{n}{k}e^{-skh}\left[\int_{-kh}^{0}f(\theta)e^{-s\theta}\,d\theta+F(s)\right]. \qquad \text{[A5.9]}$$

As for the remaining integral, a sequence of integrations by parts leads to:

$$\int_{-kh}^{0}f(\theta)e^{-s\theta}\,d\theta=\sum_{i=0}^{\infty}\left[-\left(\frac{1}{s}\right)^{i+1}f^{(i)}(\theta)e^{-s\theta}\right]_{-kh}^{0}, \qquad \text{[A5.10]}$$

or even:

$$\int_{-kh}^{0}f(\theta)e^{-s\theta}\,d\theta=-\sum_{i=0}^{\infty}\left[\left(\frac{1}{s}\right)^{i+1}f^{(i)}(0)-\left(\frac{1}{s}\right)^{i+1}e^{skh}f^{(i)}(-kh)\right]. \qquad \text{[A5.11]}$$

By putting [A5.11] into [A5.9], we can determine the equation:

$$\mathcal{L}\left[\mathrm{D}^n f(t)\right]=\frac{1}{h^n}\sum_{k=0}^{\infty}(-1)^k\binom{n}{k}\left[e^{-skh}\left(F(s)-\sum_{i=0}^{\infty}\left(\frac{1}{s}\right)^{i+1}f^{(i)}(0)\right.\right.$$
$$\left.\left.+\sum_{i=0}^{\infty}\left(\frac{1}{s}\right)^{i+1}e^{skh}f^{(i)}(-kh)\right)\right], \quad [A5.12]$$

namely:

$$\mathcal{L}\left[\mathrm{D}^n f(t)\right]=\frac{1}{h^n}\sum_{k=0}^{\infty}(-1)^k\binom{n}{k}\left[e^{-skh}\left(F(s)-\sum_{i=0}^{\infty}\left(\frac{1}{s}\right)^{i+1}f^{(i)}(0)\right)\right.$$
$$\left.+\sum_{i=0}^{\infty}\left(\frac{1}{s}\right)^{i+1}f^{(i)}(-kh)\right], \quad [A5.13]$$

or:

$$\mathcal{L}\left[\mathrm{D}^n f(t)\right]=\left(F(s)-\sum_{i=0}^{\infty}\left(\frac{1}{s}\right)^{i+1}f^{(i)}(0)\right)\frac{1}{h^n}\sum_{k=0}^{\infty}(-1)^k\binom{n}{k}e^{-skh}$$
$$+\frac{1}{h^n}\sum_{k=0}^{\infty}(-1)^k\binom{n}{k}\sum_{i=0}^{\infty}\left(\frac{1}{s}\right)^{i+1}f^{(i)}(-kh), \quad [A5.14]$$

or else, given that

$$\frac{1}{h^n}\sum_{k=0}^{\infty}(-1)^k\binom{n}{k}e^{-skh}=\frac{1}{h^n}\sum_{k=0}^{\infty}(-1)^k\binom{n}{k}\left(e^{-sh}\right)^k$$

$=\frac{1}{h^n}\left(1-e^{-sh}\right)^n=s^n$, for h infinitely small: [A5.15]

$$\mathcal{L}\left[\mathrm{D}^n f(t)\right]=s^nF(s)-\sum_{i=0}^{\infty}s^{n-i-1}f^{(i)}(0)+\frac{1}{h^n}\sum_{k=0}^{\infty}(-1)^k\binom{n}{k}\sum_{i=0}^{\infty}\left(\frac{1}{s}\right)^{i+1}f^{(i)}(-kh). \quad [A5.16]$$

The second series of [A5.16] can be rewritten in the form:

$$\frac{1}{h^n}\sum_{k=0}^{\infty}\sum_{i=0}^{\infty}(-1)^k\binom{n}{k}\left(\frac{1}{s}\right)^{i+1}f^{(i)}(-kh), \quad [A5.17]$$

namely, by changing the summation order:

$$\frac{1}{h^n}\sum_{i=0}^{\infty}\sum_{k=0}^{\infty}(-1)^k\binom{n}{k}\left(\frac{1}{s}\right)^{i+1} f^{(i)}(-kh), \qquad [A5.18]$$

or:

$$\frac{1}{h^n}\sum_{i=0}^{\infty}\left(\frac{1}{s}\right)^{i+1}\sum_{k=0}^{\infty}(-1)^k\binom{n}{k}f^{(i)}(-kh), \qquad [A5.19]$$

or even:

$$\sum_{i=0}^{\infty}\left(\frac{1}{s}\right)^{i+1}\frac{1}{h^n}\sum_{k=0}^{\infty}(-1)^k\binom{n}{k}f^{(i)}(-kh), \qquad [A5.20]$$

or else, by remarking that

$$\begin{aligned}\frac{1}{h^n}\sum_{k=0}^{\infty}(-1)^k\binom{n}{k}f^{(i)}(-kh) &= \left[\frac{1}{h^n}\sum_{k=0}^{\infty}(-1)^k\binom{n}{k}f^{(i)}(t-kh)\right]_{t=0} \\ &= \left[\mathrm{D}^n f^{(i)}(t)\right]_{t=0} = \left[f^{(n+i)}(t)\right]_{t=0} = f^{(n+i)}(0):\end{aligned} \qquad [A5.21]$$

$$\sum_{i=0}^{\infty}\left(\frac{1}{s}\right)^{i+1} f^{(n+i)}(0), \qquad [A5.22]$$

or finally:

$$\sum_{i=0}^{\infty}s^{-i-1}f^{(n+i)}(0). \qquad [A5.23]$$

This expression when put into equation [A5.16] determines the general form of the Laplace transform of $\mathrm{D}^n f(t)$, namely:

$$\mathcal{L}\left[\mathrm{D}^n f(t)\right] = s^n F(s) - \sum_{i=0}^{\infty}s^{n-i-1}f^{(i)}(0) + \sum_{i=0}^{\infty}s^{-i-1}f^{(n+i)}(0). \qquad [A5.24]$$

In the particular case when n takes integer values, the well-known formula is retrieved:

$$\mathcal{L}\left[\mathrm{D}^n f(t)\right] = s^n F(s) - \sum_{i=0}^{n-1}s^{n-i-1}f^{(i)}(0). \qquad [A5.25]$$

A5.2. Response to any input of the second generation CRONE control for non-zero initial conditions

Let us take again the free control loop of Figure 2.15, of open-loop transmittance:

$$\beta(s)=\left(\frac{1}{\tau s}\right)^n, \qquad \text{[A5.26]}$$

then let us now consider a non-zero input $E(s)$ that allows us to define a closed-loop transmittance, relative to $E(s)$, that is to say a *tracking transmittance*, namely:

$$F(s)=\frac{\beta(s)}{1+\beta(s)}=\frac{1}{1+(\tau s)^n}, \qquad \text{[A5.27]}$$

from which we draw (denoting by $S(s)$ the loop output):

$$(\tau s)^n S(s)+S(s)=E(s), \qquad \text{[A5.28]}$$

a symbolic equation that, in time domain, admits for original equation:

$$\tau^n D^n s(t)+s(t)=e(t) \text{ with } D=d/dt. \qquad \text{[A5.29]}$$

The aim of this section is to determine the solution of this differential equation for an arbitrary input $e(t)$ and in the presence of initial conditions on the output $s(t)$ and its derivatives.

Given the general expression of the Laplace transform of $D^n s(t)$ defined by relation [A5.24], equation [A5.29] admits the operational representation:

$$\tau^n s^n S(s)-\tau^n\sum_{i=0}^{\infty}s^{n-i-1}s^{(i)}(0)+\tau^n\sum_{i=0}^{\infty}s^{-i-1}s^{(n+i)}(0)+S(s)=E(s), \qquad \text{[A5.30]}$$

or, by gathering the terms in $S(s)$:

$$S(s)\left(1+\tau^n s^n\right)=E(s)-\tau^n\sum_{i=0}^{\infty}s^{-i-1}s^{(n+i)}(0)+\tau^n\sum_{i=0}^{\infty}s^{n-i-1}s^{(i)}(0), \qquad \text{[A5.31]}$$

from which we draw the transform $S(s)$:

$$S(s) = \frac{E(s)}{1+\tau^n s^n} - \tau^n \sum_{i=0}^{\infty} \frac{s^{-i-1}}{1+\tau^n s^n} s^{(n+i)}(0) + \tau^n \sum_{i=0}^{\infty} \frac{s^{n-i-1}}{1+\tau^n s^n} s^{(i)}(0). \qquad [A5.32]$$

The original $s(t)$ is written in the form:

$$\begin{aligned} s(t) = \mathcal{L}^{-1}\left[\frac{E(s)}{1+\tau^n s^n}\right] - \tau^n \sum_{i=0}^{\infty} s^{(n+i)}(0)\,\mathcal{L}^{-1}\left[\frac{s^{-i-1}}{1+\tau^n s^n}\right] \\ + \tau^n \sum_{i=0}^{\infty} s^{(i)}(0)\,\mathcal{L}^{-1}\left[\frac{s^{n-i-1}}{1+\tau^n s^n}\right]. \end{aligned} \qquad [A5.33]$$

The first term of the above equation can be written, from the Borel–Duhamel theorem, as:

$$\mathcal{L}^{-1}\left[\frac{E(s)}{1+\tau^n s^n}\right] = \mathcal{L}^{-1}\left[\frac{1}{1+\tau^n s^n}\right] * \mathcal{L}^{-1}\left[E(s)\right] = \mathcal{L}^{-1}\left[\frac{1}{1+\tau^n s^n}\right] * e(t). \qquad [A5.34]$$

The inverse transform:

$$\mathcal{L}^{-1}\left[\frac{1}{1+\tau^n s^n}\right], \qquad [A5.35]$$

is the unit impulse response of a system with explicit generalized derivative, namely:

$$\mathcal{L}^{-1}\left[\frac{1}{1+\tau^n s^n}\right] = \sum_{j=0}^{\infty} (-1)^j \frac{t^{n(j+1)-1}}{\tau^{n(j+1)}\Gamma\left(n(j+1)\right)}. \qquad [A5.36]$$

Relation [A5.34] becomes:

$$\mathcal{L}^{-1}\left[\frac{E(s)}{1+\tau^n s^n}\right] = \left[\sum_{j=0}^{\infty} (-1)^j \frac{t^{n(j+1)-1}}{\tau^{n(j+1)}\Gamma\left(n(j+1)\right)}\right] * e(t), \qquad [A5.37]$$

namely, from the convolution product definition:

$$\mathcal{L}^{-1}\left[\frac{E(s)}{1+\tau^n s^n}\right] = \int_0^t \left[\sum_{j=0}^{\infty} (-1)^j \frac{\theta^{n(j+1)-1}}{\tau^{n(j+1)}\Gamma\left(n(j+1)\right)}\right] e(t-\theta)\, d\theta. \qquad [A5.38]$$

The inverse transform that figures in the first series of [A5.33] calls the form:

$$\mathcal{L}^{-1}\left[\frac{s^{-i-1}}{1+\tau^n s^n}\right]=\left(\frac{d}{dt}\right)^{-i-1}\mathcal{L}^{-1}\left[\frac{1}{1+\tau^n s^n}\right], \qquad \text{[A5.39]}$$

namely, given [A5.36]:

$$\mathcal{L}^{-1}\left[\frac{s^{-i-1}}{1+\tau^n s^n}\right]=\sum_{j=0}^{\infty}\left(\frac{d}{dt}\right)^{-i-1}(-1)^j\frac{t^{n(j+1)-1}}{\tau^{n(j+1)}\Gamma(n(j+1))}, \qquad \text{[A5.40]}$$

or, given the generalized differentiation formula

$$\left(\frac{d}{dt}\right)^n\left[at^{\alpha}u(t)\right]=a\Gamma(\alpha+1)\frac{t^{\alpha-n}}{\Gamma(\alpha-n+1)}u(t): \qquad \text{[A5.41]}$$

$$\mathcal{L}^{-1}\left[\frac{s^{-i-1}}{1+\tau^n s^n}\right]=\sum_{j=0}^{\infty}(-1)^j\frac{t^{n(j+1)+i}}{\tau^{n(j+1)}\Gamma(n(j+1)+i+1)}. \qquad \text{[A5.42]}$$

Let us note that this result can be differently established by remarking that a differentiation of order $-i-1$ is nothing but an integration of order $i+1$, namely:

$$\begin{aligned}\mathcal{L}^{-1}\left[\frac{s^{-i-1}}{1+\tau^n s^n}\right]&=\sum_{j=0}^{\infty}\int_0^t\int_0^{t_1}\cdots\int_0^{t_i}(-1)^j\frac{t_{i+1}^{n(j+1)-1}}{\tau^{n(j+1)}\Gamma(n(j+1))}dt_{i+1}dt_i\ldots dt_1\\&=\sum_{j=0}^{\infty}(-1)^j\frac{t^{n(j+1)+i}}{\tau^{n(j+1)}\Gamma(n(j+1))(n(j+1))(n(j+1)+1)\cdots(n(j+1)+i-1)(n(j+1)+i)},\end{aligned} \qquad \text{[A5.43]}$$

or, given that

$$\begin{aligned}&\Gamma(n(j+1))(n(j+1))=\Gamma(n(j+1)+1)\\&\Gamma(n(j+1)+1)(n(j+1)+1)=\Gamma(n(j+1)+2)\\&\Gamma(n(j+1))(n(j+1))\cdots(n(j+1)+i)=\Gamma(n(j+1)+i+1):\end{aligned} \qquad \text{[A5.44]}$$

$$\mathcal{L}^{-1}\left[\frac{s^{-i-1}}{1+\tau^n s^n}\right]=\sum_{j=0}^{\infty}(-1)^j\frac{t^{n(j+1)+i}}{\tau^{n(j+1)}\Gamma(n(j+1)+i+1)}.\ \text{QED} \qquad \text{[A5.45]}$$

The inverse transform that figures in the second series of [A5.33] can be written as:

$$\mathcal{L}^{-1}\left[\frac{s^{n-i-1}}{1+\tau^n s^n}\right]=\mathcal{L}^{-1}\left[s^n\frac{s^{-i-1}}{1+\tau^n s^n}\right], \qquad \text{[A5.46]}$$

namely:

$$\mathcal{L}^{-1}\left[\frac{s^{n-i-1}}{1+\tau^n s^n}\right]=\left(\frac{d}{dt}\right)^n\mathcal{L}^{-1}\left[\frac{s^{-i-1}}{1+\tau^n s^n}\right], \qquad \text{[A5.47]}$$

or, given relation [A5.42]:

$$\mathcal{L}^{-1}\left[\frac{s^{n-i-1}}{1+\tau^n s^n}\right]=\sum_{j=0}^{\infty}(-1)^j\left(\frac{d}{dt}\right)^n\frac{t^{n(j+1)+i}}{\tau^{n(j+1)}\Gamma\left(n(j+1)+i+1\right)}, \qquad \text{[A5.48]}$$

or else, given the generalized differentiation formula [A5.41]:

$$\mathcal{L}^{-1}\left[\frac{s^{n-i-1}}{1+\tau^n s^n}\right]=\sum_{j=0}^{\infty}(-1)^j\frac{t^{nj+i}}{\tau^{n(j+1)}\Gamma\left(nj+i+1\right)}. \qquad \text{[A5.49]}$$

By putting relations [A5.38], [A5.42] and [A5.49] into [A5.33], we can determine the solution $s(t)$ of the differential equation [A5.29]:

$$\begin{aligned} s(t)=&\int_0^t\sum_{j=0}^{\infty}(-1)^j\frac{\theta^{n(j+1)-1}}{\tau^{n(j+1)}\Gamma\left(n(j+1)\right)}e(t-\theta)\,d\theta \\ &-\tau^n\sum_{i=0}^{\infty}s^{(n+i)}(0)\sum_{j=0}^{\infty}(-1)^j\frac{t^{n(j+1)+i}}{\tau^{n(j+1)}\Gamma\left(n(j+1)+i+1\right)} \\ &+\tau^n\sum_{i=0}^{\infty}s^{(i)}(0)\sum_{j=0}^{\infty}(-1)^j\frac{t^{nj+i}}{\tau^{n(j+1)}\Gamma\left(nj+i+1\right)}, \end{aligned} \qquad \text{[A5.50]}$$

namely:

$$\begin{aligned} s(t)=&\int_0^t\sum_{j=1}^{\infty}(-1)^{j-1}\frac{\theta^{nj-1}}{\tau^{nj}\Gamma\left(nj\right)}e(t-\theta)\,d\theta \\ &+\sum_{i=0}^{\infty}\left[s^{(i)}(0)\sum_{j=0}^{\infty}(-1)^j\frac{t^{nj+i}}{\tau^{nj}\Gamma\left(nj+i+1\right)}\right. \\ &\left.-s^{(n+i)}(0)\sum_{j=0}^{\infty}(-1)^j\frac{t^{n(j+1)+i}}{\tau^{nj}\Gamma\left(n(j+1)+i+1\right)}\right]. \end{aligned} \qquad \text{[A5.51]}$$

We pose:

$$X^T(0)=\left[s^{(0)}(0)\,\vdots\, s^{(1)}(0)\,\vdots\ldots\vdots\, s^{(i)}(0)\,\vdots\ldots\right]; \qquad [A5.52]$$

$$\varphi^{(0)}(n,t,\tau)=\sum_{j=0}^{\infty}(-1)^j\frac{t^{nj}}{\tau^{nj}\Gamma(nj+1)}; \qquad [A5.53]$$

$$\varphi^{(-i)}(n,t,\tau)=\sum_{j=0}^{\infty}(-1)^j\frac{t^{nj+i}}{\tau^{nj}\Gamma(nj+i+1)}, \qquad [A5.54]$$

a relation deduced from [A5.53] by applying formula [A5.41];

$$\Phi^T(n,t,\tau)=\left[\varphi^{(0)}(n,t,\tau)\,\vdots\,\varphi^{(-1)}(n,t,\tau)\,\vdots\ldots\vdots\,\varphi^{(-i)}(n,t,\tau)\,\vdots\ldots\right]. \qquad [A5.55]$$

$$\sum_{i=0}^{\infty}s^{(i)}(0)\sum_{j=0}^{\infty}(-1)^j\frac{t^{nj+i}}{\tau^{nj}\Gamma(nj+i+1)} \qquad [A5.56]$$

is then written in the form:

$$\sum_{i=0}^{\infty}s^{(i)}(0)\varphi^{(-i)}(n,t,\tau)=\varphi^{(0)}(n,t,\tau)s^{(0)}(0)+\varphi^{(-1)}(n,t,\tau)s^{(1)}(0)$$
$$+\cdots+\varphi^{(-i)}(n,t,\tau)s^{(i)}(0)+\cdots=\Phi^T(n,t,\tau)X(0). \qquad [A5.57]$$

$$\sum_{i=0}^{\infty}s^{(n+i)}(0)\sum_{j=0}^{\infty}(-1)^j\frac{t^{n(j+1)+i}}{\tau^{nj}\Gamma(n(j+1)+i+1)} \qquad [A5.58]$$

can be written as:

$$\sum_{i=0}^{\infty}s^{(n+i)}(0)\sum_{j=0}^{\infty}(-1)^j\frac{1}{\tau^{nj}\Gamma(nj+i+1)}\frac{\Gamma(nj+i+1)t^{nj+i-(-n)}}{\Gamma(nj+i-(-n)+1)}, \qquad [A5.59]$$

namely, by considering formula [A5.41]:

$$\sum_{i=0}^{\infty}s^{(n+i)}(0)\sum_{j=0}^{\infty}(-1)^j\frac{1}{\tau^{nj}\Gamma(nj+i+1)}\left(\frac{d}{dt}\right)^{-n}t^{nj+i}, \qquad [A5.60]$$

or:

$$\sum_{i=0}^{\infty} s^{(n+i)}(0)\left(\frac{d}{dt}\right)^{-n} \sum_{j=0}^{\infty}(-1)^j \frac{t^{nj+i}}{\tau^{nj}\Gamma(nj+i+1)}, \qquad \text{[A5.61]}$$

or even, given the expression of $\varphi^{(-i)}(n,t,\tau)$:

$$\sum_{i=0}^{\infty} s^{(n+i)}(0)\,\varphi^{(-n-i)}(n,t,\tau), \qquad \text{[A5.62]}$$

or else:

$$\begin{aligned}\varphi^{(-n)}(n,t,\tau)s^{(n)}(0)+\varphi^{(-n-1)}(n,t,\tau)s^{(n+1)}(0)+\cdots+ \\ \varphi^{(-n-i)}(n,t,\tau)s^{(n+i)}(0)+\cdots = \Phi^{(-n)^T}(n,t,\tau)X^{(n)}(0).\end{aligned} \qquad \text{[A5.63]}$$

Putting relations [A5.57] and [A5.63] into equation [A5.51] enables us to obtain a more condensed form of the solution $s(t)$, namely:

$$\begin{aligned}s(t) = \int_0^t \sum_{j=1}^{\infty}(-1)^{j-1}\frac{\theta^{nj-1}}{\tau^{nj}\Gamma(nj)}e(t-\theta)\,d\theta + \Phi^T(n,t,\tau)X(0) \\ -\Phi^{(-n)^T}(n,t,\tau)X^{(n)}(0)\end{aligned}. \qquad \text{[A5.64]}$$

Any of relations [A5.50], [A5.51] or [A5.64] shows that *the evolution of the output depends on the initial values of all the integer-order derivatives of* $s(t)$, *namely* $s^{(0)}(t)$, $s^{(1)}(t)$,..., $s^{(i)}(t)$,..., *and also all the integer-order derivatives of* $s^{(n)}(t)$, *namely* $s^{(n)}(t)$, $s^{(n+1)}(t)$,..., $s^{(n+i)}(t)$,... .

In the particular case where the differential equation order is an integer, namely $n=k$, mathematical simplifications are obvious. Equation [A5.51] is, indeed, written in the form:

$$\begin{aligned}s(t) = \int_0^t \sum_{j=1}^{\infty}(-1)^{j-1}\frac{\theta^{kj-1}}{\tau^{kj}(kj-1)!}e(t-\theta)\,d\theta + \sum_{i=0}^{\infty} s^{(i)}(0)\sum_{j=0}^{\infty}(-1)^j\frac{t^{kj+i}}{\tau^{kj}(kj+i)!} \\ -\sum_{i=0}^{\infty} s^{(k+i)}(0)\sum_{j=0}^{\infty}(-1)^j\frac{t^{k(j+1)+i}}{\tau^{kj}(k(j+1)+i)!},\end{aligned} \qquad \text{[A5.65]}$$

or:

$$s(t)=\int_0^t\sum_{j=1}^{\infty}(-1)^{j-1}\frac{\theta^{kj-1}}{\tau^{kj}(kj-1)!}e(t-\theta)\,d\theta+\sum_{i=0}^{\infty}s^{(i)}(0)\sum_{j=0}^{\infty}(-1)^j\frac{t^{kj+i}}{\tau^{kj}(kj+i)!}$$
$$-\sum_{i=k}^{\infty}s^{(i)}(0)\sum_{j=0}^{\infty}(-1)^j\frac{t^{kj+i}}{\tau^{kj}(kj+i)!},\qquad [A5.66]$$

namely, finally:

$$s(t)=\int_0^t\sum_{j=1}^{\infty}(-1)^{j-1}\frac{\theta^{kj-1}}{\tau^{kj}(kj-1)!}e(t-\theta)\,d\theta+\sum_{i=0}^{k-1}s^{(i)}(0)\,\varphi^{(-i)}(k,t,\tau).\qquad [A5.67]$$

This expression confirms that the solution depends only on the derivatives of orders 0 to $(k-1)$ at the initial time.

A5.3. Bibliography

[ERD 62] ERDELYI A., *Operational Calculus and Generalized Functions*, Holt, Rinehart and Winston, 1962.

[GRÜ 67] GRÜNWALD A.K., "Ueber begrenzte Derivationen und deren Anwendung", *Zeitschrift für Mathematik and Physik*, vol. 12, pp. 441–480, 1867.

[OUS 91] OUSTALOUP A., *La commande CRONE*, Hermès, Paris, 1991.

[OUS 95] OUSTALOUP A., *La dérivation non entière: théorie, synthèse et applications*, Hermès, Paris, 1995.

[OUS 99a] OUSTALOUP A., MATHIEU B., *La Commande CRONE: du scalaire au multivariable*, Hermès, Paris, 1999.

[OUS 99b] OUSTALOUP A., SABATIER J., LANUSSE P., "From fractal robustness to the CRONE control", *Fractional Calculus and Applied Analysis (FCAA): An International Journal for Theory and Applications*, vol. 2, no. 1, pp. 1–30, January 1999.

Appendix 6

Fractality and Non-integer Differentiation

A6.1. On interdependence between fractality and non-integer differentiation

In geometry, the continuity of the dimension of a figure or of the associated space implies the notion of figure or space of a *non-integer dimension* or generically of *fractality* or *fractal* (a fractal is defined as a figure whose dimension is non-integer).

In systemics, the system order continuity implies the notion of *non-integer order* system or generically of *non-integer differentiation.*

The above description leads us to say that *fractality is a matter of geometry, whereas non-integer differentiation is a matter of systemics or, more precisely, of system dynamics*. The interdependence of these two concepts can only then appear when there is a correlation between geometry and dynamics, notably when geometry conditions dynamics, or more precisely when geometry conditions a physical phenomenon governed by a differential equation. A geometry of fractal dimension then leads to a differential equation of non-integer order. The opposite, where dynamics conditions geometry, admits a characteristic illustration example in signal processing: the white noise filtering by a non-integer order filter indeed leads to a colored noise of fractal dimension (this phenomenon lies at the origin of the *fractal noise* synthesis).

Because of the interdependence so defined, the examples for which non-integer derivative is used as a modeling tool are numerous in physics. The most characteristic examples are mentioned hereafter in a non-exhaustive manner.

In physico-chemistry, in a fractal interface between a metal and an ionic medium, the current is proportional to the non-integer derivative of the voltage [LEM 83].

In electricity, in a capacitor, the current is proportional to the non-integer derivative of the voltage if at least one of its armatures is rough [OUS 86].

In mechanics, in the relaxation of water on a dyke, the flow is proportional to the non-integer derivative of the pressure at the water-dyke interface if the dyke's internal structure is porous [OUS 91]. Likewise, in a viscoelastic material, the constraint is proportional to the non-integer derivative of the deformation [BAG 83].

In acoustics, in a wind instrument, the non-integer derivative is used to model the viscothermal losses [MAT 93].

The analogy between fractality and non-integer differentiation concepts is obvious by the attribution of a continuous character to magnitudes considered so far to be discrete, as much as the analytic formalization of their interdependence is a matter of high complexity. Our contribution (in this book) is precisely inscribed in this issue. Indeed, it consists of highlighting the close link between these two concepts, notably through the change from a fractal dimension to a non-integer differentiation order. To this effect, we propose two systems as study supports, whose fractality results from roughness for one and porosity for the other: the first study system is a metal-vacuum capacitor whose at least one armature presents a *rough surface structure*; the second study system is a water-dyke interface whose dyke presents a *porous volume structure*. These two study systems are, indeed, significant to the two dual approaches inherent to the respective changes *from roughness and porosity to non-integer differentiation.*

In the framework of our developments, and on account of the recursive character of fractality, roughness and porosity are considered through their recursivity, whether of fractal origin or not. The consideration of a non-fractal recursive geometry indeed enables us to show that a fractal recursive geometry is not necessary for obtaining a non-integer differentiation, a geometry exclusively recursive being sufficient. It is true that *fractality leads to non-integer differentiation more by its recursivity than by fractality itself* [OUS 95].

The change from porosity to non-integer differentiation through a water-dyke interface as a study support has been extensively studied in this book in Chapters 1 and 2 and also in Appendix 2. This appendix constitutes a direct complement (or, more precisely, dual) in the matter, in the sense that it studies the change from roughness to non-integer differentiation through a metal-vacuum capacitor as a study support.

A6.2. A metal-vacuum capacitor as a study support to roughness/non-integer differentiation change

The considered study support is a *physical support* to study the change from roughness to non-integer differentiation.

This physical support is a capacitive device formed by two plane metallic armatures (or electrodes), of negligible resistivity, whose intern face of one armature is covered of a resistive deposit (of resistivity ρ) presenting a surface roughness of recursive or fractal nature.

The roughness is assumed to be uniformly distributed on the whole surface. The roughness profiles that various transversal cuts determine then present the same recursive motif whatever the cut.

The study principle consists of determining the admittance form of an arbitrary transversal slice of elementary thickness dz.

Given that the system is electrically equivalent to an infinity of parallel elementary slices, the global admittance comes down to the sum extended to infinity of the elementary admittances of each of the slices. Their common form then determines the one of the global admittance.

For study clarity reasons, the elementary slices are chosen in such a way that their width x_0 contains only one recursive motif. Moreover, x_0 is considered unitary as a simplifying normalization, namely $x_0 = 1$.

A last study consideration consists of assuming that the current and field lines are orthogonal to the plane defined by the system of Cartesian coordinates $x0z$. This simplifying hypothesis is at the origin of the following approximate study that explains the non-integer differentiation phenomenon recorded on the capacitors whose armatures present surface imperfections: indeed, *the current is then no longer proportional to the first derivative of the voltage, but to the non-integer derivative of the voltage, the order being between 0 and 1.*

A6.3. Roughness of recursive nature

A6.3.1. *First construction steps of a (non-fractal) recursive motif and of the equivalent electrical circuit*

A6.3.1.1. *Step 0*

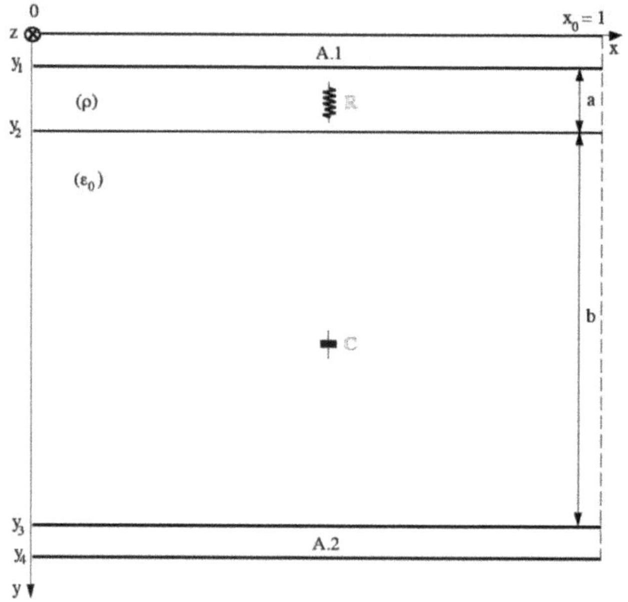

Figure A6.1. *Initial step of the recursive motif construction*

$$R = \rho \frac{a}{\mathrm{d}z} \; ; \; C = \varepsilon_0 \frac{\mathrm{d}z}{b} . \qquad \text{[A6.1]}$$

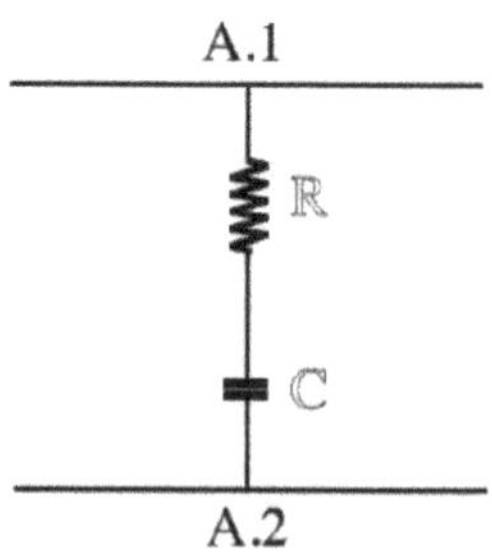

Figure A6.2. *Equivalent electric circuit with one branch*

A6.3.1.2. *Step 1*

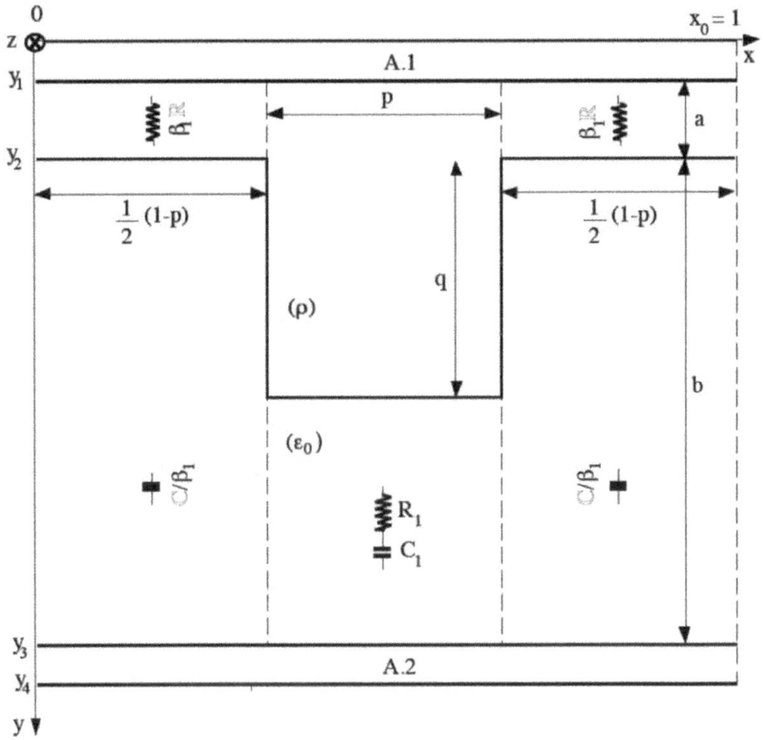

Figure A6.3. *The first step of the recursive motif construction*

$$\beta_1 R = \rho \frac{a}{\frac{1}{2}(1-p)\,\mathrm{d}z} = \frac{2}{1-p} R \text{, hence } \beta_1 = \frac{2}{1-p}; \qquad \text{[A6.2]}$$

$$\frac{C}{\beta_1} = \varepsilon_0 \frac{\frac{1}{2}(1-p)\,\mathrm{d}z}{b} = \frac{1-p}{2} C \text{, hence } \frac{1}{\beta_1} = \frac{1-p}{2}; \qquad \text{[A6.3]}$$

$$R_1 = \rho \frac{a+q}{p\,\mathrm{d}z}; \; C_1 = \varepsilon_0 \frac{p\,\mathrm{d}z}{\text{b-q}}. \qquad \text{[A6.4]}$$

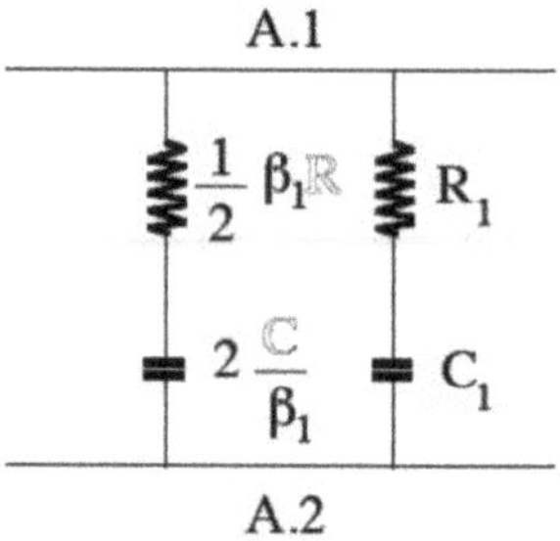

Figure A6.4. *Equivalent electric circuit with two branches*

A6.3.1.3. *Step 2*

$$\beta_2 R = \rho \frac{a}{\frac{1}{4}(1-p)^2 \,\mathrm{dz}} = \frac{4}{(1-p)^2} R \text{, therefore } \beta_2 = \frac{4}{(1-p)^2}; \qquad [A6.5]$$

$$\frac{C}{\beta_2} = \varepsilon_0 \frac{1}{4}(1-p)^2 \frac{\mathrm{dz}}{\mathrm{b}} = \frac{(1-p)^2}{4} C \text{, therefore } \frac{1}{\beta_2} = \frac{(1-p)^2}{4}; \qquad [A6.6]$$

$$R_1 = \rho \frac{a+q}{p\,\mathrm{dz}}; \; C_1 = \varepsilon_0 \frac{p\,\mathrm{dz}}{\mathrm{b\text{-}q}}; \qquad [A6.7]$$

$$R_2' = \rho \frac{a + \frac{q}{2}(1-p)}{\frac{p}{2}(1-p)\,\mathrm{dz}}; \; C_2' = \varepsilon_0 \frac{\frac{p}{2}(1-p)\mathrm{dz}}{\mathrm{b}\text{-}\frac{q}{2}(1-p)}; \qquad [A6.8]$$

$$R_2 = \frac{1}{2} R_2' = \frac{1}{2} \rho \frac{a + \frac{q}{2}(1-p)}{\frac{p}{2}(1-p)\,\mathrm{dz}}; \qquad [A6.9]$$

$$C_2 = 2C_2' = 2\varepsilon_0 \frac{\frac{p}{2}(1-p)\,\mathrm{dz}}{b - \frac{q}{2}(1-p)}. \qquad [A6.10]$$

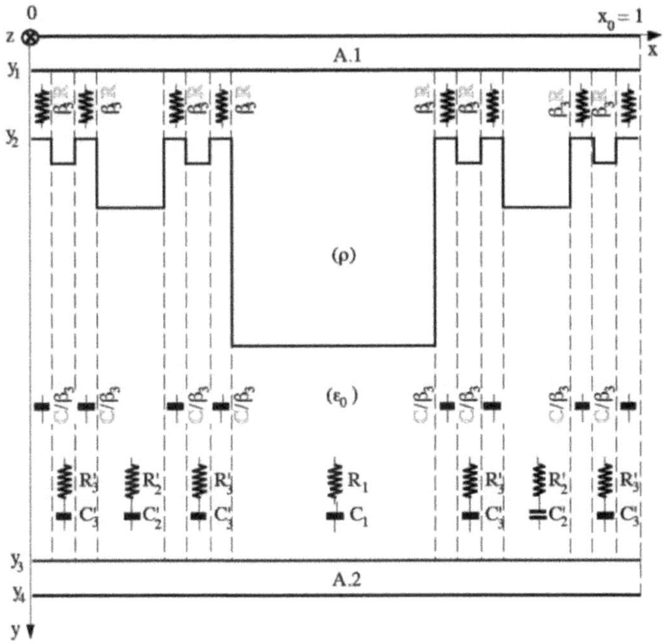

Figure A6.5. *The second step of the recursive motif construction*

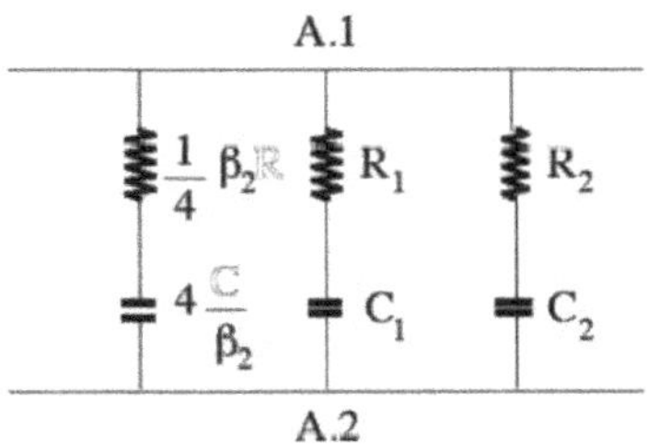

Figure A6.6. *Equivalent electric circuit with three branches*

A6.3.1.4. *Step 3*

$$\beta_3 R = \rho \frac{a}{\frac{1}{8}(1-p)^3\, dz} = \frac{8}{(1-p)^3} R \text{, hence } \beta_3 = \frac{8}{(1\text{-}p)^3}; \qquad \text{[A6.11]}$$

$$\frac{C}{\beta_3} = \varepsilon_0 \frac{1}{8}(1-p)^3 \frac{\mathrm{dz}}{\mathrm{b}} = \frac{(1-p)^3}{8} C \text{, hence } \frac{1}{\beta_3} = \frac{(1-p)^3}{8}; \qquad \text{[A6.12]}$$

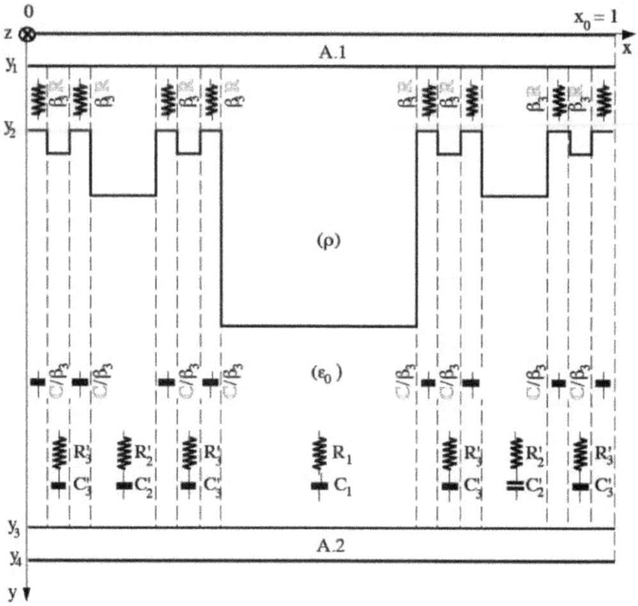

Figure A6.7. *The third step of the recursive motif construction*

$$R_1 = \rho \frac{a+q}{p\,\mathrm{dz}}\,;\; C_1 = \varepsilon_0 \frac{p\,\mathrm{dz}}{\mathrm{b\text{-}q}}\,; \qquad [A6.13]$$

$$R_2^{'} = \rho \frac{a+\frac{q}{2}(1-p)}{\frac{p}{2}(1-p)\,\mathrm{dz}}\,;\; C_2^{'} = \varepsilon_0 \frac{\frac{p}{2}(1-p)\mathrm{dz}}{\mathrm{b\text{-}}\frac{\mathrm{q}}{2}(1-p)}\,; \qquad [A6.14]$$

$$R_2 = \frac{1}{2}R_2^{'} = \frac{1}{2}\rho \frac{a+\frac{q}{2}(1-p)}{\frac{p}{2}(1-p)\,\mathrm{dz}}\,; \qquad [A6.15]$$

$$C_2 = 2C_2^{'} = 2\varepsilon_0 \frac{\frac{p}{2}(1-p)\,\mathrm{dz}}{b-\frac{q}{2}(1-p)}\,; \qquad [A6.16]$$

$$R_3' = \rho \frac{a + \frac{q}{4}(1-p)^2}{\frac{p}{4}(1-p)^2 \, \mathrm{dz}} ; \; C_3' = \varepsilon_0 \frac{\frac{p}{4}(1-p)^2 \, \mathrm{dz}}{b - \frac{q}{4}(1-p)^2} ; \quad [A6.17]$$

$$R_3 = \frac{1}{4} R_3' = \frac{1}{4} \rho \frac{a + \frac{q}{4}(1-p)^2}{\frac{p}{4}(1-p)^2 \, \mathrm{dz}} ; \quad [A6.18]$$

$$C_3 = 4C_3' = 4\varepsilon_0 \frac{\frac{p}{4}(1-p)^2 \, \mathrm{dz}}{b - \frac{q}{4}(1-p)^2} . \quad [A6.19]$$

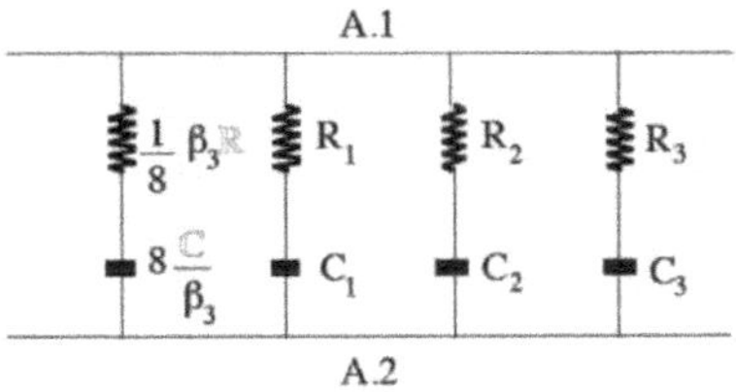

Figure A6.8. *Equivalent electric circuit with four branches*

A6.3.2. *Generalization of the equivalent electrical circuit construction*

The extension of the results obtained for the first four steps (0–3) is carried out by denoting the rank of the recursive motif construction steps by m, and then by introducing this rank in the expressions of the resistances and capacitances of each of the cells in order to generalize them.

A6.3.2.1. *Step 1 (m=1)*

$$R_1 = \frac{1}{2^0} \rho \frac{a + \frac{q}{2^0}(1-p)^0}{\frac{p}{2^0}(1-p)^0 \, \mathrm{dz}} = \frac{1}{2^{1-1}} \rho \frac{a + \frac{q}{2^{1-1}}(1-p)^{1-1}}{\frac{p}{2^{1-1}}(1-p)^{1-1} \, \mathrm{dz}} , \quad [A6.20]$$

namely:

$$R_m = \frac{1}{2^{m-1}} \rho \frac{a + \frac{q}{2^{m-1}}(1-p)^{m-1}}{\frac{p}{2^{m-1}}(1-p)^{m-1} \, \mathrm{dz}} ; \quad [A6.21]$$

$$C_1 = 2^0 \varepsilon_0 \frac{\frac{p}{2^0}(1-p)^0 \,\mathrm{dz}}{b - \frac{q}{2^0}(1-p)^0} = 2^{1-1} \varepsilon_0 \frac{\frac{p}{2^{1-1}}(1-p)^{1-1} \,\mathrm{dz}}{b - \frac{q}{2^{1-1}}(1-p)^{1-1}}, \qquad \text{[A6.22]}$$

namely:

$$C_m = 2^{m-1} \varepsilon_0 \frac{\frac{p}{2^{m-1}}(1-p)^{m-1} \,\mathrm{dz}}{b - \frac{q}{2^{m-1}}(1-p)^{m-1}}. \qquad \text{[A6.23]}$$

A6.3.2.2. *Step 2 ($m=2$)*

$$R_1 = \frac{1}{2^0} \rho \frac{a + \frac{q}{2^0}(1-p)^0}{\frac{p}{2^0}(1-p)^0 \,\mathrm{dz}} = \frac{1}{2^{2-2}} \rho \frac{a + \frac{q}{2^{2-2}}(1-p)^{2-2}}{\frac{p}{2^{2-2}}(1-p)^{2-2} \,\mathrm{dz}}, \qquad \text{[A6.24]}$$

namely:

$$R_{m-1} = \frac{1}{2^{m-2}} \rho \frac{a + \frac{q}{2^{m-2}}(1-p)^{m-2}}{\frac{p}{2^{m-2}}(1-p)^{m-2} \,\mathrm{dz}}; \qquad \text{[A6.25]}$$

$$C_1 = 2^0 \varepsilon_0 \frac{\frac{p}{2^0}(1-p)^0 \,\mathrm{dz}}{b - \frac{q}{2^0}(1-p)^0} = 2^{2-2} \varepsilon_0 \frac{\frac{p}{2^{2-2}}(1-p)^{2-2} \,\mathrm{dz}}{b - \frac{q}{2^{2-2}}(1-p)^{2-2}}, \qquad \text{[A6.26]}$$

namely:

$$C_{m-1} = 2^{m-2} \varepsilon_0 \frac{\frac{p}{2^{m-2}}(1-p)^{m-2} \,\mathrm{dz}}{b - \frac{q}{2^{m-2}}(1-p)^{m-2}}; \qquad \text{[A6.27]}$$

$$R_2 = \frac{1}{2^1} \rho \frac{a + \frac{q}{2^1}(1-p)^1}{\frac{p}{2^1}(1-p)^1 \,\mathrm{dz}} = \frac{1}{2^{2-1}} \rho \frac{a + \frac{q}{2^{2-1}}(1-p)^{2-1}}{\frac{p}{2^{2-1}}(1-p)^{2-1} \,\mathrm{dz}}, \qquad \text{[A6.28]}$$

namely:

$$R_m = \frac{1}{2^{m-1}} \rho \frac{a + \frac{q}{2^{m-1}}(1-p)^{m-1}}{\frac{p}{2^{m-1}}(1-p)^{m-1}\,\mathrm{dz}}; \quad \text{[A6.29]}$$

$$C_2 = 2^1 \varepsilon_0 \frac{\frac{p}{2^1}(1-p)^1\,\mathrm{dz}}{b - \frac{q}{2^1}(1-p)^1} = 2^{2-1}\varepsilon_0 \frac{\frac{p}{2^{2-1}}(1-p)^{2-1}\,\mathrm{dz}}{b - \frac{q}{2^{2-1}}(1-p)^{2-1}}, \quad \text{[A6.30]}$$

namely:

$$C_m = 2^{m-1}\varepsilon_0 \frac{\frac{p}{2^{m-1}}(1-p)^{m-1}\,\mathrm{dz}}{b - \frac{q}{2^{m-1}}(1-p)^{m-1}}. \quad \text{[A6.31]}$$

Let us form the ratio of the resistances of the last and its previous cells:

$$\frac{R_m}{R_{m-1}} = \frac{1}{1-p} \frac{a + \frac{q}{2^{m-1}}(1-p)^{m-1}}{a + \frac{q}{2^{m-2}}(1-p)^{m-2}}. \quad \text{[A6.32]}$$

Likewise, let us form the ratio of the capacitances of the last and its previous cells. It then becomes:

$$\frac{C_m}{C_{m-1}} = (1-p) \frac{b - \frac{q}{2^{m-2}}(1-p)^{m-2}}{b - \frac{q}{2^{m-1}}(1-p)^{m-1}}. \quad \text{[A6.33]}$$

A6.3.2.3. *Step 3* $(m = 3)$

$$R_1 = \frac{1}{2^0} \rho \frac{a + \frac{q}{2^0}(1-p)^0}{\frac{p}{2^0}(1-p)^0\,\mathrm{dz}} = \frac{1}{2^{3-3}} \rho \frac{a + \frac{q}{2^{3-3}}(1-p)^{3-3}}{\frac{p}{2^{3-3}}(1-p)^{3-3}\,\mathrm{dz}}, \quad \text{[A6.34]}$$

namely:

$$R_{m-2} = \frac{1}{2^{m-3}} \rho \frac{a + \frac{q}{2^{m-3}}(1-p)^{m-3}}{\frac{p}{2^{m-3}}(1-p)^{m-3}\,\mathrm{dz}}; \quad \text{[A6.35]}$$

$$C_1 = 2^0 \varepsilon_0 \frac{\frac{p}{2^0}(1-p)^0 \mathrm{d}z}{b - \frac{q}{2^0}(1-p)^0} = 2^{3-3} \varepsilon_0 \frac{\frac{p}{2^{3-3}}(1-p)^{3-3} \mathrm{d}z}{b - \frac{q}{2^{3-3}}(1-p)^{3-3}}, \qquad \text{[A6.36]}$$

namely:

$$C_{m-2} = 2^{m-3} \varepsilon_0 \frac{\frac{p}{2^{m-3}}(1-p)^{m-3} \mathrm{d}z}{b - \frac{q}{2^{m-3}}(1-p)^{m-3}}; \qquad \text{[A6.37]}$$

$$R_2 = \frac{1}{2^1} \rho \frac{a + \frac{q}{2^1}(1-p)^1}{\frac{p}{2^1}(1-p)^1 \mathrm{d}z} = \frac{1}{2^{3-2}} \rho \frac{a + \frac{q}{2^{3-2}}(1-p)^{3-2}}{\frac{p}{2^{3-2}}(1-p)^{3-2} \mathrm{d}z}, \qquad \text{[A6.38]}$$

namely:

$$R_{m-1} = \frac{1}{2^{m-2}} \rho \frac{a + \frac{q}{2^{m-2}}(1-p)^{m-2}}{\frac{p}{2^{m-2}}(1-p)^{m-2} \mathrm{d}z}; \qquad \text{[A6.39]}$$

$$C_2 = 2^1 \varepsilon_0 \frac{\frac{p}{2^1}(1-p)^1 \mathrm{d}z}{b - \frac{q}{2^1}(1-p)^1} = 2^{3-2} \varepsilon_0 \frac{\frac{p}{2^{3-2}}(1-p)^{3-2} \mathrm{d}z}{b - \frac{q}{2^{3-2}}(1-p)^{3-2}}, \qquad \text{[A6.40]}$$

namely:

$$C_{m-1} = 2^{m-2} \varepsilon_0 \frac{\frac{p}{2^{m-2}}(1-p)^{m-2} \mathrm{d}z}{b - \frac{q}{2^{m-2}}(1-p)^{m-2}}; \qquad \text{[A6.41]}$$

$$R_3 = \frac{1}{2^2} \rho \frac{a + \frac{q}{2^2}(1-p)^2}{\frac{p}{2^2}(1-p)^2 \mathrm{d}z} = \frac{1}{2^{3-1}} \rho \frac{a + \frac{q}{2^{3-1}}(1-p)^{3-1}}{\frac{p}{2^{3-1}}(1-p)^{3-1} \mathrm{d}z}, \qquad \text{[A6.42]}$$

namely:

$$R_m = \frac{1}{2^{m-1}} \rho \frac{a + \frac{q}{2^{m-1}}(1-p)^{m-1}}{\frac{p}{2^{m-1}}(1-p)^{m-1}\,\mathrm{d}z}; \qquad \text{[A6.43]}$$

$$C_3 = 2^2 \varepsilon_0 \frac{\frac{p}{2^2}(1-p)^2\,\mathrm{d}z}{b - \frac{q}{2^2}(1-p)^2} = 2^{3-1} \varepsilon_0 \frac{\frac{p}{2^{3-1}}(1-p)^{3-1}\,\mathrm{d}z}{b - \frac{q}{2^{3-1}}(1-p)^{3-1}}, \qquad \text{[A6.44]}$$

namely:

$$C_m = 2^{m-1} \varepsilon_0 \frac{\frac{p}{2^{m-1}}(1-p)^{m-1}\,\mathrm{d}z}{b - \frac{q}{2^{m-1}}(1-p)^{m-1}}. \qquad \text{[A6.45]}$$

The ratio of the resistances of the last and its previous cells is written as:

$$\frac{R_m}{R_{m-1}} = \frac{1}{1-p} \frac{a + \frac{q}{2^{m-1}}(1-p)^{m-1}}{a + \frac{q}{2^{m-2}}(1-p)^{m-2}}, \qquad \text{[A6.46]}$$

and the ratio of their capacitances is:

$$\frac{C_m}{C_{m-1}} = (1-p) \frac{b - \frac{q}{2^{m-2}}(1-p)^{m-2}}{b - \frac{q}{2^{m-1}}(1-p)^{m-1}}. \qquad \text{[A6.47]}$$

A6.3.3. *Change from recurrent laws to recursive laws*

Establishing the electric circuits corresponding to different construction steps leads us to a *recurrent parallel arrangement of series RC cells*. It is true that the identity of relations [A6.32] and [A6.46] then [A6.33] and [A6.47] obtained, respectively, for two well distinct steps, $m = 2$ and $m = 3$, shows that there exist two recurrence relations between the resistances and the capacitances of the last two cells, whatever the step rank m, namely:

$$\frac{R_m}{R_{m-1}} = \frac{1}{1-p} \frac{a + \frac{q}{2^{m-1}}(1-p)^{m-1}}{a + \frac{q}{2^{m-2}}(1-p)^{m-2}} \qquad \text{[A6.48]}$$

and

$$\frac{C_m}{C_{m-1}} = (1-p)\frac{b - \frac{q}{2^{m-2}}(1-p)^{m-2}}{b - \frac{q}{2^{m-1}}(1-p)^{m-1}}. \qquad [A6.49]$$

Dependent on the rank, these recurrence laws express the *non-recursivity* of the resistance and capacitance distribution.

On the contrary, considering a high number of construction steps (m being sufficiently large) makes these recurrent laws independent of the rank, namely:

$$\frac{R_m}{R_{m-1}} = \frac{1}{1-p} \qquad [A6.50]$$

and

$$\frac{C_m}{C_{m-1}} = 1-p. \qquad [A6.51]$$

Furthermore, considering a nil enables us to reduce the recurrent law [A6.48] to a recursive law without having to formulate any condition on the step number, namely:

$$\frac{R_m}{R_{m-1}} = \frac{1}{2}. \qquad [A6.52]$$

A6.3.4. *Capacitor model*

Finally, the recursive laws [A6.50]–[A6.52] express the *recursivity* of the parallel arrangement of series RC cells. Their form suggests the introduction of *recursive factors* α and η (Figure A6.9) such that:

$$\alpha^{-1} = \frac{R_m}{R_{m-1}} = \begin{cases} \dfrac{1}{1-p} & \text{for } a \neq 0 \ \ (\text{namely } \alpha = 1-p) \\[2mm] \dfrac{1}{2} & \text{for } a = 0 \ \ (\text{namely } \alpha = 2) \end{cases} \qquad [A6.53]$$

and

$$\eta^{-1} = \frac{C_m}{C_{m-1}} = 1-p \quad \left(\text{namely } \eta = \frac{1}{1-p}\right). \qquad [A6.54]$$

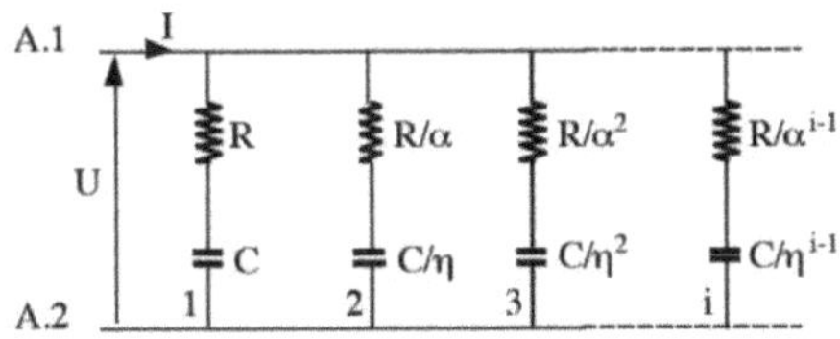

Figure A6.9. *Recursive parallel arrangement of series RC cells*

We are thus in the presence of a *recursive parallel arrangement of series RC cells* such as that covered in Chapters 2 and 6 and for which the differentiation non-integer order is given by:

$$n = \frac{\log \alpha}{\log(\alpha\eta)} \text{ with } 0 < n < 1. \qquad [A6.55]$$

A6.4. Roughness of fractal nature

A6.4.1. *First construction steps of a fractal recursive motif and of the equivalent electrical circuit*

A6.4.1.1. *Step 0*

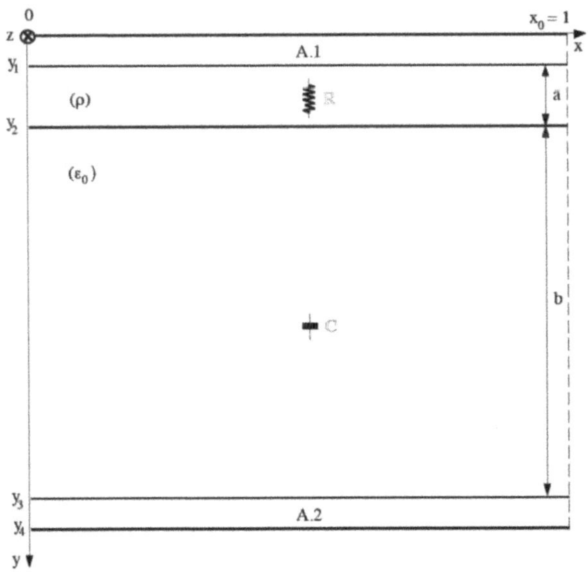

Figure A6.10. *Initial step of the fractal motif construction*

$$R = \rho \frac{a}{\mathrm{d}z}\,;\; C = \varepsilon_0 \frac{\mathrm{d}z}{b}\,. \qquad [A6.56]$$

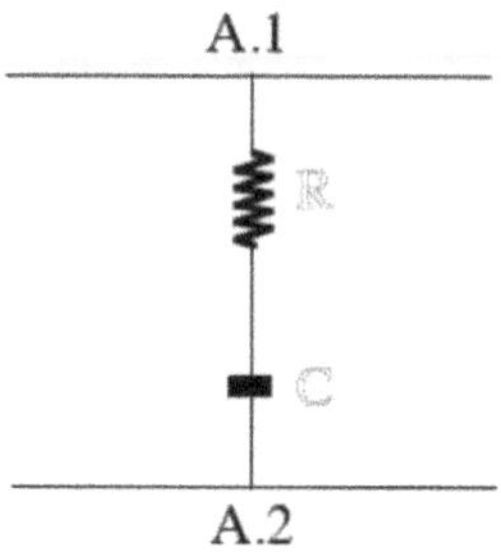

Figure A6.11. *Equivalent electric circuit with one branch*

A6.4.1.2. *Step 1*

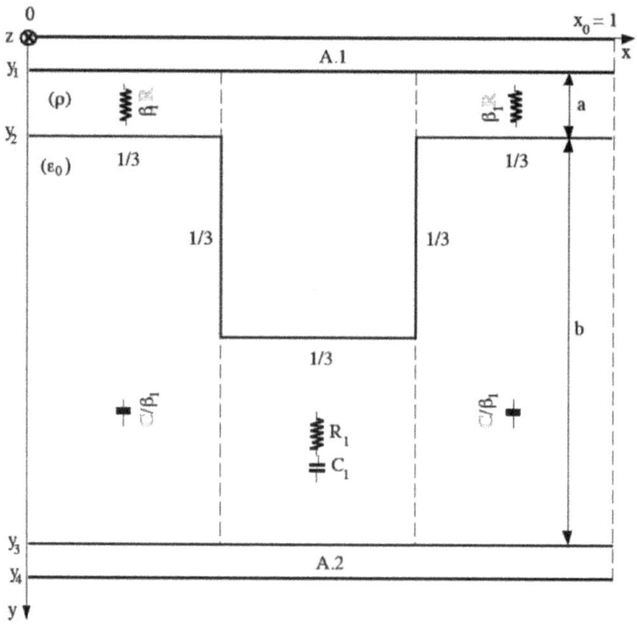

Figure A6.12. *The first step of the fractal motif construction*

$$\beta_1 R = \rho \frac{a}{\frac{1}{3}\mathrm{d}z} = 3R \;\rightarrow\; \beta_1 = 3\,; \qquad [A6.57]$$

$$\frac{C}{\beta_1} = \varepsilon_0 \frac{\frac{1}{3}\mathrm{dz}}{b} = \frac{1}{3}C \rightarrow \frac{1}{\beta_1} = \frac{1}{3}; \qquad [A6.58]$$

$$R_1 = \rho \frac{a + \frac{1}{3}}{\frac{1}{3}\mathrm{dz}}; \; C_1 = \varepsilon_0 \frac{\frac{1}{3}\mathrm{dz}}{\mathrm{b} - \frac{1}{3}}. \qquad [A6.59]$$

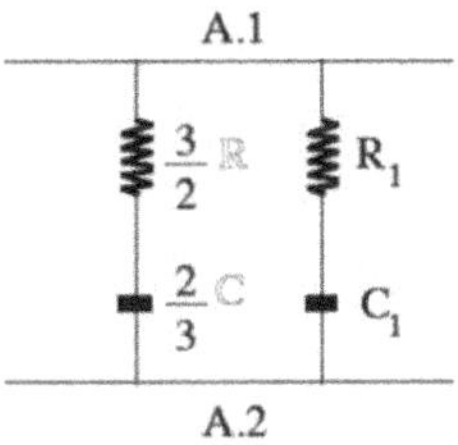

Figure A6.13. *Equivalent electric circuit with two branches*

A6.4.1.3. *Step 2*

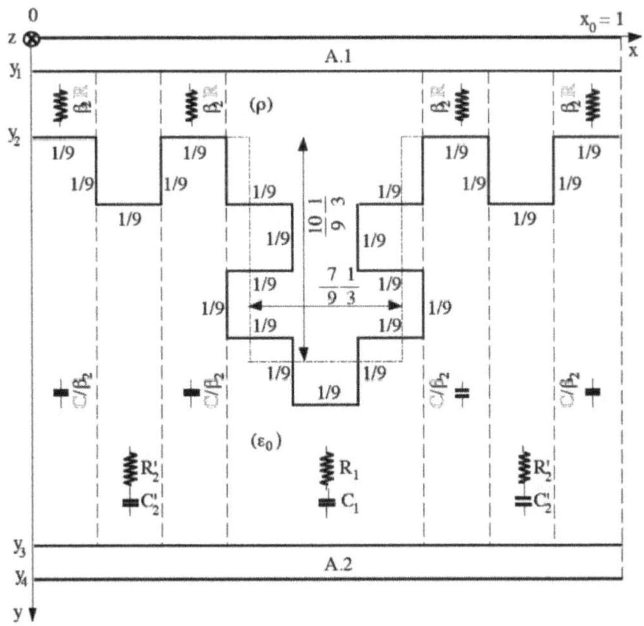

Figure A6.14. *The second step of the fractal motif construction*

$$\beta_2 R = \rho \frac{a}{\frac{1}{9}\mathrm{dz}} = 9R \rightarrow \beta_2 = 9; \qquad [A6.60]$$

$$\frac{C}{\beta_2} = \varepsilon_0 \frac{\frac{1}{9}\mathrm{dz}}{b} = \frac{1}{9} C \rightarrow \frac{1}{\beta_2} = \frac{1}{9}. \qquad [A6.61]$$

The rigorous determination of R_1 and C_1 is extremely difficult. So, we propose an approximate calculation which, as it seems, is sufficiently significant to render the transformation that step 2 defines (Figure A6.15).

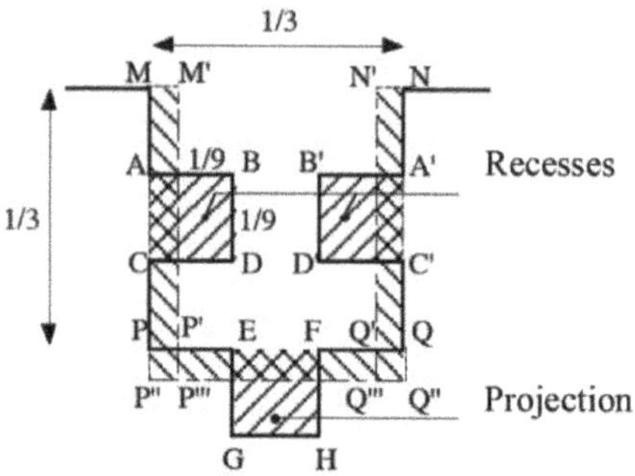

Figure A6.15. *Overhang modification*

The notches (or recesses) ABCD and A'B'C'D' and the projection $EFGH$, respectively, applied on the overhang MNPQ have specific effects. Indeed, we intuitively feel that:

– the notches tend to thin down the overhang;

– the projection tends to lengthen it.

In order to quantify these effects, it is proposed (Figure A6.15) to spread out:

– the surface lacks ABCD and A'B'C'D' along the segments MP and NQ;

– the surface excess $EFGH$ along the segment PQ. The surface identities are translated into:

$$\mathrm{MM'}\frac{1}{3} = \left(\frac{1}{9}\right)^2 \rightarrow \mathrm{MM'} = \frac{1}{3}\frac{1}{9} \qquad [A6.62]$$

$$\mathrm{NN'}\frac{1}{3}=\left(\frac{1}{9}\right)^2 \rightarrow \mathrm{NN'}=\frac{1}{3}\frac{1}{9}$$ [A6.63]

$$\mathrm{PP''}\frac{1}{3}=\left(\frac{1}{9}\right)^2 \rightarrow \mathrm{PP''}=\frac{1}{3}\frac{1}{9}$$ [A6.64]

$$\mathrm{M'N'}=\frac{1}{3}-\frac{2}{3}\frac{1}{9}=\frac{7}{9}\frac{1}{3}$$ [A6.65]

$$\mathrm{M'P'''}=\frac{1}{3}+\frac{1}{3}\frac{1}{9}=\frac{10}{9}\frac{1}{3}.$$ [A6.66]

Bringing the overhang $MNPQ$ back to the overhang $M'N'P'''Q'''$ amounts to neglecting the surface of the squares $PP'P''P'''$ and $QQ'Q''Q'''$, whose value which is equal to $\left(\frac{1}{3}\frac{1}{9}\right)^2=\frac{1}{243}$ is effectively negligible.

$$R_1=\rho\frac{a}{\frac{1}{3}\mathrm{dz}}+\rho\frac{\frac{10}{9}\frac{1}{3}}{\frac{7}{9}\frac{1}{3}\mathrm{dz}}=\frac{\rho}{\mathrm{dz}}\left[\frac{a}{\frac{1}{3}}+\frac{10}{7}\right];$$ [A6.67]

$$C_1-\varepsilon_0\frac{\frac{7}{9}\frac{1}{3}\mathrm{dz}}{b-\frac{10}{9}\frac{1}{3}};$$ [A6.68]

$$R_2'=\rho\frac{a+\frac{1}{9}}{\frac{1}{9}\mathrm{dz}};\ C_2'=\varepsilon_0\frac{\frac{1}{9}\mathrm{dz}}{b-\frac{1}{9}};$$ [A6.69]

$$R_2=\frac{1}{2}R_2'=\frac{1}{2}\rho\frac{a+\frac{1}{9}}{\frac{1}{9}\mathrm{dz}};$$ [A6.70]

$$C_2=2C_2'=2\varepsilon_0\frac{\frac{1}{9}\mathrm{dz}}{b-\frac{1}{9}}.$$ [A6.71]

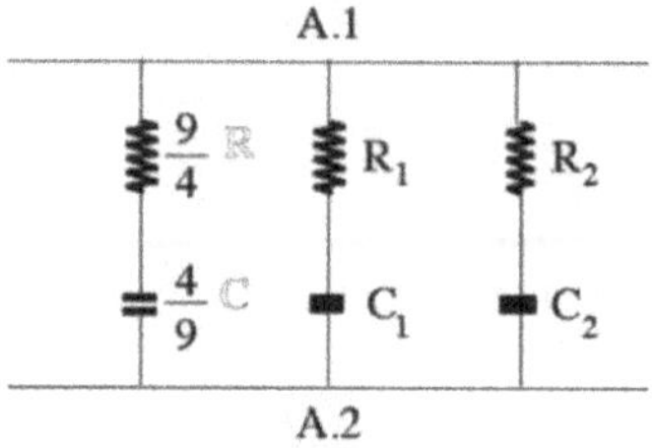

Figure A6.16. *Equivalent electric circuit with three branches*

A6.4.1.4. *Step 3*

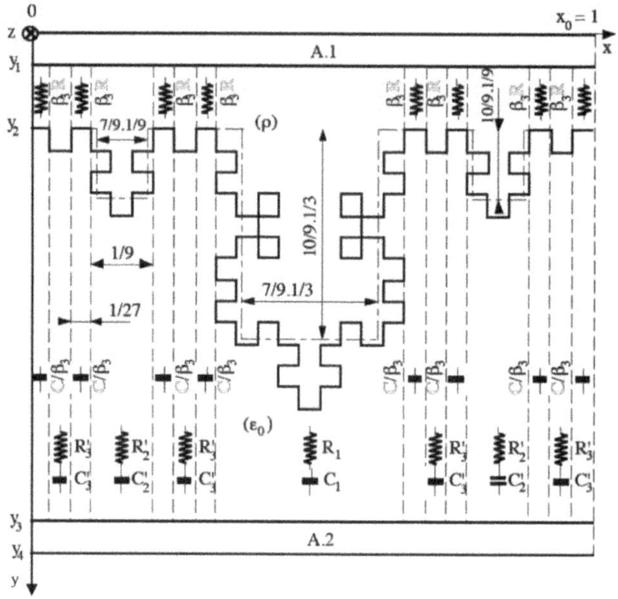

Figure A6.17. *The third step of the fractal motif construction*

$$\beta_3 R = \rho \frac{a}{\frac{1}{27}\mathrm{d}z} = 27R \quad \rightarrow \quad \beta_3 = 27\,; \qquad [A6.72]$$

$$\frac{C}{\beta_3} = \varepsilon_0 \frac{\frac{1}{27}\mathrm{d}z}{b} = \frac{1}{27}C \rightarrow \quad \frac{1}{\beta_3} = \frac{1}{27}. \qquad [A6.73]$$

If the values of R_1 and C_1 are reasonably affected by the transformation corresponding to step 2 (first-order modification), it is possible to consider that they are practically no more, starting from the transformation that step 3 (second-order modification) defines; therefore:

$$R_1 = \frac{\rho}{\mathrm{dz}}\left[\frac{a}{\frac{1}{3}} + \frac{10}{7}\right] \tag{A6.74}$$

and

$$C_1 = \varepsilon_0 \frac{\frac{7}{9}\frac{1}{3}\mathrm{dz}}{b - \frac{10}{9}\frac{1}{3}}. \tag{A6.75}$$

$$R_2' = \rho\frac{a}{\frac{1}{9}\mathrm{dz}} + \rho\frac{\frac{10}{9}\frac{1}{9}}{\frac{7}{9}\frac{1}{9}\mathrm{dz}} = \frac{\rho}{\mathrm{dz}}\left[\frac{a}{\frac{1}{9}} + \frac{10}{7}\right]; \tag{A6.76}$$

$$C_2' = \varepsilon_0 \frac{\frac{7}{9}\frac{1}{9}\mathrm{dz}}{b - \frac{10}{9}\frac{1}{9}}; \tag{A6.77}$$

$$R_2 = \frac{1}{2}R_2' = \frac{1}{2}\frac{\rho}{\mathrm{dz}}\left[\frac{a}{\frac{1}{9}} + \frac{10}{7}\right]; \tag{A6.78}$$

$$C_2 = 2C_2' = 2\varepsilon_0 \frac{\frac{7}{9}\frac{1}{9}\mathrm{dz}}{\mathrm{b} - \frac{10}{9}\frac{1}{9}}; \tag{A6.79}$$

$$R_3' = \rho\frac{a + \frac{1}{27}}{\frac{1}{27}\mathrm{dz}};\; C_3' = \varepsilon_0\frac{\frac{1}{27}dz}{b - \frac{1}{27}}; \tag{A6.80}$$

$$R_3 = \frac{1}{4} R_3' = \frac{1}{4} \rho \frac{a + \frac{1}{27}}{\frac{1}{27} \mathrm{d}z}; \qquad [A6.81]$$

$$C_3 = 4C_3' = 4\varepsilon_0 \frac{\frac{1}{27} dz}{b - \frac{1}{27}}. \qquad [A6.82]$$

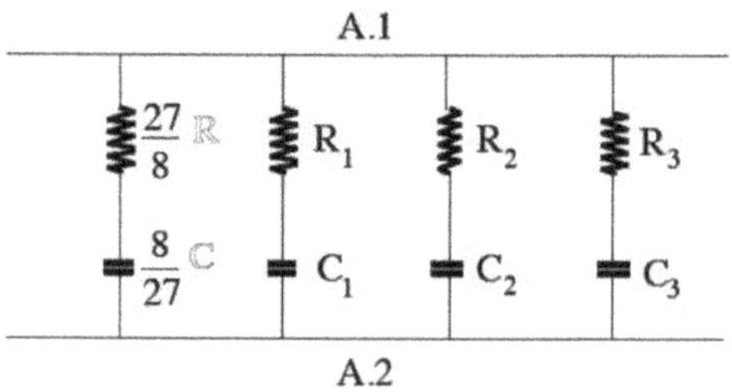

Figure A6.18. *Equivalent electric circuit with four branches*

A6.4.2. *Generalization of the equivalent electrical circuit construction*

In this section, the aim is to go back over the resistance and capacitance expressions by taking care of the introduction of the rank m of the fractal motif construction steps.

A6.4.2.1. *Step 1* $(m = 1)$

$$R_1 = \frac{1}{2^0} \rho \frac{a + \frac{1}{3^1}}{\frac{1}{3^1} \mathrm{d}z} = \frac{1}{2^{1-1}} \rho \frac{a + \frac{1}{3^1}}{\frac{1}{3^1} \mathrm{d}z}, \qquad [A6.83]$$

namely:

$$R_m = \frac{1}{2^{m-1}} \rho \frac{a + \frac{1}{3^m}}{\frac{1}{3^m} \mathrm{d}z}; \qquad [A6.84]$$

$$C_1 = 2^0 \varepsilon_0 \frac{\frac{1}{3^1}\mathrm{d}z}{b-\frac{1}{3^1}} = 2^{1-1}\varepsilon_0 \frac{\frac{1}{3^1}\mathrm{d}z}{b-\frac{1}{3^1}}, \qquad [A6.85]$$

namely:

$$C_m = 2^{m-1}\varepsilon_0 \frac{\frac{1}{3^m}\mathrm{d}z}{b-\frac{1}{3^m}}. \qquad [A6.86]$$

A6.4.2.2. *Step 2 ($m = 2$)*

$$R_1 = \frac{1}{2^0}\frac{\rho}{\mathrm{d}z}\left[\frac{a}{\frac{1}{3^1}} + \frac{10}{7}\right] = \frac{1}{2^{2-2}}\frac{\rho}{\mathrm{d}z}\left[\frac{a}{\frac{1}{3^{2-1}}} + \frac{10}{7}\right], \qquad [A6.87]$$

namely:

$$R_{m-1} = \frac{1}{2^{m-2}}\frac{\rho}{\mathrm{d}z}\left[\frac{a}{\frac{1}{3^{m-1}}} + \frac{10}{7}\right]; \qquad [A6.88]$$

$$C_1 = 2^0 \varepsilon_0 \frac{\frac{7}{9}\frac{1}{3^1}\mathrm{d}z}{b-\frac{10}{9}\frac{1}{3^1}} = 2^{2-2}\varepsilon_0 \frac{\frac{7}{9}\frac{1}{3^{2-1}}\mathrm{d}z}{b-\frac{10}{9}\frac{1}{3^{2-1}}}, \qquad [A6.89]$$

namely:

$$C_{m-1} = 2^{m-2}\varepsilon_0 \frac{\frac{7}{9}\frac{1}{3^{m-1}}\mathrm{d}z}{b-\frac{10}{9}\frac{1}{3^{m-1}}}; \qquad [A6.90]$$

$$R_2 = \frac{1}{2^1}\rho\frac{a+\frac{1}{3^2}}{\frac{1}{3^2}\mathrm{d}z} = \frac{1}{2^{2-1}}\rho\frac{a+\frac{1}{3^2}}{\frac{1}{3^2}\mathrm{d}z}, \qquad [A6.91]$$

namely:

$$R_m = \frac{1}{2^{m-1}} \rho \frac{a + \frac{1}{3^m}}{\frac{1}{3^m} \mathrm{dz}}; \qquad \text{[A6.92]}$$

$$C_2 = 2^1 \varepsilon_0 \frac{\frac{1}{3^2} \mathrm{dz}}{b - \frac{1}{3^2}} = 2^{2-1} \varepsilon_0 \frac{\frac{1}{3^2} \mathrm{dz}}{b - \frac{1}{3^2}}, \qquad \text{[A6.93]}$$

namely:

$$C_m = 2^{m-1} \varepsilon_0 \frac{\frac{1}{3^m} \mathrm{dz}}{b - \frac{1}{3^m}}. \qquad \text{[A6.94]}$$

The ratios of the resistances and capacitances of the last two cells are:

$$\frac{R_m}{R_{m-1}} = \frac{1}{2} \frac{3^m a + 1}{3^{m-1} a + \frac{10}{7}} \qquad \text{[A6.95]}$$

and

$$\frac{C_m}{C_{m-1}} = 2 \frac{\frac{1}{3^m}}{b - \frac{1}{3^m}} \frac{b - \frac{10}{9} \frac{1}{3^{m-1}}}{\frac{7}{9} \frac{1}{3^{m-1}}}. \qquad \text{[A6.96]}$$

A6.4.2.3. *Step 3* $(m = 3)$

$$R_1 = \frac{1}{2^0} \frac{\rho}{\mathrm{dz}} \left[\frac{a}{\frac{1}{3^1}} + \frac{10}{7} \right] = \frac{1}{2^{3-3}} \frac{\rho}{\mathrm{dz}} \left[\frac{a}{\frac{1}{3^{3-2}}} + \frac{10}{7} \right], \qquad \text{[A6.97]}$$

namely:

$$R_{m-2} = \frac{1}{2^{m-3}} \frac{\rho}{\mathrm{dz}} \left[\frac{a}{\frac{1}{3^{m-2}}} + \frac{10}{7} \right]; \qquad \text{[A6.98]}$$

$$C_1 = 2^0 \varepsilon_0 \frac{\frac{7}{9}\frac{1}{3^1}\mathrm{d}z}{b - \frac{10}{9}\frac{1}{3^1}} = 2^{3-3} \varepsilon_0 \frac{\frac{7}{9}\frac{1}{3^{3-2}}\mathrm{d}z}{b - \frac{10}{9}\frac{1}{3^{3-2}}}, \qquad \text{[A6.99]}$$

namely:

$$C_{m-2} = 2^{m-3} \varepsilon_0 \frac{\frac{7}{9}\frac{1}{3^{m-2}}\mathrm{d}z}{b - \frac{10}{9}\frac{1}{3^{m-2}}}; \qquad \text{[A6.100]}$$

$$R_2 = \frac{1}{2^1}\frac{\rho}{\mathrm{d}z}\left[\frac{a}{\frac{1}{3^2}} + \frac{10}{7}\right] = \frac{1}{2^{3-2}}\frac{\rho}{\mathrm{d}z}\left[\frac{a}{\frac{1}{3^{3-1}}} + \frac{10}{7}\right], \qquad \text{[A6.101]}$$

namely:

$$R_{m-1} = \frac{1}{2^{m-2}}\frac{\rho}{\mathrm{d}z}\left[\frac{a}{\frac{1}{3^{m-1}}} + \frac{10}{7}\right]; \qquad \text{[A6.102]}$$

$$C_2 = 2^1 \varepsilon_0 \frac{\frac{7}{9}\frac{1}{3^2}\mathrm{d}z}{b - \frac{10}{9}\frac{1}{3^2}} = 2^{3-2} \varepsilon_0 \frac{\frac{7}{9}\frac{1}{3^{3-1}}\mathrm{d}z}{b - \frac{10}{9}\frac{1}{3^{3-1}}}, \qquad \text{[A6.103]}$$

namely:

$$C_{m-1} = 2^{m-2} \varepsilon_0 \frac{\frac{7}{9}\frac{1}{3^{m-1}}\mathrm{d}z}{b - \frac{10}{9}\frac{1}{3^{m-1}}}; \qquad \text{[A6.104]}$$

$$R_3 = \frac{1}{2^2}\rho\frac{a + \frac{1}{3^3}}{\frac{1}{3^3}\mathrm{d}z} = \frac{1}{2^{3-1}}\rho\frac{a + \frac{1}{3^3}}{\frac{1}{3^3}\mathrm{d}z}, \qquad \text{[A6.105]}$$

namely:

$$R_m = \frac{1}{2^{m-1}} \rho \frac{a + \frac{1}{3^m}}{\frac{1}{3^m} \mathrm{dz}}; \qquad [A6.106]$$

$$C_3 = 2^2 \varepsilon_0 \frac{\frac{1}{3^3} dz}{b - \frac{1}{3^3}} = 2^{3-1} \varepsilon_0 \frac{\frac{1}{3^3} dz}{b - \frac{1}{3^3}}, \qquad [A6.107]$$

namely:

$$C_m = 2^{m-1} \varepsilon_0 \frac{\frac{1}{3^m} \mathrm{dz}}{b - \frac{1}{3^m}}. \qquad [A6.108]$$

The ratios of the resistances and capacitances of the last two cells, respectively, admit for expressions:

$$\frac{R_m}{R_{m-1}} = \frac{1}{2} \frac{3^m a + 1}{3^{m-1} a + \frac{10}{7}} \qquad [A6.109]$$

and

$$\frac{C_m}{C_{m-1}} = 2 \frac{\frac{1}{3^m}}{b - \frac{1}{3^m}} \frac{b - \frac{10}{9} \frac{1}{3^{m-1}}}{\frac{7}{9} \frac{1}{3^{m-1}}}. \qquad [A6.110]$$

A6.4.3. *Change from recurrent laws to recursive laws*

Establishing the electric circuits obtained for the different construction steps leads us to a *recurrent parallel arrangement of series RC cells*, the corresponding recurrence laws being defined by the relations (see identities [A6.95] and [A6.109] then [A6.96] and [A6.110]):

$$\frac{R_m}{R_{m-1}} = \frac{1}{2} \frac{3^m a + 1}{3^{m-1} a + \frac{10}{7}} \qquad [A6.111]$$

and

$$\frac{C_m}{C_{m-1}} = 2\frac{\frac{1}{3^m}}{b-\frac{1}{3^m}} \frac{b-\frac{10}{9}\frac{1}{3^{m-1}}}{\frac{7}{9}\frac{1}{3^{m-1}}}. \qquad [A6.112]$$

Dependent on the rank, these laws express the *non-recursivity* of the resistance and capacitance distribution.

On the contrary, for sufficiently large m (consideration of a high number of construction steps), these laws lose their dependency on the rank and thus become recursive, namely:

$$\frac{R_m}{R_{m-1}} = \frac{3}{2} \qquad [A6.113]$$

and

$$\frac{C_m}{C_{m-1}} = \frac{6}{7}. \qquad [A6.114]$$

Furthermore, considering a nil enables us to make the recurrent law [A6.111] recursive without imposing any condition on the step number, namely:

$$\frac{R_m}{R_{m-1}} = \frac{7}{20}. \qquad [A6.115]$$

A6.4.4. *Capacitor model*

Finally, the recursive laws [A6.113]–[A6.115] express the *recursivity* of the parallel arrangement of series RC cells. Their form suggests the introduction of *recursive factors* α and η such that:

$$\alpha^{-1} = \frac{R_m}{R_{m-1}} = \begin{cases} \frac{3}{2} & \text{for } a \neq 0 \quad \left(\text{namely } \alpha = \frac{2}{3} = 0,666\right) \\ \frac{7}{20} & \text{for } a = 0 \quad \left(\text{namely } \alpha = \frac{20}{7} = 2,857\right) \end{cases} \qquad [A6.116]$$

and

$$\eta^{-1} = \frac{C_m}{C_{m-1}} = \frac{6}{7} \quad \left(\text{namely } \eta = \frac{7}{6} = 1,166\right). \qquad [A6.117]$$

As in section A6.3.4, we thus obtain a *recursive parallel arrangement of series RC cells* such as that discussed in Chapters 2 and 6 and for which the differentiation non-integer order is given by:

$$n = \frac{\log \alpha}{\log(\alpha\eta)} \text{ with } 0 < n < 1. \quad [A6.118]$$

Given the numerical values of the recursive factors notably obtained for $a = 0$, namely:

$$\alpha = 2{,}857 \text{ and } \eta = 1{,}166, \quad [A6.119]$$

the differentiation order so defined is particularized in conformity with the non-integer value:

$$n = 0{,}872. \quad [A6.120]$$

A6.5. Bibliography

[BAG 83] BAGLEY R.L., TORVICK P.J., "A theoretical basis for the application of fractional calculus to viscoelasticity", *Journal of Rheology*, vol. 27, pp. 201–210, June 1983.

[LE 83] LE MEHAUTE A., GREPY G., *Introduction to Transfer and Motion in Fractal Media: The Geometry of Kinetics*, Solid State Ionics 9 & 10, North-Holland Publishing Company, pp. 17–30, 1983.

[LEM 90] LE MEHAUTE A., *Les géométries fractales*, Hermès, Paris, 1990.

[LEM 98] LE MEHAUTE A., NIGMATULLIN R.R., NIVANEN L., *Flèches du temps et géométrie fractale*, Hermès, Paris, 1998.

[MAN 75] MANDELBROT B., *Les Fractals*, Flammarion, 1975.

[MAT 93] MATIGNON D., D'ANDREA-NOVEL B., DEPALLE P., *et al.*, "Viscothermal losses in wind instruments: a non integer model", *International Symposium on the Mathematical Theory of Networks and Systems (MTNS '93)*, Regensburg, Germany, 2–6 August 1993.

[OUS 86] OUSTALOUP A., *Colloque des fractales en Mathématiques et en Physique*, Marseille, France, 13–18 January 1986.

[OUS 91] OUSTALOUP A., *La commande CRONE*, Hermès, Paris, 1991.

[OUS 95] OUSTALOUP A., *La dérivation non entière: théorie, synthèse et applications*, Hermès, Paris, 1995.

Index

D

E, F

G

P

Q, R

S

V

W, Z